Aquaculture

New Horizons

Aquaculture

New Horizons

Hiralal Chaudhuri
A. B. Chaudhuri

2024
Daya Publishing House®
A Division of
Astral International Pvt. Ltd.
New Delhi – 110 002

First Published, 2009

Reprinted, 2024

ISBN: 978-93-90259-84-7 (HB)

Published by : **Daya Publishing House®**
A Division of
Astral International Pvt. Ltd.
– ISO 9001:2015 Certified Company –
4736/23, Ansari Road, Darya Ganj
New Delhi-110 002
Ph. 011-43549197, 23278134
E-mail: info@astralint.com
Website: www.astralint.com

Printed at : **Replika Press Pvt. Ltd.**

Acknowledgements

The editor acknowledges the sincere advice extended by Dr. R.D. Chakraborty, Aquaculture expert from time to time in the preparation of the document. Thanks are due to the organizations of Central Inland Fisheries Institute and "Fishings Chimes" for allowing some of their research findings and picture to be utilized by the undersigned. Dr. H. Srinivasa Rao, Prof. H.S. Swingle and Dr. Himadri Mookerjee (all deceased) had great influence on Prof. Chaudhuri's research successes; to them I, on behalf of Prof. Chaudhuri, offer my heart felt gratitude and respect.

A.B. Chaudhuri

Dedication

This treatise could be shaped on the suggestions, inspiration and financial assistance extended by Late Mukul Chaudhuri, wife of the Professor Chaudhuri; this sacred work is dedicated to her memory on behalf of millions of poor fishermen and maritime poor population and on my behalf as she had been an inseparable associate in the research success of Professor Chaudhuri in bringing "Protein Revolution" to provide nutrition to the poor and especially in countries where Prof. Chaudhuri worked. Her philanthropic activities for the cause of poor fishermen families of Southeast Asia had a divine touch.

A.B. Chaudhuri

Foreword

Getting together work of a great scientist spread over decades is an enormous task. This further compounded if the field of work though inter-related is diverse.

Dr. Hiralal Chaudhuri the pre-eminent fisheries scientist of India as also a leading light in his field in the world occupies a niche of his own.

Modern aquaculture, as we know it today is much indebted to his unitiring work, the crowning glory of which is his successful work 'induced fish breeding' of commercial carp species in confined waters. His outstanding work, has opened the door and made possible the undertaking of planned fish farming.

This has somewhat obscured his other great achievements of developing the technology of attaining high fish production in aquaculture. He is that rare scientist who has contributed the most to both the main fields in culture fisheries.

In going through his work compiled in these volumes one would realise how much he has made possible successful aquaculture.

As a co-worker of this great scientist for some years, the writer has privileged to observe the meticulous planning and the thoroughness in his work. This reflected in his publications, written in lucid and easy to follow language born out of clear understanding of the subject.

A teacher of outstanding ability, which is legion of students spread over the world would testify, later in his life he was invited to join as professor at the University at Los Banos in the Phillippines.

Dr. Chaudhuri's publications provide practical guidance in aquaculture and should inspire further studies in this sphere. Dr. Chaudhuri, as would be revealed by study of his work was a pioneer and provided encouragement and inspiration to other and this should lead to further work in aquaculture.

R. D. Chakraborty

Retd. Head of Division (Pond Culture)
Central Inland Fisheries Research Station,
Cuttack, Bangalore

Professor Chaudhuri's Consent for Publication

Although I have always been averse to name, fame and public appreciation for whatever little I have achieved in my Scientific endeavour, I rather reluctantly permit Shri A. B. Chaudhuri of 131, N.S.C. Bose Road, Block–10, Flat-4, Kolkata–700 040 to edit my research documents, lecture notes, reports etc. relating to aquaculture, limnology and ichthyology either in part or full or modify them for suitable publications.

All these papers were widely discussed, readout, quoted and presented in national and international Scientific Workshops, Symposia and Congregations. As such no accent is necessary for the present publication from any body or any organization.

H. Chaudhuri

Formerly Chief Technical Advisor, FAO/UNDP and Regional Aquaculture Coordinator South East Asian Fisheries Development Centre, Manila

Preface

Aquaculture and food production are synonymous. World population is increasing at an exponential rate and consequently the demand for food is also increasing. The green revolution in Agriculture has contributed substantially to solve the food problem, but the nutritional standard of the vast majority of the people of the developing countries of the world is still much below the average level. There is acute shortage of animal protein in the diet of those poverty-stricken people. The situation would become worse as according to the prediction of U.N. demographers (1986), the world population is expected to rise from 4 billions by 2025 A.D. To feed these additional mouths there would be hardly any arable land left for expansion of agricultural area. Hence, in the future, mankind would have to develop more and more reliance on the oceans for food thus shifting the main source of food from Agriculture to Aquaculture.

Fish is highly nutritious food and a very important and cheap source of high quantity protein. Over 60 per cent people of the developing countries of the world derive 40 per cent or even 70-75 per cent of animal protein intake in their diet from fish. Till the middle of the present century, oceans and seas were the main source of fish supply for human consumption. The belief that the resources of seas are inexhaustible has been proved to be a myth as the fish catch instead of increasing with the increasing demands of the increased population, dropped down. The decline in the fish yield was mainly due to over-exploitation and several other factors such as, high cost of fuel, lack of foreign exchange, implementation of Exclusive Economic Zone Law, pollution of coastal waters etc. Aquaculture researchers have been able to bridge this gaps substantially.

It's my great pleasure to compile this most important publication. The book contains several most important papers on Aquaculture either presented in FAO or other international seminars or published as a technical paper. These should be read in conjunction with other unpublished work and also with the three manuals of the Professor Chaudhuri to be published soon. All these must constitute a comprehensive review of work in the development and analysis of potentials and problems in the field of food production. Most of the presentations are in Chaudhuri's words.

To some countries aquaculture is something of a novelty. A group of people view aquaculture as a way to make further contributions to the supply of animal protein food, while others consider it as a means to raise high-priced aquatic species for profit.

It should be deciphered from this presentations that aquaculture means many diverse things ranging from modest improvement in the environment in which aquatic animals live, through extensive methods of fish farming to highly sophisticated systems of intensive culture. A great number of species of plants and animals are involved in the study. (Modified from F.A.O. Aquaculture Conference, 1976).

J. Sia *et al.*, mention (vol. v No. 4 Oct-Dec, 2004, in "Aquaculture" beyond 2000–going global) that "over the past three decades, aquaculture has developed to become the fastest growing food producing sector in the world. Aquaculture has expanded, diversified, intensified and technologically advanced and its contribution to aquatic food production has increased significally".

This editor however, thinks that Professor Chaudhuri vowed in mid fifties or even earlier, much before the "Green Revolution" in Agriculture struck a success, to initiate a "Blue Revolution" or a "Protein Revolution" to create a vast protein resource on a sustained basis to feed the millions of poor people suffering from malnutrition and protein difficiency. His success in Induced Breeding and introduction of other techniques in aquaculture development testify to his practical and realistic approach to applied research and humanitarian services. As such, it should be proper to say that over the past five decades aquaculture has developed to become the fastest growing food producing sector.

J. Sia observes–"Aquaculture is highly diverse and consists of a broad spectrum of systems, practices and operations, ranging from backyard small, household pond system to large-scale, highly intensive products. A large proportion of aquaculture production comes from small-scale producers in developing and low income food deficit countries and the sector contributes to food security, poverty alleviation and social well-being in many countries contributions of aquaculture to trade, both local and international have increased over the past decades, and its share in the generation of income and employment for economic development has increased in many countries."

In fact Professor Chaudhuri clearly indicated in many of his deliberations that in the event of growing human population aquaculture has to play a major role than agriculture. Most of the papers Prof. Chaudhuri presented in the treatise clearly show various facets of development of aquaculture.

Aquaculture has made enormous progress in India, largely driven by markets and specifies innovations in last 50 years but in this process externalities, such as the environment, feed quality and management and social impacts could not taken care of.

Multiple technological innovations discovered in 1950s and 1960s made revolutionary changes in fisheries.

Indian Fisheries scientists established in the past an enviable record in the development of improved fish production technologies and their transfer to end users.

The success is due to the following :

1. Induced breeding of fishes.
2. Composite fish culture.
3. Polyculture.
4. Marine shrimp farming.
5. Fresh water prawn farming.

Aquaculture has registered as the fastest growing food production sector in the world to contribute to fast production. This sector creates job, revitalise communities suffering from the collapse of traditional fisheries stocks, utilizes advanced technologies to resolve natural resource confluicts, increases fish export and enhances domestic production of fin fish and shell fish. FAO Resources Chief (Fishing Chimes, Feb, 2007)

Serge Garcia, Director, FAO Fishery Resources Division said, "Fishing as we understand it today is a lot more complicated than it used to be". He said that complete fishery systems were affected not only by fishing but, also, by a wide range of simingly unrelated human activities, shifts in global consumption trends, climate change and more. As a consequence, good management was not just about dealing exclusively with targeted fish species any more, according to him.

He gives examples of factors that also need to be assessed, including :

1. Over fishing of one popular species leading to the collapse of others further up the food chain.
2. Pesticide run off from farms creating dead zones where fish cannot survives.
3. Turtles and Dolphins accidentally entangling themselves in nets and dying, as do fish not intentionally targeted for capture.
4. Erosion.
5. Dumping of ballast water by cargo ship etc.

Serge Garcia further says, "There are no easy answers to this complexity of interactions but we were to look at fisheries from a broader perspective and strive to create fisheries management plans that look at the multiple effects of fishing and how other human activities and the environmental effect fish stocks." Experts call this ecosystem approach to fisheries management.

Hope this treatise will meet some of these gaps.

The readers must keep in mind that Professor Chaudhuri is in no way involved in the compilations of this treatise as he merely permitted, though reluctantly, to edit his publications. So the editor shoulders entire responsibility if there be errors, mistakes and misrepresentation of facts in this treatise. Considering brevity of the size of the book many charts, graphs and diagrams have been excluded; yet most of the presentations are in his words.

A.B. Chaudhuri

131, N.S.C. Bose Road,
Block–10, Flat–4,
Kolkata–700 040

Contents

Introduction

Professor Hiralal Chaudhuri the doyen among the aquaculture, biologists of the world achieved monumental success in his research venture all through his service tenure. Chapter-I records some of the successes and the felicitations he received from various Luminaries of the world.

The professor did not ever think to collate all his works together and publish in a book form. The people interested in Aquaculture, Limnology and Ichthyology were also oblivious of the useful work done by him in absence of a consolidated work.

On such a background the vision and ingenuity of M/s Daya Publishing House in requesting this editor their sincere desire to publish monumental finding of professor Hiralal Chaudhuri not only testifies a constructive step, but also a remarkable gesture to make public aware of his successes in fishery research which ensures aquaculture to lead a premiere role than agriculture in food production of the country.

But unfortunately, most of professor's reprints, reports and unpublished works were already donated to the library of the University of the Philippines, and therefore, only handful documents available for compilation.

However, concerted effort of this editor resulted in retrieval of a few documents and two compilations had been contemplated which were:

1. Manuals of Aquaculture, Limnology and Ichthyology, and the second one
2. Aquaculture: Beyond 2000–New Horizons (documentation of technical papers).

The subject matter of this treatise no (2) highlights the following as may be found in the contents some of which are:

Induced Breeding, Fish Seed Production, Introduction of exotics, Fresh water aquaculture, Some selected consultancy projects, Modern Aquaculture, Fish culture–Management technique, Biotechnology to enhance Aquaculture production, Short deliberations of Prof. Chaudhuri, Some contributions on miscelleneous success and several others.

These are only a few amongst Professor's extensive work. The success could be evaluated from the observations made by eminent scientists some extracts of whose letters have been presented in chapter–1.

R.D. Chakraborty, the renowned aquaculturist who worked with professor says–"His published papers record his language and as such selected ones on induced breeding are last words on the subject."

In the field of aquaculture I was privileged to have worked with him in Cuttack when the zenith in the field was reached. The two papers on high fish production based on work done by him and guided throughout are:

1. Record fish production with intensive culture of Indian and exotic carps (current since May, 20, 1974 by Chaudhuri *et al.*) and,
2. A new high in fish production in India with record yields by composite fish culture in fresh water ponds (Aquaculture 6 (1976) by Chaudhuri *et al.*).

To me these present the high standards achieved in freshwater aquaculture in India.

Dr. Chakraborty considers the consultancy report on Burma was an outstanding contribution. He further says,–"Dr. Chaudhuri, the guide and teacher, stands apart and these qualities based on his great knowledge of his subject and executed has made him a truly great teacher. I feel his extension work of demonstrating and teaching aquaculture has contributed largely to the spread of it in our part of the world and he remains the Guru par excellence."

The depth of success the professor attained in the first few years of his research has been acknowledged by several top aquaculturist and fishery biologists of the world (refer Chapter 1); further success achieved during his service tenure may be read from Dr. P. K. Samanta's life sketch of the professor.

All these, though very briefly mentioned, present professor's research successes but only few of his work could be exposed in this treatise. The editor sincerely hopes to document further on the professor's work in his next treatise.

A.B. Chaudhuri

131, N.S.C. Bose Road,
Block–10, Flat–4,
Kolkata–700 040

Chapter 1

A Prelude

Why Honour Professor Hiralal Chaudhuri: An Observation Relating "Induced Breeding"

This treatise presents only a fraction of Prof. Chaudhuri's technical publications. Most of his research documents, reports etc. were donated to the libraries of the University of the Philippines and International Centre for Limnology and Aquaculture Centre and could not be considered for presentation in this treatise.

In order to appreciate the depth of the professor's vast knowledge and experience over the subject the editor does not find any citation better and appropriate than Dr. P. K. Samanta's record made in vol. 24, No. 10, January, 2005 issue of "Fishing Chimes". As such his entire write up is reproduced below with little modification:

"Prof. Chaudhuri's scientific research work in Aquatic sciences is multi-dimensional, with a focus on fish farming and fish breeding. His achievements in these spheres have been acclaimed the world over. He is a teacher of rare qualities and he taught and trained hundreds of students. His dedicated and pioneering work generated the technology of induced breeding of fishes which totally transformed the course of Indian fish culture sector and showered prosperity among fish farmers and on other related stake holders. The unique and economically potent sequel of his achievement brought spontaneous laurels from various scientific bodies and academies both at home and abroad. These are many but some of them are mentioned hereunder.

Prof. Hiralal Chaudhuri is a towering technocrat who has inspired and beneficially influenced the careers of many including mine in a remarkable way. The profound impact he made on fisheries scientists and farmers of India of South and South-East Asia and other parts of the globe through his lectures and training classes has been a major highlight of his professional career on which he can look back with pride. He has the rare distiction of emerging as a world class expert in fish breeding, fish seed production and fish farming. Prof. Chaudhuri is acclaimed as the foremost authority in the field of induced breeding of fishes like Indian Major Carps, Milk Fish, Mullet, Cat Fishes etc. His other unique contributions include Fish seed transportation technology intensive composite and mixed fish farming and hybridization of Indian Carps. These contributions revolutionized inland fish culture in India and in several other countries.

The results of his many outstanding studies and research works are vibrant in many research papers, reviews, technical manuals, monographs and other publications. His brilliant academic phase and professional activities were chronicled in many publications.

In the year 1943 itself Prof. Chaudhuri embarked on his professional career. After brief stints of work as lecturer in zoology in Sylhet and as Research Assistant at the Indian Veterinary Research Institute, Izatnagar in Uttar Pradesh, a potentially useful shift that gave a new direction to his career took place in 1948. He could secure a posting at the Central Inland Fisheries research Institute, Barrackpore, West Bengal in this year. This was a key development in his career, as there was no looking back after this.

After an exciting duration of investigation work on riverine fishries of Hooghly, Bhagirathi, Dhamodar and Ganga where he used to study spawn of major carps like Catla and also spawn of Hilsa from Nimta, Dhulian Ganga and Shabebganj river banks, he got the opportunity of diversifying into culture fishery research line, which was to lead him later to name and fame that have deservedly come to him. The foundation for his later successes was laid by the research work he did under Prof. Himadri Kr. Mukherjee, Dept. of Zoology, Science College University of Calcutta as part of his work at CIFRI. During this period he conducted an experiment on artificial fertilization of fish eggs through stripping of Goby (*Gobiopturus chuno*) pair. The scientific report on this, as presented by him, was published in the Proceedings of National Institute of Sciences in India (1951).

In 1949, Chaudhuri's scene of work temporarily shifted to the CIFRI's Sub-Station at Cuttack for the purpose of working out suitable remedial measures to stem the on going heavy morality of spawn of Rohu and Catla in Orissa State, at the behest of Gajendra Nath Mitra, the then The Director of Fisheries of the State. After necessary investigations, Prof. Chaudhuri successfully devised a technique to reduce the spawn mortality. Within a very short time he also devised a method of producing micro-plankton to serve as food of fish hatchlings. As a result, the survival of hatchlings increased to the extent of 90 per cent. This crowning success earned him a regular transfer to the Cuttack Sub-Station of CIFRI as Junior Research Officer. This outstanding contribution became part of a breakthrough he later achieved in culture fishery history of India. This was the technology of induced breeding of major carps.

In early part of 1954 under US TCM programme, the Govt. of India sent Chaudhuri on deputation to the USA under a scheme of Fish and Wild Life Service. The Chief of US Dept. of Interior arranged for the training of Prof. Chaudhuri under the world famous fishery scientist, Prof. H. S. Swingle in Auburn University. That single year's stay at this university was, according to him a memorable chapter in his life. Within a short period of 9 months, Chaudhuri could get his M.S. degree in fish management (in 1955), securing A grade certificate and he was awarded the membership of Gama-Sigma-Delta Society of Aquaculture for higher academic success and Research. Simultaneously he was awarded a Certificate of Merit by Fish and Wild Life Service, Dept. of Interior, US Govt., for successfully completing the training in Fishery Biology and Fish Culture. He was also awarded a Certificate of Merit by Foreign Operation Administration of Govt. of USA in recognition of his successful participation in the Technical Assistance Programme.

During the course of his long and illustrious service at the CIFRI, Cuttack, for over three decades from 1948 to 1975, Chaudhuri's outstanding contributions had been multidimensional, far reaching and revolutionary. These consisted of Induced Carp breeding, selective breeding and hybridization, Carp seed production, propagation and culture of exotic carps, intensive carp culture, composite culture and poluculture of Indian and Chinese Carps, integrated fish-crop livestock farming system. At the CIFRI's Sub-Station at Cuttack under the guidance of Dr. K. H. Alikunhi he succeeded initially in the induced breeding of a Carp minnow Esomus danricus by hormone injection. Then he succeeded in the breeding of riverine catfish by pituitary hormone injection. These successes paved the way for his acclaimed later success in the induced breeding of major carps.

Dr. G. N. Mitra, Director of Fisheries, Orissa, extended additional facilities for Chaudhuri's research at Angul Fish Farm of the State Fisheries Dept. On 10th July, 1957, Chaudhuri achieved a success in inducing the Asiatic Carp *Cirrhinus reba* in an aquarium through hypophysation and within 18 hours spawning took place. Thus the first success was achieved at Angul Fish Farm by Prof. Chaudhuri with support from Alikunhi.

This was not all. His main achievement, however, was his outstanding success in inducing Rohu and Mrigal to breed tough administration of extract of fish pituitary gland. The news of Prof. Chaudhuri's success spread all over India as well as abroad in no time. Thus 10th July 1957, the day on which the breakthrough was achieved, became a Red Letter Day for Indian farmers and this success opened up other avenues such as selective breeding of carps and their hybridization. The technology of induced breeding of carps, gifted by Prof. Chaudhuri, has now taken roots in the country. Much later, after four decades and over, at the initiative of Dr. Nita Chaudhury, the then Union Joint Secretary (Fisheries), Mr. Nitish Kumar the then Minister of Agriculture, Govt. of India gave recognition to the unique significance of 10th July in the history of culture fishery development of India by declaring it as the date for celebration of the Special Fish Farmers Day, which is a great tribute to the achievement of Hiralal Chaudhuri.

On other laurel that came to him later was his success in 1958 in the hybridizing of carps by crossing major and minor carps of the general *Labeo cirrhinus* and *Catla*. At this important stage of his career, he had to shift his activities to Calcutta to join as Fisheries Extensions Officer at Barrackpore. Chaudhuri, however, could go back to CIFE's Centre at Cuttack soon after and resume his work.

In 1961, Chaudhuri obtained his Ph.D from the University of Calcutta. Later in 1962, the National Institute of Science in India awarded 'Chandra Kala Hora Memorial Gold Medal' to Prof.

Chaudhuri mainly for succeeding for the first time in the world in the breeding of Asiatic carps and developing viable technique for large scale production of their seeds. Chaudhuri also received the prestigious 'Rafi Ahmed Kidwai Memorial Gold Medal' in recongnition of his valuable contributions to the developments of Aquaculture in India.

In the year, 1964, Hiralal Chaudhuri took charge of the Cuttack Centre of CIFRI. In the same year, Food and Agricultural Organisation (FAO) offered an appointment in Taiwan to him as Fishery Adviser. In 1968 he was invited to the USSR to deliver a lecture on Genetic Selection and Hybridization of cultivated fishes. From 1967 to 1970 he was on deputation to Burma (now Myanmar) as Fishery Advisor and from 1971 to 1975 he acted as Senior Scientist and Officer in Charge of CIFRI and Head of Fish Culture Division. This time he succeeded in producing fish at 10 t/ha/yr through polyculture of Indian and Chinese Carps.

In 1972, a request came from FAO, to function as Aquaculture Specialist for a period of 3 years in Bogota, Columbia. Prof. Chaudhuri was not released on time by India Govt. for which the consultancy went to other country. In 1974, again at the request of FAO, Chaudhuri succeeded in Milk Fish Breeding in Indonesia. Thereafter, as appointed by the International Development Research Centre (IRDC) he worked as Fishery Adviser at Research Institute in Madri, Malaysia.

During his stay at Cuttack Research Station, the instructional film on Induced Fish Breeding produced by the Films Division of the Govt. of India under Prof. Chaudhuri's technical guidance received National Honour and also the award for the best film at FAO sponsored Symposium on Aquaculture in Kyoto, Japan in 1976. It also received the award of Silver Bear in Germany and the President's Gold Medal in India.

In his research career, he came in contact with Dr. H. Srinivasa Rao who also helped him immensely to carry out his Research. Dr. Rao left the mortal world in 1973.

At Pacific Science Congress held at Vancouver in 1975, Dr. Kuronuma, an eminent Fishery Scientist of Japan welcomed Chaudhuri as *Father of Induced Breeding* and since then this epithet became an accepted fact. Chaudhuri's work received focal attention of the biologists all over the world. He as graded as an outstanding researcher and top class scientist as early as late 1950s. Hiralal Chaudhuri took voluntary retirement from Governmental service in February, 1976.

After retirement, Prof. Chaudhuri travelled widely as an FAO expert through Myanmar, Philippines, Malaysia, Laos, Fiji and Sudan. He advised the local fisheries authorities on various professional aspects of fish culture (1976-1993). In fact, Chaudhuri was offered high value consultancies/assignments by the FAO, IRDC, ESCAP, World Bank, SEARCA and SEAFDEC (South East Asian Fisheries Development Centre), Philippines to work on the development of Fisheries and Aquaculture and to train personnel in a number of developing countries.

Chaudhuri served as FAO/UNDP Fisheries Expert for about 10 years in Burma and Laos PDR. He acted as the Senior Technical Adviser and Visiting Scientist under SEAFDEC, Philippines from 1976-1979 and also as Chief Technical Adviser in Vietnam and Laos from 1979 to 1984. He made significant contributions in the Breeding of Milk Fish (Artificially fertilizing stripped milk fish eggs with stripped melt of the fish and studying the development and rearing upto post larval stage) besides training a host of visiting Fisheries Officials from different countries. In 1988, he joined University of Philippines at Los Banos as visiting Professor and he was in charge of the Institute of Biological sciences. After hectic career of 50 years of dedicated professional work, Chaudhuri decided to relax. He return back to Calcutta where he lives now.

During the period from 1994-2002 he was the recipient of various awards in India, in California and Las Vegas, USA. In 1994, a unique recognition of his professional stature came by way of institution of a chair in fisheries in his name at the University of Calcutta.

The Inland Fishery Society of India conferred its Fellowship on Prof. Chaudhuri, Association of Aquaculturists, CIFA, Kausalyaganga, Bhubaneswar, Orissa, felicitated Prof. Chaudhuri, on 27th August, 1996. The World Aquaculture Society awarded its "Life membership" to Prof. Chaudhuri in February 1998 in recognition of his life devotion to aquaculture research and training. In Sep 2000, the Zoological Society, Calcutta bestowed the Honorary National Fellowship of the Society for Advancement of Aquaculture and members of Fisheries Fraternity, Bangalore, felicitated Prof. Chaudhuri. In December, 2000, Fishing Chimes Jaysree Charitable

Trust instituted and annual Best Fish Farmer Gold Medal Award in his name in 1999. On 10th July 2001 he was presented with a citation by the Dept. of Fisheries, Aquaculture, Aquatic Research and Fishing Harbours, Govt. of West Bengal. The Central Institute of Fisheries Education, Mumbai, has set up a Foundation in the name of Prof. Chaudhuri on 15th July, 2000, to promote activities in the field of Fisheries, Aquaculture and allied sciences. On 21 September, 2001, the Department of Zoology, Calcutta University, has also established "Dr. Hiralal Chaudhuri Academy" for the development Fisheries and Aquaculture.

Prof. Chaudhuri is of the firm view that aquaculture development around the globe has reached a stage of making significant contribution to the supply of fish protein to millions of people around the World suffering malnutrition. He is hopeful that with further intensive research it will be possible to obtain the desired level of production of animal protein by composite and integrated fish culture and polyculture for the benefit of mankind". [Fishing Chimes Vol. 24 No. 10, January, 2005]

This journal "Fishing Chimes" also records the professor as follows:

- ☆ The major carp culture sector owes its present flourishing status to Hiralal Chaudhuri who functioned at CIFRI's erstwhile Centre at Cuttack under the guidance of K. H. Alikunhi. He ushered in the system of induced breeding of major carps through administration of carp pituitary extract into the body of brooder major carps and also later standardised the system.
- ☆ It was on 10th July, 1957, the first success in induced breeding was achieved at Angul in Orissa by Chaudhuri with support from Alikunhi. The spawn that was produced out of this experiment was successfully raised into fry and later into fingerlings. Thus, 10th July, 1957 is a red letter day of Indian fish farmers. In fact, 10th July every year should be a day of celebration by all fisheries departments of the country and fish farmers. This unique success has generated an unprecedented fish culture activity over the entire country leading to prosperity among fish farmers. In course of time, this success opened up other avenues such as selective breeding of carps and their hybridisation.

☆ The technology of induced breeding has now taken roots. The first ever workshop on the dissemination of the technology of induced breeding was accomplished through a FAO/UNDP training programme in induced breeding of carps held at CIFRI shortly after the standardisation of the technology became part of the working practices of all in the line.

☆ Chaudhuri does not claim that the total credit for the work accomplished goes to him. He says that the use of pituitary hormons as an additional stimulus to induce fishes to breed was attempted as early as 1930. What was accomplished by him at CIFRI'S centre was the result of a further quantum of devoted inputs taking the cue from the past work. He is extremely happy that in the course of 40 years after the first success, the induced breeding technique has been improved and refined in India as well as in quite a few other countries of the world. It gives him immense joy and satisfaction that other hypophyseal hormones, HCG, neuro-hormones and quite a few other hormones came into vogue for the same purpose. The latest developments by way of introduction of products based on peptization marketed under the trade name such as Ovaprim, Ovatide, Ovapel etc. gives him happiness that his work has eventually peaked into such a high level of development that it has become so convenient to fish breeders."

Rare International Honour for Dr. Hiralal Chaudhuri, Indian Fishery Scientist

☆ Dr. Hiralal Chaudhuri, the eminent Fisheries Scientist, was recently honoured in U.S.A for his outstanding contributions in the field of finfish aquaculture. In September, 1994 a colloquium was held at the Bodega Bay Marine Laboratory in California, U.S.A., on the 'Applications of Endocrinology to Pacific Rim Aquaculture'. It was organised by the Universities of California and Tokyo with particular efforts made Dr. Howard Bern, world renowned endocrinologist. Thirty two distinguished endocrinologists/aquaculturists of the Pacific Rim countries were invited to participate and present papers. The plaque presented to Dr. Chaudhuri read 'Colloquium of the Bodega Marine Laboratory of the University of California, Davis, on Applications of Endocrinology to Pacific Rim Aquaculture' is pleased to honour Professor Hiralal Chaudhuri by dedicating its proceedings to him in recognition of his pioneering efforts in the use of reproductive hormones in finfish aquaculture and for training of generations of students in India, Laos and the Philippines. We commend him to his sustained international contributions over five decades and for his talent in applying the result of basic research to the solution of practical problems'. The papers presented at this seminar are to be published in a special volume of the renowned journal 'Aquaculture' and dedicated to Dr. Chaudhuri.

☆ This well-deserved rare honour to Dr. Chaudhuri is the crowning glory of an outstanding scientist whose contributions have made possible the planned freshwater aquaculture of today in respect of induced breeding of carps by the administration of fish pituitary hormones, the most important scientific methodology developed in the field of culture fisheries. This pioneering research work done by him in the fifties made possible to obtain seed of good quality instead of the earlier method of collecting mixed seed (desirable and undesirable fish species together) from natural sources. His work made possible the raising of unmixed fish seed in farm ponds and subsequent stocking of waters with kind of fish seed desired.

☆ Dr. Chaudhuri did not rest with the development of this scientific method of breeding fish but carried on to take this to its logical conclusion *i.e.,* optimising production from fish ponds. He developed high yielding carp culture techniques in the late sixties, the first outstanding results obtained in Burma while serving as a FAO/UNDP Inland Fishery Biologist; he achieved high production (10 tonnes/ha) of carps in fish ponds. Later, on return to India he demonstrated the practicability of achieving high fish yields in ponds by carp polyculture (composite fish culture) through applied research work in Cuttack at the

Central Inland Fisheries Research Institute, the same Institute where he had developed the induced fish breeding technique.

He served the FAO/UNDP as a fisheries expert for about 10 years working in Burma and Lao PDR. He was the Regional Aquaculture co-ordinator of the South Eas Asian Fisheries Development Centre (SEAFDEC), Philippines, as also its Senior Technical Adviser and visiting scientist for about 7 years and made significant contributions towards the success in breeding of the milkfish, *Chamos,* besides training a host of visiting fisheries officials from different countries.

A unique appreciation came his way in 1994 when the University of Calcutta, his alma mater, instituted a chair in Fisheries in Chaudhuri's name.

Inland aquaculture owes a permanent debt to Dr. Hiralal Chaudhuri, for making possible through his painstaking brilliant efforts, production of quality fish seed and how best to culture them through scientific farming to obtain high yields of food fish. (Source: Fishing Chimes, May, 1995)

Induced Breeding

Felicitations for Monumental Successes

It remains a paradox–not known why–while dead silence prevailed among the intelligentsia and beauracrats in India about the monumental success in Induced Breeding and in other associated fields brought by the scientists of Central Fisheries, while many top biologists of the world showered accolades and felicitations on Professor Chaudhuri and his associates; but for their recognition the Protein revolution in India would have remained unknown.

Glimpses of Views of a Few World Experts

On Professor Chaudhuri's Success

☆ "The best student ever to enroll in Fisher in the Institute."
(Dr. Swingle, Alabama Institute, USA, 1957)

☆ "Congratulation on your success in induced Spawning."
(Dr. Stilman Wright, USA Dept. of Interior 195

☆ "You are making Fish Cultural History; You are making an important contribution to economics."

(Dr. J. W. Atz, U.S.A., 1958)

☆ "If you succeeded in Induced breeding of Milk Fish it might be a big revolution in Biology."

(Dr. S. Sarig, Israel, 1958)

☆ "You furthered my education very much and I offer my congratulations on your success."

(Dr. J.G. Crawford, IBRD, Adviser, Governor World Bank, 1974)

☆ "Your intensive fish culture has been monumental. These works remain as glorious as a Golden Age in the annals of Fishery research."

(Dr. C. V. Kulkarni, India 1976)

☆ "Congratulations on your appointment in SEAFDEC".

(Dr. G. R. Fish, Newzealand, 1976)

☆ "It is not surprise to me that you should have been invited to join SEAFDEC and I warmly wish you well in that new responsibility–much as you will be missed in India."

(Dr. J. G. Crawford, Chairman, Technical Advisory Council Research)

☆ "I warmly welcome your appointment in SEAFDEC. India will miss you."

(Sir John Crawford, Chairman, Technical Advisory Council, Aquaculture, 1976)

☆ "Chaudhuri, if you succeed in induced breeding of Carps, it will be comparable/similar to getting the Noble Prize."

(Dr. B. R. Bhatia, Deputy Fisheries Adviser, Govt. of India)

☆ "Your appointment in SEAFDEC is a cause of considerable pleasure, to others in this country as well as myself."

(Dr. G. R. Fish, Fishery Research Lab. Newzealand)

☆ "Your services can be provided for coordinating the stimulated development of aquaculture"... "Your duties with SEAFDEC" may help catalize on harmonious development activities in South-East Asia."

(Dr. W.H.L. Allsop, Canadian Director, Aquaculture, 1976)

☆ "If anybody could accomplish this difficult task (of induced breeding of Milk Fish) Dr. Chaudhuri would be one."

(Dr. D. D. Moss, USA Aquaculturist)

☆ "I think you are doing wonders with inadequate equipment you have to work with, and you set an example for us."

(Dr. C. F. Hickling, Director, Trop. Fish Culture Institute, Malacca & Eminent Fish Culturist Research Institute)

☆ "Your letter of December, 1957 was very exciting to me. For several year since my visit to the Far East I am waiting for the day when Scientists from that part of the world will succeed to induced spawning of some of the important of that region.

It seems to me that if you succeeded with Indian carps then, why not follow with other species like Chinese carps, Mullets and *Chanos chanos*? This might be a big revolution Biology and fish culture as well. So, I wish you much success and progress. Meanwhile I am waiting for your publications on these experiments."

Sincerely yours,

Sd/- S. Sarig

Editor: 'BAMIDGEH'

(Israel Journal of Fisheries and Aquaculture)

Nir-David, D.N. ISRAEL. December, 27, 1957

[N.B.: *Dr. Shumuel Sarig is one of the few TOP Fishery Scientists of Israel and has earned World-wide recognition as an Aquaculturist and Fish Pathologist as well an Editor.*]

☆ ....I want to thank you for all your kindness and help to me during my visit. And that goes for your colleagues as well I think you are doing wonders with inadequate equipment you have to work with, and you set an example for us."

Yours sincerely,

Sd/- C. F. Hickling
Eminent Fish Culturist of U.K.
British Fish Culture Specialist
Director Tropical Fish Culture Research Institute
Batu Berendum, Malacca, Malaysia.
(London, August 29, 1965)

☆ I wanted to tell you how much we enjoyed our visit to your fisheries research station. It was an eye opener to us all, amply confirmed when we repeated our visit in West Bengal.

As Chairman of the Technical Advisory Council I was very interested because of the attention we have been giving to aquaculture and also as a Governor of IDRC I was doubly interested. You furthered my education very much and I offer my warm congratulations on your success.

With good wishes.

Yours sincerely

Sd/- J. G. Crawford
Agriculture Adviser,
World International Bank for Reconstruction and Development
November 25, 1974

☆ I wish to congratulate you on your new position with the Southeast Asian Fisheries Development Centre. I am sure you will find this a challenging experience, and I wish you every success.

Sd/- Lauren R. Donaldson
World Famous Fishery Scientist, Professor Emeritus of Fisheries
University of Washington,
May 20, 1976

[N.B.: *Prof. Donaldson became world famous because of his work on Pacific Salmon. He was awarded Hatai Medal for his contribution to science in the Pacific in 1975.*]

☆ I was very pleased to hear from you and I would like to extend my congratulations to you on your new appointment in SEAFDEC in the Philippines.

Your appointment is cause of considerable pleasure to others in this country as well as myself............

☆ I have sent a copy of the full report that our Freshwater Sciences Committee made to the Pacific Science Association to you under separate cover. I am most grateful to you and to the other members of the Committee for the help I received during our term of office.

Sd/- Dr. G.R. Fish
Fisheries Research Laboratory,
P.O. Nagapuna, Rotoura, New Zealand
May 6, 1976

[N.B.: *Dr. Fish was one of the topmost Fisheries Scientists of New Zealand. He was Chairman of the Freshwater Sciences committee, Pacific Science Association of which Dr. Chaudhuri was a member.*]

☆ I was shocked to learn from your letter that you had opted voluntary retirement. Your association with important researchers in Inland Fisheries and especially your contribution in respect of induced breeding of carps and also intensive fish culture has undoubtedly been quite monumental. These works will remain as glorious as a Golden Page in the annals in the Inland Fisheries research in India. You were very rightly awarded the covetous Rafi Ahmed Memorial Prize. I one again congratulate on that meritorious achievement. Nevertheless, you have retired at a time when our country needed you most. I do not know whether there are chances of rethinking. However, I trust that you will continue the work in your beloved field of Inland Fisheries. I earnestly wish you furthei laurels and happy long life.

Sd/- Dr. C. V. Kulkarni
Indian Fisheries Scientist and Administrator
Director of Fisheries of Maharashtra (Retired),
Bombay, India
March 16, 1976

[N.B.: *Dr. Kulkarni is renowned Fisheries Scientist and an administrator in India. He retired as the Director of Fisheries, Bombay and held many high positions. This letter was written after hearing from Dr. Chaudhuri about his voluntary retirement from ICAR Service.*]

☆ I was delighted to hear of your receipt of the Rafi Ahmed Kidwai Memorial Prize, which you thoroughly deserve. I continue to sing your praises of the very fine work done by you and your colleagues in developing aquaculture.

It is no surprise to me that you should have been invited to join the Southeast Asian Fisheries Develpment Centre and I warmly wish you well in that new responsibility–much as you will be missed in India.

I am sure our paths will cross again and you will be interested to know that TAC be again taking up the question of research in Aquaculture at its October meeting this year."

Sd/- Sir John G. Crawford

[N.B.: *As Chairman of the Technical Advisory Council (TAC) Crawford advocated priority for Aquaculture Research*]

☆ The intention of SEAFDEC together with the Southeast Asian Centre for Graduate Training and Research (SEARCA) is that through their joint auspices, and the partial funding from the Milkfish Project (IDRC), your services can be provided for coordinating the stimulated development of aquaculture programs in the countries affiliated to SEAFDEC and SEARCA.

May I indicate a few comments in regard to how we anticipate your duties with SEAFDEC may help catalyze on aquaculture program. It is anticipated that your long and harmonious association with aquaculture research and development activities in Southeast Asia (including, I believe, Burma, Fiji, Malaysia, Indonesia, Bangladesh besides your home country of India), will enable you to assistance of SEAFDEC and SEARCA.............

You have already earned the respects and compliments of many projects in different countries, now you have a bigger fish pond to swim in.

Yours sincerely

Sd/- W.H.L. Allsopp

Dr. W.H.L. Allsopp
Canadian Aquaculturist,
Head: IDRC Fisheries
Associate Director (Fisheries)
International Development Research Centre
Otawa, Canada
March 29, 1976

[N.B.: *The above letter was written just after Dr. Chaudhuri's voluntary retirement, when the offer came to join the position as the Regional Aquaculture Coordinator. Although he was appointed by SEAFDEC, Dr. Chaudhuri was intended to monitor also the projects to SEARCA and IDRC Aquaculture Projects.*]

✩ We have been keeping up with your progress in breeding the milkfish. With some visitors the other day, Dr. Shell* made a comment that if anybody could accomplish this difficult task that one Dr. Chaudhuri would be the one. Please do keep us informed of your milkfish particularly when you have been successful in not only producing the larvae..... In reference to publishing your work on the milk fish I would suggest that you contact two journals:

(1) Aquaculture (2) Fishfarming International

Sd/- D. D. Moss
Dr. Donovan D. Moss
Famous U.S. Aquaculturist
Assistant Director, International Centre for Aquaculture
Department of Fisheries and Allied Aquaculture,
Auburn University, Auburn, AL 36849-5419
June 30, 1977

[N.B.: * *Dr. E.W. Shell is the Prof. and Head, Department of Fisheries and Allied Aquacultures and Director, International Centre for Aquaculture.*

The paper "Observations on artificial fertilization of eggs and the embryonic and larval development of milkfish Chanos chanos (Forskal)" by H. Chaudhuri et al. was published in 'AQUACULTURE', 13 (1978), 95-113, (Printed in Netherlands)]

✩ It is very nice that the eggs of Chanos which were collected undoubtedly under confinement have been grown to post larval stage. This fact itself, the world record in strict sense, is a significant step towards the goal. I am very glad to find this record established in the Philippines before Hawaii and Taiwan.

.... As to publication of your valuable contribution in early date, I strongly suggest it done by SEAFDEC............. SEAFDEC is an International Organization whose contribution should not be kept under national level.

Sexual dimorphism of *Chanos* through your hand and Biology was made clear.

Sd/- Katsuzo Kuronuma
Dr. Kuronuma,
Eminent Ichthyologist and Fisheries Scientist of Japan
June 29, 1977

[N.B.: *The letter was written after Dr. Chaudhuri's success in artificially fertilizing milkfish (Chanos) eggs caught from the sea and studying the development and rearing upto the postlarval stage. He published the paper in the Journal 'Aquaculture' from Netherlands. He had also succeeded in determining the sexes of milkfish (Identifying charactors).*]

Another letter for Dr. Kuronuma:

First of all please accept my sincere appreciation for your everlasting efforts contributing to the development of Asian countries in many ways.

Sd/- Katzsuzo Kuronuma

[N.B.: *Dr. Katsuzo Kuronuma, one of the most eminent Ichthyologists in the world, was the President of Tokyo University of Fisheries and President of the Ichthyological Society of Japan. He was also a renowned fisheries personnel in Japan. He welcomed Dr. Chaudhuri as the "Father of Induced Breeding" when Chaudhuri first met him at the Pacific Science Congress held at Vancouver in 1975. Dr. Kuronuma expired a few months back.*]

☆ Dr. Chaudhuri's letter Mr. J. Kamp

[N.B.: *Mr. Jan Kamp was the Resident Representative, United Nations Development Programme (UNDP) at Vientiane, LAO, PDR.*]

"My project was the most successful and could demonstrate a production of 3.5 to 4 tons/ha/yr by integrated farming and polyculture of various species by almost without any or with very little inputs, Lao PDR is a landlocked country (Least Development Country) and was not able to import costly fertilizers or fish feeds. Thus the integrated fish-crop-livestock farming systems demonstrated near Vientiane was recommended to be a model fish farm to be practised throughout the country.

FAO, UNDP and the Government of Lao recommended for my extension for one more year beyond the retirement age of 62 years".

Excerpt from the Memo. from the FAO/UNDP [LAO PDR] Fisheries Department, FAO, Rome May 4, 1982

Mr. Kamp stopped over at FAO, HQ. in conjunction with his home leave travel.

☆ In this brief account of the general situation of the country, Mr. Kamp pointed out various difficulties on the part of the Government in the implementation of the projects; in particular there are shortage of trained counterpart personnel, lack of fuel, lack of funds and inadequate transport and communication facilities.

He rated the Fisheries Project (LAO/78/014) as one of the most successful project in spite of the slow progress caused by the above-mentioned difficulties.

.... Please accept my warm congratulations on your success in winning the awards for your films on Aquaculture.

Sd/- J. G. Crawford

[N.B.: *The film "INDUCED BREEDING" was awarded (6th September, 1976) First Prize among 72 films exhibited at the FAO World Conference in Aquaculture at Kyoto, Japan in 1976. The film was produced by the Information Ministry, Govt. of India under the technical supervision of Dr. Chaudhuri. Except the photography and music, Dr. Chaudhuri had other contributions including writing of the dialogue and editing. Another film equally good, was also produced along with the above film on "Composite Fish Culture" under direct supervision from Institute at Cuttack*]

Another letter from Sir John Crawford

August 11, 1978

☆ .... I am sorry you were unable to take up the offer from the World Bank, but I am sure there will be other opportunities.

I remain interested and whenever you can report some major progress in the field of aquaculture do please let me know.

Sd/- J. G. Crawford

[*At this time his address was: Research School of Pacific Studies, Australian National University*]

☆ Congratulations to you on your successful experiments in inducing spawning in carps by means of hormone injection. I trust that you will be able to transfer the results of your laboratory work to the practical application on carps throughout the country, and thus solve the problem presented by failure to spawn naturally.

I take this opportunitity to wish you a Happy New Year in 1958, and continued success in your work. You will be interested to know that I had a recent conversation with Dr. Swingle and that he was extremely complementary of your work at the Alabama Polytechnic Institute. He told me that you were one of the best students ever to enroll in fishery work at the Institute. This statement is gratifying to me as I am sure it would be to you.

Sincerely yours,

Sd/- Dr. Stillman Wright,
Cheif Office of Foreign Activities
United States Department of Interior Office of Foreign Activities,
Fish and Wildlife Service
Washington 25, D.C.
December 27, 1957

☆ First of all, I want to congratulate you on your success in inducing freshwater food-fishes of India to spawn in captivity. You are certainly making fish cultural history, and in view of the tremendous importance of fish have in animal providing animal protein food, the item in man's diet most critically in demand, you are also making an important contribution to economics.

I should be very happy if Dr. Pickford and I could speak of your success in our forthcoming article in ENDEAVOUR. As you undoubtedly know, this journal has a very wide distribution and I am sure it would make your fine work known to many people who might otherwise remain ignorant of it................

Sincerely yours,
James W. Atz
New York Zoological Society
The Aquarium: Zoological Park
July 23, 1958

☆ From 'ENDEAVOUR'

By J. W. Atz and G. E. Pickford

'THE USE OF PITUITARY HORMONES IN FISH CULTURE'

July, 1959

Promising development have recently occurred in India. The principal problem in Indian fish culture has been obtaining seed fish regularly... H. Chaudhuri of Central Fisheries Cuttack, has now been able to induced spawning in six species of carps...." (Dr. Pickford and Dr. Atz are world famous fish Physiologists. They are the authors of the famous book "Physiology of Pituitary Gland of Fishes".

Chapter 2

Modern Aquaculture: Professor's Vision and Modern Concept

It's Contribution to Human Nutrition

Aquaculture plays a very important role in food production. World population is increasing at an exponential rate and consequently the demand for food is also increasing. The green revolution in Agriculture has contributed substantially to solve the food problems, but the nutritional standard of the vast majority of people of the developing countries of the world is still much below the average level. There is acute shortage of animal protein in the diet of those poverty-stricken people. The situation would become worse as according to the prediction of U.N. demographers (1986), the world population is expected to rise from 4 billions to 6 billions by the end of the century, which again would exceed 8 billions by 2025 A.D. To feed these additional mouths there would be hardly any arable land left for the expansion of agricultural areas. Hence, in the future, mankind would have to depend more and more on the oceans for food, thus shifting their reliance on the main source of food from AGRICULTURE to AQUACULTURE*.

---

* Aquaculture is the aquatic counterpart of agriculture–an all embracing term including culture of any aquatic organism (Simple definition). The most common aquatic organisms cultured are the finfishes, shell fishes (Crustaceans: shrimps, prawns, crabs etc. Molluscans: oyesters, mussels, clams, cockles etc.) and seaweed and unicellular algae.

Fish is highly nutritious food and a very important and cheap source of high quality protein. Over 60 per cent people of the developing countries of the world derive 40 per cent of even as high as 70-75 per cent of animal protein intake in their diet from fish. Till the middle of the present century, oceans and seas were the main source of fish supply for human consumption. The belief that the resources of seas are inexhaustible has been proved to be a myth as the marine fish catch instead of increasing with the increasing demands of the increased population, dropped down. The decline in fish yield was mainly due to over-exploitation and several other factors, such as, high cost of fuel, lack of foreign exchange, implementation of Exclusive Economic Zone law, pollution of coastal waters etc.

Aquaculture has been identified as one of the major strategies to increase world fish production to supplement depleted production from capture fisheries. Till the end of second world war, aquaculture production was quite low since it was mostly dependent on the traditional type of fish culture practices. The average production of fish was 60-100 kg/ha/yr to the maximum of about 500-800 kg/ha/yr. But with the modernization of aquaculture techniques during the last few decades, global production through aquaculture have incresaed significantly from meager 2 per cent of the total world production in 1960's to about 15 per cent in 1980's and it is expected that the contribution of aquaculture would incresae to about one-third of total global production by the turn of the century. During 1970's when the marine production was about 50-60 million MT (metric ton) per year, the share of aquaculture was only about 10-15 per cent. It has been estimated that while capture of fish from the seas even with improved technologies and greater efficiency in catching effort, it can only reach a maximum of about double the present level and would be around 100-120 m. MT in one year. On the contrary, however, aquaculture has the potential to increase the production 10 to 12 times or up to 5005 m. MT by making all round technological advancements and innovations of most recent scientific approaches.

In this lecture, attempts would be to give an idea as to how the evolution of aquaculture systems has taken place to increase fish yield from meager 500 kg to as high 50-200 MT/ha/yr or even upto 2000-4000 MT/ha/yr that could be achieved. Modern aquaculture, therefore, is expected to play a major role in supplementing fish protein in the diet of the protein-deficient poorer masses of the Third world countries.

A gradual incease in the production of fish through aquaculture was primarily effected through the implementation of the following four programmes: (Four 'I's')

1. Increasing area of culture: (expansion of culture area; reclamation of derelict water areas / swamps; construction of modern fish farms and hatcheries)
2. Improvement of Technology: (from traditional to scientific fish farming)
3. Intensification of culture: (from extensive to sem-intensive and intensive type of culture)
4. Introduction of Non-conventional methods: (fishpen/cage culture; running water culture; raceways; integrated fish-crop-livestock farming systems).

Introduction of improved technology at the various phases of aquaculture management including pond fertilization, stock manipulation, supplementary feeding and control of fish parasites and diseases and others resulted in enhancement of growth rate, higher survival and much higher yield of fish. Modernization of the method of transport of fish spawn, fry, fingerlings and adult fishes have reduced mortality to the minimum during and also immediately after transport.

The greatest breakthrough in aquaculture, however, was achieved when success was attained in inducing the cultivated fish species to breed in captivity. One of the main constraints in the development and expansion of the aquaculture industry had been the acute shortage of seed of the commercially important cultured species. Majority of these group of fishes either do not attain full maturity or fail to reproduce in captivity when reared in ponds or in other enclosed waters. These fishes include the difficult-to-spawn Asiatic carps, the popular euryhaline species, namely, the mullets, milkfish, seabass and others, the catadromous eels and majority of the marine commercial fishes. The only source of their seed was their natural habitats where they naturally breed at will, for obtaining adequate quantities of their seed for culture, was felt for decades. Finally, the monumental success was first achieved in mid-fifties' with the Indian carps and thereafter in subsequent decades majority of the important species could be successfully induced-breed. Fully matured fishes were induced to spawn by injection of fish pituitary hormones and the technique was developed to produce millions of quality seed of these valuable fishes for culture. The success has helped expansion of culture of these species to remote areas. Also, studies of genetic selection and hybridization could be initiated in these species for developing strains superior to the present species. Availability of abundant seed has also made intensification of their culture possible for obtaining very high yields by intensive and super-intensive culture. Further research on induced breeding programme has led to some improvement as well as sophistication of the technique. Besides piscine pituitary hormones, other hormones such as, HCG, synthetic LH-RH and its analogues, salmon gonadotrophin-releasing hormone (sGnRH-A) combined with domperidone (a dopamine antagonist) having the trade name 'Ovaprim' are being used in the induced breeding programme for seed production. In addition, steroids, thyroid hormone and also non-hormonal substances have also been experimented with.

Polyculture or mixed-species culture is another improved method of aquaculture for enhancing fish production. In polyculture, a number of quick-growing fish species of different feeding habits are cultured together so that there is very little competition for food among the different species. Their proper combinations are able to utilize all available resources in a pond to produce fish. By this management technique the yield of fish could be enhanced from around 1-2 MT/ha/yr to 10-20 MT/ha/yr. Also by introducing suitable exotic fishes of superior qualities, yield of fish could be enhanced considerably.

Introduction of African Nile tilapia (*Oreochromis niloticus*) in recent years to several countries, resulted in the yields of 600-800 MT/ha/yr by development of fast growing strains by genetic research and intensive farming technology. A high-yielding red hybrid strain is considered as the aquatic chicken. It is now feasible to project an annual world aquaculture production of 2 million MT by tilapia alone. Mono-sex culture of tilapia (only males) is yet another improved technology (Sex-reversal by feeding male sex-hormones to fry) to obtain very high production of the species.

Of the non-conventional methods introduced for enhancing aquaculture production, a few are mentioned below:–The modernization of the age-old S.E Asian cage and pen culture practices has been responsbile for extremely high production of fresh, brackish and marine fishes. Japan produces million tons of yellowtail and seabream in marine cages every year. Running water fish culture characterized by abundant supply of oxygen-rich water, effective elimination of metabolities, optimum water temperature, intensive stocking and feeding ad libitum, is capable of giving unbelievably high rate of yields. Running water culture of rainbow trout Idaho in USA and common carp in Japan have produced over 2000 MT and 4100 MT/ha/yr (estimated production) respectively. High yields of fishes are also obtained in silos, raceways and in recirculating systems.

Yet another high yielding aquaculture system is the integrated fish-crop-livestock farming systems where aquaculture is combined with crop and vegetable farming together with livestock raising. In this system there is mutual benefit among the different commodities, bye-products and wastes of one being utilized by another. Fish farming enjoys most of the benefits since the wastes and bye-products are utilized for enhancing production of fish thereby saving expenditures on fertilizers and feeds. The culture of Asiatic carps and tilapia help production of fishfood organisms by recycling of animal wastes and converting those to fishflesh. The production cost is very low and from 5-10 MT/ha/yr yield could be obtained by recycling animal wastes alone. This type of culture is encouraged for production of low-cost fish for the low-income section of the consumers.

Aquaculture of oyesters, mussels and seaweeds in the shallow areas of coastal waters mostly in the inter-tidal zone, are giving millions of tons of production. This 'Three dimensional aquaculture' doesn't need (fertilizers or feeds. Also, green algae such as, *Chlorella* and blue-green *Spirulina* (rich in (50-70 per cent) protein, vitamins and minerals) are produced at a mass scale as feeds in the hatcheries and in rearing ponds. These species are regarded as the potential food for mankind in distant future.

We have, so far given more importance for the production of low-cost fishes meant especially for the low-income group of consumers. But aquaculture is also an important source of earning foreign exchange as well as for providing good income and employment. In recent years, shrimp culture has become very popular all over the world. In many developing countries shrimps are produced mainly for export to USA, Japan and Europe for earning dollars. The tiger shrimp (*Penaeus monodon*) is the most important species for export purposes. Mostly intensive or even super-intensive culture is practiced to get an increased production of 10 to over 40 MT of marketable shrimps per ha per crop. The latest production figures showed that the world's shrimp farmers have produced 712,000 MT of shrimps in 1995 (valued at US$ 6000 million approx.) which is 27 per cent of the global shrimp production.

At the end, it may be safely concluded that aquaculture development has reached a stage that it is now capable of making a significant contribution to the supply of fish protein to millions of poor people of the world suffering from malnutrition due to deficiency of animal protein in their diets. Further intensive research on the biological, endocrinological, nutritional, environmental and socio-economical aspects of aquaculture supported by development of appropriate technical manpower, institutions, and infrastructures, combined with massive financial investments and suitable legislations, it will be possible to obtain unbelievable production of this animal protein for the benefit of mankind.

Recent Advances in Global Aquaculture

Aquaculture has been recognized in recent year, as one of the most important strategies to accelerate global fish production. Fish, as a group is the final result of a complex biological cycle in an aquatic ecosystem. Fish culture had been an age-old practice in Asia, especially, in China and India, but the average production had been generally low because of the traditional type of culture. After the second world war, the aquaculture technology was gradually improved and more and more scientific methods were introduced. As a result, the fish production was increased from a meager 500-600 kg/ha/yr to 2-4 mt/ha/yr. However, during the last few decades, all round development in the aquaculture industry had taken place. Consequently, the yield of fish has gone up considerably due to intensification and expansion of modern aquaculture throughout the world. Such remarkable achievements could be possible mainly because of the application of the latest technologies in the two very important aspects of aquaculture, namely, (A) Fish Seed Production and (B) Food-fish Production.

(A) Fish Seed Production

The availability of adequate quantity of quality seed of the cultivated fish species, is an essential pre-requisite for obtaining large-scale success in the culture of any fish. Aquaculture industry in the world suffered tremendously due to the extreme dearth of seed for culture, since the majority of the major cultivable fishes either fail to reproduce in capacity, or do not attain full maturity when reared in ponds or other enclosed waters. They generally breed in their natural habitats under optimum environmental conditions. The Asiatic carps (Indian and Chinese carps), cultured for centuries in ponds, belong to this category, as also the important euryhaline species such as, mullets, milkfish and sea-bass, the catadromous freshwater eels and majority of the commercially important marine fishes. For decades in the past, concerted efforts were made by fishery scientists to breed fishes under artificial conditions. The success was finally achieved in late mid 1950's when members of Asiatic carps were induced to breed by injection of fish pituitary hormones. A dependable technique was subsequently developed to produce millions of seeds of these economic varieties of Indian and Chinese carps and induced fish breeding became a routine procedure in the various countries of Asia and Europe to produce quality seed to meet the demands of seed of these food-fishes for culture. Following this success, in subsequent decades, it was possible to induce breed the popular grey mullets, milkfish, sea-bass, several catfishes and a few other species. More recently, a number of marine commercial species have been induced to spawn by administration of fish pituitary and other hormones. Hormones of mammalian origin, both natural and synthetic, as well as a few other hormones are progressively being used in recent years for inducing spawning in fishes. A few important hormones used in regular seed production programme, are mentioned below:

1. HCG (Human Chorionic Gonadotropin obtained from human pregnancy urine) alone or in combination with FPE (Fish Pituitary extract) gave varying success in breeding fishes and obtaining their seed.
2. Luteinizing Hormone-Releasing Hormone (LG-RH) and its analogues. Here attempts had been made to intervene a hypothalamic-hypophysis interface in the endocrinal chain instead of injecting exogenous gonadotropins to stimulate hypophysis directly.

3. The mammalian (mGnRH) and/or piscine (sGnRH) gonadotropin releasing hormone analogue combined with dopamine antagonists, is a futher improvement over the previous one.
4. Salmon gonadotropin-releasing hormone analogue (sGnRH-A) in combination with domperiodone, known as "Linpe Method" is largely used in China in the large-scale seed production of Chinese carps. 'Ovaprim' its trade name, prepared by Syndel Lab., Canada, is used for induced breeding in several countries.

However, the poor fish-farmers of majority of the developing countries of the world still prefer to use the whole pituitary gland extract mainly because of the lower cost, simpler technique and ready availability of the hormones.

The induced breeding technique has also opened a new line of research on the genetic selection and hybridization in these fishes for producing superior strains of fishes.

(B) Food-Fish Production

With the expansion of cultural areas and intensification of culture, the overall production of fish through aquaculture had increased considerably. Besides, by the introduction of scientific cultural techology including improved management practices, such as, stock manipulation, pond fertilization, supplementary feeding, disease control etc., the fish production could be increased from a meager 500-600 kg/ha/yr to about 2 to 4 mt/ha/yr.

Further, with the introduction of exotic species of superior qualities and faster growing Chinese bighead, silver carp, grass carp etc., an elaborate system of culture is being practiced, known as poly-culture or mixed culture of a number of fast growing compatible species of fishes having different feeding habits and different feeding zones. By poly-culture of a large number of species, it is possible that the different varieties of food that are produced and are available in the pond environment, are fully utilized and converted to fish flesh. By mixed-fish forming the production could be increased from 8-10 tons to as high as 20 mt/ha/yr.

Also, introduction of a faster growing species of tilapia (*Oreochromis niloticus*) in many countries of Asia and the Pacific, the fish production had been increased tremendously. Further by genetical research and intensive forming technology, a red hybrid strain has been produced which could give a production as high as from 600-800 mt/ha/yr.

The fish production could further be enhanced considerably by following Non-conventional Methods of Culture, such as (*a*) Cage and Pen Culture (*b*) Integrated fish-livestock farming systems and Recycling of Animal, and (*c*) Running water fish Culture.

1. *Cage and Pen Culture*: Cage and Pen culture, although an age-old practice in a few countries of South-east Asia, the modernization of cultural techniques have in recent years given high yields in fresh-brackishwater as well as through mariculture. In Japan, millions of metric tons of yellow-tail and sunbreams are produced every year in marine cages.
2. *Integrated fish-livestock-crop Farming Systems*: This is a very high yielding aquaculture system where aquaculture is combined with livestock raising and vegetable/crop farming. In this system there is mutual benefit among the different commodities and the wastes and bye-products of livestocks and crops thereby saving expenditures on feeds and fertilizers. Recycling of animal wastes is very very important since it converts the animal wastes to fish flesh by fertilizing the pond waters which otherwise would have been a nuisance and a health hazard. By simple recycling of wastes 5-10 mt/ha/yr fish can be produced without using inorganic fertilizers or supplementary feeding.
3. *Running Water Fish Culture*: The principle behind running water fish culture is that the flowing water constantly removes the nitrogenous and other metabolites from the culture waters, thereby eliminating these toxic products which when accumulated inhibit fish growth. Running water culture can give unbelievable production, such as, over 4000 mt/ha/yr (estimated) of common carps in Japan over 2000 mt/ha/yr of rainbow trouts in Idaho, USA. Such high production is possible because of intensive stocking, elimination of metabolites and feeding ad libitum in oxygen-rich flowing water.

To sum up, it may be confidently said that recent advances in aquaculture have shown that the industry has reached such a stage that is could provide a substantial portion of fish protein to the diet of protein-deficient poorer masses of the third world countries.

Fish Culture: Management Technique (Demonstration) in Aquaculture Programme

(A compilation from Prof. Chaudhuri's lectures and FAO/UNDP Programme)

Fish culture is practised traditionally in eastern India, where fish constitutes an essential part of the diet. However, growing fish for sale is largely confined to West Bengal.

Pisciculture is a vocation of communities considered backward socially and lacking in financial resources. This together with lack of management know-how and dearth of fish seed of commercial varieties are mainly responsible for the slow development of fish culture in India.

The Central Inland Fisheries Research Institute's Freshwater Fish Culture Division at Cuttack, through epoch-making research have developed effective technology of fish breeding and fish culture. The former makes possible production at will of seed of quality fish in large quantities. Average productions of over 8000 kg/ha/yr of marketable fish have been obtained by intensive fish culture in still-water dug-out ponds. Similarly well developed management measures have been developed in raising fish fry in nursery ponds and fingerlings from rearing ponds.

For effective dissemination of the advancements made in fish culture it is necessary that the methods developed are demonstrated, competently by agencies such as extension personnel, members of fish farmers development agencies trained at the Central Inland Fisheries Research Institute. The Institute through its schemes on demonstrating pisciculture in rural areas, the one launched with financial assistance from International Development Research Centre, Canada and the operational research projects, is already engaged in method demonstrations.

Method demonstrations to benefit Bank employees, members of fishermen Co-operative Societies, Village Panchayats and representatives of trade associated with fish culture are also necessary as is also the creating of awareness and appreciation of the advancement made by fisheries research, in the minds of the administrator.

The need for rapid development of fish culture in India, so richly endowed with resources, with the help of well tried and profitable methods evolved, should be brought to bear upon all concerned so that production of this nutritionally rich food, can be greatly augmented, wealth generated and employment created thereby ushering in prosperity.

Observations

Pisculture has age-old traditions in eastern India where homestead fish culture is fairly common. Fish is an essential part of the diet of majority of the population in this part of the country about 80 per cent people of West Bengal falling into this category. The home pond being stocked annually by carp fry purchased from sellers of this commodity and fishing by angling and during ceremonial occasions removing greater quantities of fish by netting through parties hired for the purpose are the type of culture practices, if it can at all be called so, more widely prevalent. Growing of fish in ponds for commercial sale of the produce is confined largely to West Bengal, where the management practices though empirical are better than those indulged in elsewhere. The so-called pond fish that reach the markets in these areas are quite often from beels and jheels (large natural impoundments) that are stocked naturally when during the rains the rivers get connected to these. Capture of the fish that grow entirely on the food resources of such impoundments, is effected by engaging fishing parties by the lessee.

Fish apart from being a nutritious and tasty item is one that is readily saleable. The fish production potential of our warm waters is high and the country is abounding in resources. Despite all these the quantity of fish produced in the country is far less than that required, with result that per capital consumption of fish in India is one of the lowest in the World.

Pisciculture as a vocation has for generations been carried out by communities considered backward socially. Economically too, except for a few prosperous fish farmers, majority of people engaged in this trade are poor and still largely dependent on financial aid and assistance in the form of fishing implements, to middlemen who quite often may not belong to their community. The progress of fish culture in India has due to various reasons been slow. Prominent among which is the lack of knowledge of proper management of fish ponds, paucity of fish seed of desirable varieties of fish, poor financial resources or support and last but not the least the awareness of fish producing capacity of our impounded waters.

Field Scenario

Inland water resources of India are vast and varied. Of these the cultivable freshwater resources are important, as the type of fish that can be cultured in these are of a kind that is liked by inhabitants of inland areas where sea-fish is fresh condition may not reach in a marketable stage due to inadequate refrigeration facilities.

The jheels, beels, ponds, irrigation tanks and reclaimable swamps together form a large water area and to this may be added the saline swamps, brackishwater impoundments etc. All these together constitute the vast inland resources which if properly utilised could augment fish production considerably.

Works by Central Fisheries: Various Facets

The Central Inland Fisheries Research Institute established by the Government of India has done pioneering work in the field of inland fisheries. Researchers carried out at the Institute's Freshwater Fish Culture Division at Cuttack in the fields of fish breeding and fish culture are epoch making and tends to bring freshwater aquaculture to the level of a worthwhile and viable industry.

The technique of hypophysation of the Indian major carps–excellent food fishes–evolved at the Cuttack Division, makes possible production of fish seed of quality fish in confined waters in contrast to traditional methods of collection of mixed (both desirable and undesirable varieties) fish seed from natural sources like rivers. The method has been so simplified that a farmer with the help of easily procurable inexpensive equipment can breed fish in his honds and raise pure seed. It is also possible to breed the same fish twice in the same breeding season, if properly cared. The fecundity of the indigenous carps is so high that even if a few fish can be bred millions of post-larvae or spawn can be produced.

Development of the method of preservation of the pituitary extract in glycerine and ampuling the same has also made possible easy storage of the injection material for future use. Transport of such ready-made extract can also be easily performed.

In the field of fish culture similar great strides have been made. Through experimental studies conducted at Cuttack it has been possible to get average production of as high as 8000 kg/ha/yr of marketable fish, by intensive fish culture in dug-out ponds without recirculation of water.

Of the essential preprequisites for successful fish culture is the raising of healthy fish fry (young fish measuring 25-40 mm). The management measures that are to be adopted in nursery ponds, where the carp hatchlings are stocked, have been well developed by research at Cuttack. The result is that raising a crop of over 6 million fry from a hectare water area in about a couple of week's time during the season, is distinctly possible. Proper pond preparation, stocking the same adequately to tap its productive potential, feeding the fry with artificial feed mixtures are some of the important management procedures adopted in this mono-culture of young fish fry.

Techniques that could ensure productions around 0.2 millions of healthy carp fingerlings of 10-15 cm size in about 3 months have been developed. It is these fingerlings that form the stocking material for culture of table-sized fish in large ponds. Weight-wise production of these can be over 3000 kg/ha/3 months.

Multi-species Culture

Multi-species culture techniques of raising together indigenous and exotic carps to table size has been developed in a decade of research at Cuttack to result in high fish yields. The management measures have been raised to the stage of a well defined technology which has come to be known as composite fish culture. Average productions of 8000 kg/ha/yr of marketable fish have been raised for a number of ponds. This is more than 10-12 times higher than the average fish yield in the country from freshwater ponds. Emphasis is on the need for foddering fish with available nutritious agricultural by-products to achieve high productions of fish of desirable health and size.

Lacunae and Suggestions

The pacc of inland fisheries development is slow despite the fact that our warm-waters have high fish productive potential.

Among other things as mentioned earlier is the factor that could be considered as responsible for a such state of affairs is the unawareness of the fish culturists, fish-farmers or fish-breeders of the tremendous advancements made in these branches by fisheries research.

The most effective method of dissemination of such usable information would be to let people see for themselves, the methods developed, at work. As is known, seeing is believing. The techniques adopted, the inputs required, the quantum of these used and their method of application, post-stocking care and management necessary in fish ponds are required to be known. Solving problems that might arise during the actual culture can all be witnessed and absorbed by the trainees fish culturists. Since based on the agro-climatic conditions variations are made in the management methods, such demonstration of well-tried and established methods to be effective for large areas of the country, needs to be adopted in different zones.

The flow of information in such demonstration is not necessarily one way as the trainee fish-culturists also could offer helpful suggestions, indicate locally available substitutes like in fish toxins and feeds or labour and cost saving devices or ways. Experience gained in different zones could be utilised to effect modificiations that would make possible better use of the technology, to suit local needs. The field experiences gained while applying the research developed advancements are invaluable as after all the measures are production oriented *i.e.* these are useful to the extent these can be translated under field conditions to produce more fish.

Agencies that may be entrusted with the undertaking of such demonstrations are those that are capable of effective demonstrations. Since for effective publicity the methodology should reach all corners of the country up to the village level it would perhaps be desirable to have the demonstrations carried out in selected villages rather than allowing the technology to trickle down to the village in the course of time, from the urban areas.

The competent demonstrators are the fisheries extension workers both of the Central Fisheries Department and State Fisheries Departments who have themselves been trained at CIFRI, the Institute that is conducting research achieve higher per hectare production of fish from impounded waters. As a matter of fact, a scheme has already been launched with financial assistance from IDRC, Canada, in demonstrating fish culture in selected rural areas in the eastern states of West Bengal and Orissa, aimed at development of pisiculture in rural areas. The CIFRI is the executing agency. With gradual training, the panchayat and block officials (village level workers) could take up such work in their own areas.

A Fish Farmer's Development Agency has recently been created in some of the States with Central financial assistance. Officials attached to the agency in different zones as also the farmer-member of the agency could receive instructions at CIFRI on recent fish culture and breeding techniques, whereafter such methods could be tried in their own areas.

Operational research projects for promotion of aquaculture in rural areas are also in essence method demonstration schemes for trading fish culturists and would-be fish farmers recently been started are manned by staff of CIFRI and as such can be considered as an extension activity of the Institute. The projects are being taken up in areas where agricultural farming is also undertaken, so that the farmers are employed all the year and have additional income through fish culture. Of course, with greater efforts fish farming could be the main source of income.

As advancements in the technology of fish breeding and fish breeding and fish culture are being made through applied fisheries research it would be necessary to have short term refresher courses for personnel entrusted with method demonstrations. The training programmes no doubt will have to organised by CIFRI being the premier Research Centre in the country. Such training should cover both theoretical lectures and practical demonstrations of work in progress and may have to be arranged during the proper season. The programmes should be field oriented (aimed at boosting the fish production) and as such should confine itself to matter directly concerning aquaculture, rather than a crowded programme including lectures on subject the trainee will have no opportunity to apply in the field. Short period participation by the trainee in the fish culture and breeding methods demonstrated would prove useful.

Among the reasons of the low level of development in aquaculture in India, despite the great strides made by research in this field is the lack of true fish culturists unlike the state of affairs in agriculture where well developed methods are used by professional farmers in their fields.

At the core of this appears to be lack of awareness of the tremendous potential of the waters under adequate management not only by the so called fish culturist as has been explained earlier but also lack of application by members of agencies that could aid in rapid development of this profitable venture, by men and finance. Without the help of the latter little progress can be made.

As such, there is as much a need to arrange method demonstrations for the benefit of Bank Employees, members of fishermen cooperative societies, village panchayats and also perhaps representative of trade to be associated with fish culture like businessmen dealing in piscicides, fish feeds inorganic fertilizers and weedicides etc. The responsibility of creating awareness and appreciation in the minds of administrative personnel rests with the fisheries officials of the State served by administrators which needs to be pursued in right earnest.

The fact that well tried, high yielding, highly profitable methods both in the field of fish breeding and fish culture has been evolved and that rapid fisheries development needs to be done with the aid of these, effective tests should be brought to bear upon members of the society at all levels. The administrators, the fish farmers, the loan-giving agencies working in concert by proper application of the fish culture methods displayed through method demonstrations can only accelerate the pace of the development fish culture, both in freshwaters and brackishwaters in the country. With this the production of the nutritionally rich food can be augmented providing gainful employment to all connected with the aquaculture industry. The wealth generated, in addition to the contribution of the produce (fish) towards filling up the protein gap, is a gain to the nation.

[N.B.: *This is a joint presentation by H. Chaudhuri and R. D. Chakraborty; paper presented at the 13th Pacific Science Congress, August 18-30, 1975*]

Aquaculture Extension: Training Programme

Technology Transfer Continuum

The discovery or invention, evolution and transfer of appropriate aquaculture technology must be viewed as a continuous process and not occurring in compartmentalized stages. Thus, all aquaculture research "travelling" along a technology transfer continuum.

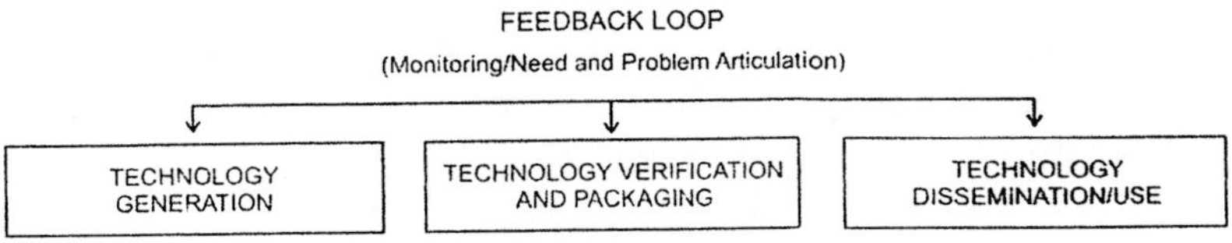

The Technology Transfer Continuum

The process has three interrelated components: technology generation, technology verification and packaging, and technology dissemination and use. As a process, technology and packaging involves a part of the research sector and a segment of the technology dissemination sector with the action happening in the users' milieu. Technology dissemination and use involves the information processors, extensions workers and the end users. A monitoring mechanism is necessarily built into the continuum to provide the forward and backward linkages between the various components.

Technology Generation

Effective aquaculture research has to be oriented toward developing feasible management and cultural practices. The projected production increases in the developing world are expected to be obtained by the application of known technlogies so that adaptive research, to enable transfer of technologies or the improvement of the techniques involved, will have the highest priority. Such research must be of a multi-disciplinary nature and systems-oriented because problems are closely the gaps in our knowledge of the systems have to be filled and problems in their application solved through team effort.

At the technology generation level, basis and applied studies will be conducted on the basis of existing aquaculture technology gaps in the region and the present and predicted problems of the industry. Technology generation will be primarily done at, but not cofined to the regional research center. Research priorities however will be based on an integrated regional aquaculture research and development program priority to be developed by the region's leading aquaculture planners and technical experts and revised periodically.

Technology Verification/Packaging

The prevailing situation in the world particularly among developing countries is the tendency to provide strong emphasis on technology generation and to bypass or conduct, in an unorganized manner, some aspects of technology verification. Research results, whether indigenous or borrowed, are often recommended for adoption without the benefit of verification as to their "workability" in a given country, production setting, or farming system.

A scheme for technology verification and packaging will be developed and installed in the regional research network. The technology verification mechanism will provide for inter-disciplinary teams of discipline and commodity-oriented researchers/subject matter specialists to conduct adaptive trials of specific aquaculture commodities taking into consideration the ecology, situation, location, and socio-economic conditions of a given aquaculture production setting. Production systems in a given area or country will be of primary consideration in these verification studies.

Results of technology verification will then be put together into a one-story package of recommendations for producers with provision for periodic updating of recommended techniques or systems. These are then fed into the extension and media pipelines.

Verification studies will take care of the great diversity of aquaculture species, production systems, ecological and socio-economic conditions prevailing in the Asian region.

Technology Dissemination/Utilization

A crucial stage in the technology transfer continuum is the stage when the producer make a decision on whether to use or reject the technology. While verification studies will considerably reduce the decision risks, it is still necessary to find out the factors that motivate farmers to adopt innovations. It is therefore important to integrate socio-economic studies in the research program. Communication studies will also be needed to develop effective ways of bringing technology packages to the fish farmer's doorstep and of reinforcing his decision, if he has to decide, to adopt. Various extension strategies and techniques could be formulated and implemented from these studies.

The nature of facilities and expertise required for undertaking inter-disciplinary research and technology verification and packaging schemes render it extremely difficult, at least for the present, to organize these on a national basis. On the other hand, centralizing these activities in one international center would be counterproductive because of the differences in species cultured and the need to study the various aspects in the set of environmental conditions in which the particular species occurs and those in which it has to be reared.

Close linkages then will have to be established between the regional and national centers so as to form effective networks of research institutions which could also organize training of core personnel for aquaculture development.

Assistance is urgently needed for the establishment and operation of such a network. Since studies are to be carried out by inter-disciplinary teams, it should be possible to barness the services of specialists in relevant disciplines rathar than aquaculture scientists per se if the overall guidance of such personnel can be assured. Also, aquaculture research calls for contribution from allied sciences, animal nutrition and feed technology, etc. Thus technical assistance from countries even not too advanced in aquaculture research could very well be utilized for regional research activities in developing regions.

It will be advantageous to combine regional research and training activities. If existing institutions and facilities are enlarged and strengthened, considerably savings can be effected.

Information

A free flow of information is vital to the development of aquaculture since the major increases in production in the near future are expected to be achieved through transfer of technology. Information flow is needed for research, training and extension, which form the major supporting services to be established. Although the types of information needed for each of these activities differ, a unified comprehensive gathering, processing and dissemination system is necessary to meet the needs of aquaculture development.

Three main stages have to be considered in an aquaculture information system–information collection, processing, and dissemination. The efficiency of information collection would depend on easy access to sources, particularly with respect to unconventional sources. By the very nature of the operation, it has to be carried out on a decentralized basis. A regional aquaculture research and development network could be utilized effectively for this purpose.

[Source: H. Chaudhuri on his UNDP/FAO/SEAFDEC–Training Programme]

Chapter 3

Induced Breeding and its Application to Aquaculture

Induced Breeding of Carps

Professor Chaudhuri who has been given the epithet "Father of Induced Breeding" has contributed many technical write ups in various journals. But only a few of such presentations from in this treatise were available to edit.

(A Few Papers Selected for Presentation)

Several technical papers of Professor Chaudhuri has been documented in this group. Of all these, the editor considers the following two papers summarises the salient features of "Induced Breeding" the papers are:

1. Induced Breeding of Carps by Hiralal Chaudhuri and S. B. Singh (ICAR Publication, 1984).
2. History of "Induced Breeding" on fish and its application to aquaculture. (Proc. Zoo. Sec. Kolkata, 1994) by H. Chaudhuri No. (*i*) above covers 82 pages, while No. (*ii*) covers 22 pages. Some selected portions of the publication No. (*i*) find place in this treatise while full text of the article No. (*ii*) has been presented.

Besides these two, the following have been presented in other chapters:

- ☆ Use of hormones in aquaculture practices.
- ☆ Efficacy of partially purified Salmon pituitary.
- ☆ Gonadotrophic in inducing spawning of Indian Carp.
- ☆ Preliminary observation on the use of Trichloroacetic acid for extraction of gonadotrophic hormone in induced spawning of carps.
- ☆ Induced breeding and hatchery management in carps.
- ☆ Induced spawning of Indian carps.

Induced Breeding Carp

Development of the Hypophysation Technique

The significant role of the pituitary gland in reproduction in animals was realized long back, but the application of this technique in inducing fish to breed is rather recent. In the beginning efforts to spawn fishes by administering mammalian hormones met with limited success. Houssay of Argentina is the pioneer in introducing the concept, in 1930, that fish pituitary can be used for spawning fishes. However, this concept was successfully applied for the first time in Brazil in 1934 by Von Inhering (1937) and his collaborators. The Russian succeeded in breeding sturgeons by 1937, (Gerbilskii, 1938).

In India, the first attempt to induce spawning of carps by administering mammalian pituitary extract was made in 1937 by Khan (1938) but without success, and unfortunately no further efforts were put in this direction till 1955. Interest in this field was reviewed in 1955 at the Cuttack Substation of the Central Inland Fisheries Research Institute when the carp minnow *Esomus danricus* and the catfish *Pseudeutropius atherinoides* were induced bred by pituitary hormone injection (Chaudhuri, 1960). By 1957 the difficult-to-spawn major Indian carps were successfully induced to spawn by injecting fish pituitary hormones (Chaudhuri and Alikunkhi, 1957). After this major break though, the breeding and hatching techniques were further improved at this Research Substation and in the following years these reached the takeoff stage at fisheries extension level (Chaudhuri, 1963). Later, by adopting the same technique, the difficult-to-spawn Chinese silver and grass carps, grown in ponds, were also successfully bred (Alikunhi *et al.* 1963), and large-scale breeding on commercial scale was done (Chaudhuri *et al.*, 1966; Singh *et al.*, 1970). With the development of this technique in India, the induced breeding method has spread all over the world and even replaced the age-old practice of collection of carp seed from natural habitats (the rivers).

Simultaneously with the development and propagation of the hypophysation technique for breeding the major Indian and Chinese carps to obtain quality seed, efforts were also made towards breeding other fishes, *viz.*, mullets, catfishes, and milkfish, by using fish pituitary hormones. Besides, efforts were being made to purify and isolate the fish pituitary hormones and to find out its substitutes such as synthetic and mammalian hormones, chemicals and steroids. (Chaudhuri and Singh)

Induced Breeding of Carps by Hypophysation

Stocking and Upkeep of Brood Fish

An adequate stock of good brood fish is a basic pre-requisite for a successful and large-scale induced fish-breeding operation by hypophysation. The brood fish may be grown in the farm or collected from outside sources and stocked in proper brood-fish ponds. Healthy ones should be selected to ensure production of healthy progenies. Transporting brood fishes from distant places for induced breeding involves difficulties such as transport cost, injury and subsequent mortality of brood fishes. Therefore it is advisable to raise them at the composite fish-seed farm itself. The size of the brood fish pond may be around 0.2 to 0.5 ha retaining a minimum of 1.2 to 1.5 m average depth of water during summer season. The brood fish ponds have to be suitably manured to produce natural food. The cultivated carps (major Indian and Chinese carps) may be stocked together at 1,500–2,000 kg/ha. The common carp, for which no hypophysation is necessary, may be grown separately in the farm itself as a donor fish for pituitary glands.

The brood fish stocked have to be adequately fed with supplementary feed (groundnut–cake + rich polish or wheat bran in 1:1 ratio) daily, at 1–2 per cent of the body weight to start with. For grass carp suitable water weeds or fodder glass are to be provided. The quantity of feed may be roughly estimated around 25-30 per cent of the body weight once or twice a week. However, it has to be adjusted according to the extend of utilization of the feed, health of the brood stock and the water quality to avoid pollution of water, or development of any thick algal bloom in the pond.

Opinions differ regarding the utility of supplementary feeding of brood fish. Under optimum feeding condition the fishes reach, proper maturity and produce large number of eggs. Surplus feeding may lead to deposition of fat in the body cavity affecting maturity adversely. Some of the research workers opine that brood stock for hypophysation should be selected from those reared on natural food alone. However, our experience at the Cuttack Research Station suggests that both natural food and proper supplementary feeding are necessary to raise a good stock of brood fish. The care of the brood stock is to start nearly 5-6 months earlier to the breeding season.

The selected brood fish should be examined periodically for their condition and for progress in maturity and freedom from parasitic infection. In case the condition of the brood fish is not satisfactory and they are emaciated, the stock should be thinned out or the rate of supplementary feeding increased suitably. Sometimes gonads fail to develop properly due to factors such as inadequacy of proper food, unfavorable ecological conditions or genetic defects. If even after ensuring adequate availability of proper food the gonadal development is unsatisfactory, it is advisable to collect brood fish from such localities where fishes have been maturing normally. Filling in the brood-fish ponds with fresh water, and some replacement, re-circulation and aeration of the standing water periodically, preferably during summer months, is highly desirable. In the event of any serious parasitic infection the brood stock has to be suitably treated or replaced. For diseases like fin-rot, caused by water moulds or bacteria, the fishes may be disinfected by giving them bath in a solution of 10 ppm of potassium permanganate for 10 to 15 minutes, or till the fishes start showing symptoms of distress.

In the selection of brood fish, both the age and size of the brood fish are important considerations. Carps 2-4 years old and weighing 2-5 kg are the best for hypophysation. Both the fecundity and ease in handling of fishes during induced breeding operation are optimum at the above age and size.

General care of brood fish during transport is equally important. Best results are obtained by transporting the brood fish individually in separate hand–nets (a bag-net with handle) or in bags made of old pieces of fry net and then put inside a fish carrier. Otherwise, they are likely to get badly injured. Use of special tanks for brood fish transport such as splashless tanks (Mammen, 1962) or other improvised ones (Patro, 1969) and that of an aesthetics have solved this problem satisfactorily. The brood fish after transport should be disinfected with 10 ppm solution of potassium permanganate or 1 ppm of 'acriflavin' before releasing in the ponds.

Characteristics of Ripe Brood Fish and their Selection of Hypophysation

The success of the entire induced breeding operation depends upon proper selection of the brood fish, particularly the females. Selection of males is usually not difficult, but the testing has to be done carefully. The fish should be held gently with the ventral side facing upwards. In the case of males when the abdomen near the vent region is pressed slightly, milt would ooze out easily but only the least loss of milt should be allowed during selection. Females with soft and bulging abdomen and slightly swollen and pinkish vent are considered to be good. However, it may be deceptive sometimes, particularly so in glass carp because of excessive feeding and in catla due to considerable deposition of abdominal fat within the body cavity. The use of a simple catheter, described earlier, is of considerable help in providing a 'field hint' in judging a good female. In a ripe female the eggs are slightly loose inside the ovary and so the catheter when inserted through the vent would get in smoothly, and on slight suction a sample of ovarian eggs is easily obtained. The eggs of uniform shape and size suggest that the female is quite ripe. In grass and silver carps the good ovarian eggs are mostly copper red and blue, respectively, though the colour varies in various ecological zones and also with age of the female. At Cuttack even pinkish eggs from mature yearlings of silver carp have yielded healthy spawn on fertilization.

For selection of brood fish for hypophysation, good females and males are kept inside separate testing hapas, but not too many fish together. The number of males selected is twice that of females, but a few more may be necessary particularly in the case of silver carp where the release of milt in males is sometimes insufficient.

Formation of a Breeding Set

Usually a breeding set consists of two males and one female. Weights of all the three fishes are recorded individually inside a hand–net by a suitable spring balance and the actual weight of the fish is determined by deducting the weight of the hand–net alone. This is necessary to calculate the dose of the pituitary hormone required for injection. One such set is usually introduced inside a breeding hapa after injection. Sometimes the release of milt from males, particularly in silver carp, may be inadequate. Therefore it is necessary to put in an additional male in the some hapa or to keep some extra selected males in a separate hapa which could be utilized at the time of stripping, if necessary. In male silver carp sometimes testes are either not fully formed or are deformed. These could be deformed due to the ecological condition of the habitat, resulting in poor yield of milt.

Collection and Preservation of Glands and Preparation of Pituitary Extracts

Particulars of Donor Fish

Species of Donor Fish

The carps themselves would serve as proper donor fishes. The common carp *(Cyprinus carpio)* would be most preferable because it is easy to raise them and is mature for most part of the year on the plains of India.

Freshness of Donor Fish

This is one of the most important considerations while collecting the pituitary glands for hypophysation from live or freshly killed donor fish. The Russian workers observed that glands from dead fishes kept at 1°–3°C for 24 hours are effective. At the Cuttack Research Station pituitary glands collected in Kolkota fish markets from well-preserved iced donor fishes were found fully potent for induced fish breeding work, though freshly killed fishes are always preferred.

Stage of Maturity of Donor Fish

This factor is very vital. Pituitary gland should be collected from donor fish which are mature or nearing full maturity. Pituitary gland is most potent in ripe gravid fish. In seasonal breeders, the genital complex attains maximal functional capacity for limited period only during the breeding season, and thereafter regresses to almost a state as in juveniles. The gonadotrophic potency of pituitary glands in such fishes increases with the advancement of maturity, reaches the maximum at the time of breeding and is lost after spawning. Therefore glands from fully ripe donors only are used and those from the immature fishes and those where regression of gonads has started, or which have already spawned earlier are discarded. However, glands from induced bred carps immediately after spawning are fully potent and are used for further hypophysation work.

The suitable period for collection of the glands is usually the breeding season of the donor fish, and this period is therefore longer for fishes which breed more than once in year, *viz.*, common carp, on our plains.

Collection of Glands

Pituitary gland can be removed from fish either after severing the head from the body or keeping the fish intact. In the latter case, a portion of cranium roof is chopped off to expose the brain. The fatty substances over and around the brain are removed and cleaned with cotton or washed with water. At first the olfactory nerves are detached and a portion of the ophthalmic nerves are separated interiorly. The brain is then lifted up and laid back. The pituitary gland is usually located below the brain just posterior to the optic nerve crossing. In some fishes the gland remains attached to the brain and is therefore lifted along with it, whereas in the major Indian and Chinese carps the gland is invariably left behind on the floor of the brain box. The membrane (duramater) covering the gland is gently detached exposing the gland fully and carefully taken out intact without causing any damage to it. If damaged the potency gets reduced.

The glands can also be collected in fish markets from severed fish heads. To avoid any mutilation of the fish heads which would otherwise reduce its market value, the pituitary gland can easily be collected by making an opening in the region of the foramen magnum on the posterior side of the brain. Through this opening the brain is either pushed forward or is pulled out leaving behind the pituitary gland exposed which can be easily taken out in a bent needle. After some experience is gained, a worker can very easily collect glands from the fish heads without either causing any injury to the gland or mutilating the fish head.

A broad saw (head-operated or electric) or an electric drill is used to obtain a care from the head region containing the pituitary gland. In view of the simplicity of operation and feasibility the methods adopted in India are quite suitable for our work.

Preservation of Glands and Storage

The pituitary glands collected for hypophysation can be preserved and stored quite in advance of the fish-breeding season. Glands from major Indian and Chinese carps are a highly potential source of viable pituitary glands for the purpose. Besides they can be preserved and stored safely for at least one year under proper care without any perceptible loss of potency. All these have made it possible to draw plans to establish 'pituitary banks' comparable to the 'blood banks' to enable widespread application of the hypophysation technique at international levels.

Some of the commonly adopted methods for the preservation and storage of pituitary glands are preservation in alcohol, preservation in acetone and freezing. These are described in the following paragraphs.

Preservation in Absolute Alcohol

This method is most commonly used in India. The absolute alcohol acts as a dehydrating and defattening agent. In this method the glands are first removed into absolute alcohol immediately, later transferred to air-tiger phials and 24 hours later are transferred to other air-tiger phials containing fresh alcohol. These phials are stored inside dark air-tight bottles containing absolute alcohol at either room temperature inside a desiccators containing anhydrous calcium chloride in a cool place or preferably the desiccators itself is kept inside a refrigerator. It is important to mention that the pituitary hormones are readily soluble in water and not in pure alcohol; therefore the alcohol should be absolute. During storage it is advisable to change the alcohol once in 3 months or the phials maintained air-tight. The potency of the glands is lost at high temperatures and on boiling. However, precise information is not yet available as regards the loss of potency of gland at high temperatures. The experience gathered at the Cuttack Research Station has shown that the glands stored at room temperature, where the higher limits have been recorded around 40°C in the course of a day during summer months, do not lose the potency. This information is of considerable significance, because refrigeration facilities are as yet beyond the reach of a common fisherman in the remote villages where the technique is to be finally taken to.

This method is highly popular with the Brazilian pisciculturists too. The potency of the glands preserved in absolute alcohol is reported not lost for 2 years, and according to them the glands remain effective even up to 5 years.

Preservation in Acetone (Acetone-Dried)

This method largely practiced in the USSR and the USA, and is highly satisfactory. The fish pituitary glands are preserved in acetone which is also a dehydrating and defattening agent, and are put inside screw capped phials for 36 hours inside a refrigerator. During this period acetone is changed 2-3 times. The phials are kept full with acetone so that there is no condensation of atmospheric moisture on the inside walls of the phials because this would result in diluting the acetone and thus causing some loss of potency of the glands. After 36 hours of storage, the phials are taken out and kept as such for sufficient time so that the acetone warms up to the room temperature. The glands are now removed from the phials and kept on a filter paper to dry at room temperature. These dried glands are stored inside air-tight sterile phials and kept inside a desiccators containing anhydrous calcium chloride which is placed inside a refrigerator.

Though this method is widely used, there is some doubt regarding the retention of hormonal potency in the acetone dried glands. Some workers reported that the dried glands lose potency after 6 months, some reported that there is no loss of potency even after 2 years of storage, and some reported that these glands are active up to 10 years.

Freezing of Glands

The pituitary glands soon after removal are stored inside a refrigerator, but this method is not commonly adopted.

Determination of Dosage

Determination of proper dosage in fish pituitary treatment is very important and various workers have tried their own methods of standardization of the same. However, the problems involved in standardization are of a very complex nature, because the dose of gonadotropin world depend upon a variety of factors including the size and stage of sexual maturity of the recipient fish, potency of the glands according to the time of its collection, freshness, stage of maturity of the donor fish, and preservation of the glands and also the ecological factors of the breeding environment at the time of taking up the induced fish breeding operation.

The Russian workers have, however, made some notable progress in this direction and have evolved a method of biological assay to express the potency of the pituitary extracts in terms of biological units. These researchers have used the European weather fish (*Misgurnus fossilis*), also known as Vy'un, as the test animal for biologically assaying fish gonadotropin. They have termed the unit as Vy'un Unit (V.U) and it is defined as "The quantity of gonadotropin which, after a single intramuscular injection, causes extra-seasonal maturation of the ovaries and ovulation in 40-50 hours at 18°-20° in 3 out of 5 female Vy'un weighing 35-45 g". Russian investigators have also calculated that a single common carp pituitary contains 10-15 V.U. during spawning season, and the most effective dose for inducing their carps to spawn at low temperature is around 10-15 V. U for the males and 25-30V. U. for the females. The Russian workers (Stronganov and Alpatov, 1951) proposed a "Frog Unit" and defined it as that quantity of hormone which causes the appearance of spermatozoa in the cloaca of 50 per cent male frogs or toads and exhibit the presence of mature spermatozoa in their sperm duct.

The Brazilians though using the hypophysation technique from 1934 have been considering an individual pituitary as the unit of measurement. They recommended repeated injections of small doses of pituitary gland in an increasing order. Spawning usually takes place after four successive injections at an interval of six hours each. To obtain successful results 2½ to 5 glands are required. Of course, a number of workers have suggested several 'fish units' provisionally in order to standardize the dosage.

In India, the Cuttack Research Station is the pioneer in the field of hypophysation of cultivatable carps. Initially an individual pituitary gland was used as an unit. Although considerable success was attained in inducing spawning in carps this way, yet the need to evolve a standardized dose for hormone injection was keenly felt, particularly because the size of pituitary glands varied considerably with the size and species of the donor fish. Expecting some comparable relationship between the size and hormone content of the pituitary glands from the ripe donors of various species, it was considered reasonable to relate the dose of the pituitary hormones with the weight of the gland and weight of the recipient fish. Hence, accurate determination of the weights of the pituitary gland is very important and the dose of the hormones is expressed as weight of the pituitary gland in mg/kg of body weight of the recipient fish (mg/kg body weight). This expression of dosage is a very convenient one and is an original approach in standardizing the hormone dose.

Das and Khan (1962) advocated the use of a 'Magur Unit' (M.G. Unit) following the Russian approach, and defined it as that amount of pituitary hormones which would induce ovulation in 50 per cent of mature magur (*Clarias batrachus*). One M. G. Unit, according to them, was equivalent to 1/10 to 1/15 of Indian major carp unit. However, this seemed to have no special advantage over the highly practicable method of expressing the dose as mg of pituitary gland/ kg body weight of the recipient fish.

The main problem in standardization of pituitary dosage for hypophysation, particularly in the difficult to breed major Indian and Chinese carps, is the difficulty in assessing the exact stage of maturity of the female fish by external examination and appearance. Therefore, in India, the dose of pituitary hormones for hypophysation of carps is determined according to the method evolved by the Cuttack Research Station.

Calculation of Doses and Dilution

Weighing of Glands

An uniform procedure is adopted for weighing the pituitary glands. The dehydrated and defattened glands are carefully taken out in a small petri-dish or water glass and picked up with a pair of forceps and placed on a filter paper. Now a clean glass phial of known weight with cork is taken and the gland is placed inside it and weighed on an electric balance or chemical balance to find out the weight of the gland alone. The total duration of drying on a filter paper *i.e.* the time after taking the gland out of alcohol and weighing it, is kept constant throughout to arrive at an uniform procedure.

From experience a fixed interval of 2 minutes has been considered to be adequate. After weighing, the individual glands are put back in absolute alcohol in separate, properly labeled phials and respective entries are made in a 'gland register'.

The above procedure is good enough for well-organized research and development–oriented large scale fish breeding programmes. To ensure easy and adequate availability of pituitary glands to the fishermen in remote areas and also for convenience of determining the doses of injection: (*i*) 'fish pituitary banks' should be established at national and international level, and (*ii*) fish pituitary extracts of known concentration, preserved in glycerine, could also be ampouled (the method of preservation and ampouling is given separately).

In India, the majority of fish culturists may not be able to afford an electrical or chemical balance to weigh the glands. For them the following simple guidelines should serve the purpose. Usually 1-3 whole glands from the ripe donor fish (common carp, rohu, catla, mrigal and silver carp) of nearly the same weight as that of recipient fish would yield successful results.

Earlier we give a general idea about the expression and determination of dosage of pituitary extract for hypophysation of carps. A more precise account dealing with it is given here separately for the major Indian and Chinese carps respectively.

1. *For Major Indian Carps*: The ripe female fish of *Catla catla, Labeo rohita and Cirrhinus mrigala* are injected twice at an interval of six hours during the conducive weather conditions. The first dose is of 2-3 mg/kg body weight and the second dose is around 5-8 mg/kg body weight of the recipient female. The male is administered 2-3 mg/kg body weight of the recipient fish at the time of second injection to the female, and even higher dose (4-5 mg/kg) may be given if the yield of milt is not quite free during testing.
2. *For Chinese Grass and Silver carps*: The above procedure is quite efficient for exotic carps too, excepting that the dose of hormone administered is usually higher than that for the Indian carps. The female is usually given the first dose

of 3-4 mg/kg body weight and the second dose of 7-10 mg/kg body weight of the recipient fish. The males are injected at the time of second injection to the female, the dose being 3-6 mg/kg body weight of the recipient fish, which is usually higher of the males of silver carp.

Apart from the above procedure, the females are also injected with a single 'knock out' dose, particularly during late in the season when the recipient fish are found to be losing condition. The 'knock out' doses are usually 10-14 mg/kg body weight for females of major Indian carps and 15-20 mg/kg body weight of the ripe females of grass and silver carps. During favorable weather conditions, the Chinese carp females are given sometimes three injections in a gradually increasing order at an interval of three hours, the total dose being only 10-14 mg/kg body weight of the recipient. The two doses are administered at an interval of six hours.

While the dose for a particular brood fishes is easily deterınined on the basis of its weight, the question as regards the extent of dilution of the hormone extract with water (distilled water) remains to be decided upon by the worker. No investigations have been carried out in India or abroad so far prescribing any particular optimum concentration of hormone or any definite volume of the extract to be injected to a fish of a given body weight. However, a concentration of 2-4 mg in 0.1 ml of the extract and a dilution of 0.2 ml/kg body weight of recipient fish have been found effective and recommended by the first author. The volume of the extract is mostly determined arbitrarily, but the size of the recipient fish, convenience of injection (not too little volume or too much) and calculation of the dosage are the important criteria. Through the above considerations are important, yet only minimum quantity of any extraneous matter (distilled water) should be injected into the fish to avoid any physiological reactions. From the practice followed at the Cuttack Research Station, the dilution of the extract is arbitrarily decided to range between 0.2 and 1.0 ml for a recipient fish depending upon its size. Usually 0.2 to 0.6 ml of extract is considered to be adequate for brood fishes ranging between 1 and 4 kg, and it could be increased up to 1.0 ml for fishes weighing up to 10 kg. For still larger brood fishes, though not considered good for hypophysation, the dilution is decided according to the quantity of glands to be used for the purpose, *e.g.* roughly 2.0 ml of extract for a 20 kg fish.

It may be useful to give a simple example below for the convenience of field workers.

Recipient Fish-Rohu

Sex	*Weight of Recipient*	*Dose (mg/kg)*	*Quantity of Gland*	*Volume of Extract*
Females	2 kg	1st injection 2	4 mg	0.4 ml
		2nd injection 6	12 mg	0.4 ml
Males	2 kg	3	6 mg	0.4 ml
	2 kg	3	6 mg	0.4 ml

In the above example, the first dose to the female would require only 4 mg of gland and it could be diluted to 0.4 ml for the convenience of preparation and injection of the extract, roughly trying to keep the dilution around 0.2 ml/kg body weight, as far as possible. At the time of the second injection to the female, the gland requirement is 12 mg, and for the two males which are also injected at the same time is 6 mg each. Hence, all the 24 mg of glands could be taken together for the preparation of the extract and diluted first 0.8 ml of water (concentration of the dose being 3 mg/0.1 ml of extract), of which 0.4 ml of extract containing 12 mg of gland is taken out and injected to the female, and the remaining 0.4 ml could be further diluted to 0.8 ml (gland concentration being 1.5 mg/0.1 ml of extract) and injected to the males.

While the above example gives some idea about determining the dose and dilution of pituitary extract for one breeding set, in actual practice when large number of sets are taken together hormone extracts of known strength for all the sets are prepared together. Extracts of known strength (known concentration) should be prepared in advance of the fish-breeding season, preserved in glycerin and stored either as such under refrigeration or preferably ampouled for large scale distribution. At the time of actual injection, the ampouled extract is again suitably diluted with distilled water.

In the light of the above, it is essential that the supply of pituitary material though the proposed 'Pituitary Banks' as glands or ampouled extracts is ensured for proper planning and propagation of the hypophysation technique all over the country.

It is always advisable to take a little more of the pituitary gland than required because some quantity of the extract may be lost while injecting a number of fishes.

Preparation of Pituitary Extract

The pituitary extract is prepared after the breeding sets are formed, weight of the recipient fishes are recorded and the dosages for injection are decided.

For making an extract, required quantity of pituitary glands are taken out of absolute alcohol and placed on a piece of filter paper and where the alcohol has evaporated the glands are put inside a tissue homogenizer of suitable size (2-50 ml capacity) and then macerated with a little distilled water or physiological salt solution*. The homogenized glands are then further diluted to make up the desired volume. Some workers prefer to use the physiological salt solution but in India only distilled water is being used with success.

The triturated (macerated) gland suspensions are injected either after or without centrifuging. The usual practice in intra peritoneal injection is to administer the entire gland suspension without centrifuging it. For intramuscular injection the gland suspension is mostly centrifuged, and the supernatant liquid alone is used for injection. According to the practice adopted in India, the gland tissue is discarded to avoid introduction of foreign matter which may cause some physiological reactions in the recipient fish. However, in other countries the entire gland suspensions are injected intramuscularly without centrifuging, and no obvious ill effects have been reported.

At the Cuttack Research Station, too, entire glands were crushed by mortar and pestle, and the suspensions without centrifuging have been used with success. This would be helpful in developing a simplified hypophysation technique for the fish farmers. The usual procedure followed at the Cuttack Research Station is still to inject the supernatant pituitary extracts after centrifugation by a hand operated or electric centrifuge. Weight of alcohol preserved pituitary glands is taken either singly or a few together and kept inside separate glass phials. These phials are serially labelled and the details of phial number, number and weight of glands, and donor species are entered into a register (Gland Register), so that at the time of preparing the extracts glands of known weight can be sorted out and wastage of the extract avoided. For centrifuging the gland extract it is convenient to use graduated centrifuge tubes, and after centrifugation the extract is showly decanted into a small 25-ml beaker from which it is easy to take it into the hypodermic syringe for injection.

Preservation of Fish Pituitary Extract

The procedure for preparing the glycerine extract is simple (Ibrahim and Chaudhuri, 1966). Glands of known weight are thoroughly macerated inside a homogeniser. The desired concentration of the extract is now decided and accordingly one third of the total volume is made up by adding distilled water in which the hormone extract is obtained after maceration. This solution is kept for 24 hours inside a refrigerator, and later glycerine is added to make up the remaining two thirds of the volume, the ratio of water to glycerine being 1:2. The hormone syspensionis again stored inside a refrigerator for another 24 hours, and afterwards it is taken out and centrifuged. The supernatant solution is filtered and stored in air-tight bottles inside a refrigerator or sealed in ampoules. The ampoules are properly labelled giving the details of hormone concentration and date of preparation. Usually a concentration of 40 mg of pituitary/ml of extract is convenient for use, though it can be altered to any desired concentration. Also the extract can be further diluted to obtain the exact dosage for injection. These have been found effective for one year and more, but it is advisable to keep them ready about a month or two in advance of the fish breeding season.

Methods of Injection

Intraperitoneal Injection

The injection is given in the peritoneal cavity of the fish. In fishes, especially in carps, there are soft regions at the base of both the pectoral and pelvic fins though which the injection needle is inserted quite easily. Usually such an injection is given at the base of the pelvic fin. The tip of the needle is inserted under a scale and is pushed though the abdominal wall into the body cavity and then directed parallel to the ventral surface. In this method there is the danger of damaging the internal organs and the injected fish may even die out of shock. This method is mostly adopted in the USA.

Intramuscular Injection

This method is very effective and convenient, and is popular in Brazil and India. The injection is given in the back muscles, caudal peduncle of the muscles in the humeral region. The Russians who were earlier using intracranial method have switched over to this method finding is more convenient and effective.

Intramuscular injection is ordinarily administered in the region of the caudal peduncle of the recipient fish. In case more than one injection is to be given, the pricks are made on either side (right and left) alternately. While injecting, the needle is first inserted under a scale parallel to the body of the fish and then the muscle is pierced through quickly at an angle of 45°; absolutely no damage is caused to any of the scales. In case the needle is inserted at an angle from the very beginning, it will be very difficult to pierce through the muscles because the needle would be obstructed by the scales arranged in imbricate fashion. A 2-ml hypodermic syringe with 0.1 ml graduations, preferably with a locking arrangement (B. D. Yale type), is convenient for use. The thickness and the length of the needle depend on the size of the recipient fish. Generally B. D. H. No. 22 needle is used for fishes weighing between 1 and 3 kg and no. 19 for the larger ones. For smaller fishes (much below 1 kg) No. 24 needle is considered adequate.

Injecting the Brood Fishes

The recipient fish are taken out from the breeding hapas one by one inside hand-nets and are gently put on a form rubber cushion of suitable size. While injecting, one person places his hands on the head and the other person holds the caudal peduncle with one hand, gently but firmly, and with the other hand the hormone is injected in the region of the caudal peduncle quite quickly. The whole operation is finished as quickly as possible. The injected fishes are soon put back inside the breeding hapas and the mouth of the hapa is securely closed to prevent escape of the injected brood fishes into the pond.

Breeding Environments

The induced carp breeding operation in India is mostly carried out in stagnant ponds which have received sufficient fresh rain water, do not have algal blooms of any sort, and also are without any crab, tilapia and common carp. In such ponds only breeding hapas are fixed. Flowing water is not essential for hypophysation of carps. Hapas can also be fixed in rivers if suitable locations are available. It may be worth mentioning here that in mainland China, Taiwan and also the USSR., the breeding environment is usually specially constructed channel, spawning tank or tub, maintaining a gentle flow of water (preferably clear, rich in oxygen and cool). This is of course desirable if facilities exist.

In India, the immediate environment for induced breeding is considered to be the breeding hapa, inside which the injected brood fishes of both the sexes are released to spawn. A description of breeding hapa is as follows:

It is a box shaped cloth container, like an inverted mosquito net with a cover. It has an opening on one side (width-wise) provided with cloth loops and button holes to enable proper closing. Through the loops a thin bamboo stick is passed to close the opening securely after releasing the injected brood fishes inside. The breeding hapa is fixed with the help of bamboo poles of suitable thickness and length. The cloth usually used is strong; fine meshed and mosquito net type, quite cheap and durable. Hapa would not suitable for hypophysation work if the cloth is thick. The mesh size of net cloth should be such that the ovarian eggs do not pass out of the breeding hapa when released by the injected female. The hapa should be tanned and washed thoroughly before use. Also proper maintenance of the hapa is very important.

The size of the breeding hapas varies: (1) 3.6 m x 1.8 m x 1.0 m for brood fishes weighing over 4 kg; (2) 2.4 m x 1.2 m x 1.0 m for brood fishes weighing between 1.5 and 4 kg; and (3) 1.8 m x 1.0 m x 1.0 m for brood fishes under 1.5 kg.

When fixed the bottom of the hapa should not touch the muddy pond bottom, and at least 15-20 cm of the hapa should remain above the water level. It should be fixed at a depth where it is easily accessible to the fishermen.

The brood fishes after a few hours of final injection spawn naturally inside the breeding hapa, mostly in the case of Indian carps and sometimes in grass and silver carps too.

Spawning and Fertilization

Ordinarily the major Indian carps spawn within 6-8 hours after the final injection to females and males. During extremely conducive weather conditions it could take place even much earlier, i. e. within 3-4 hours. Chinese carps, too, are ready for strippii.g in about 6-8 hours, and much earlier under favourable ecological conditions. On a few occasions when ecological conditions were ideal, the females of grass and silver carps were ready for stripping even after the first initial low dose and the eggs released were successfully fertilized by milt from uninjected males. The above experiences are valuable for the purpose of deciding the timings for injection. In case a single dose of injection is considered adequate, this could be done late in the evening, and spawning could be over by early hours of morning. In the even of two injections, which is usually the case, the first injection could be given around 4 pm and the second injection at 10 pm so that spawning is over by early morning, an ideal time for it. These timings are not quite suitable when stripping has to be done. In this case the first injection is given at about 6-7 pm and the second injection 6 hours later so that the fishes are ready for stripping in the morning hours. Such an injection schedule is ideal for the fish to spawn, but is somewhat tedious.

The injected fishes, particularly the major Indian carps, spawn naturally inside breeding hapa, and fertilized eggs are mostly obtained as the female and male spawning is synchronized. Natural spawning and fertilization of released eggs have been recorded in Chinese carps, sometimes in grass carp, and rarely in silver carp.

If the injected female does not spawn within 8 hours a third injection of higher dose may be administered, but it is usually not favoured. Such fishes are mostly not in good and ripe conditions, and even when these are forced to spawn under the influence of excess of hormone yield of fertilizable eggs is poor.

Natural Spawning After Injection

The ripe brood fishes of major Indian carps (rohu, catla, mrigal and calbasu) from ponds spawn naturally inside breeding hapas fixed in ponds on hypophysation. Often sex play between the female and male is noticed a few hours after the final injection of hormone extract, and mostly fertilized eggs are obtained inside the hapa. Good fertilized eggs exhibit blastodisc and following developmental stages. In bad eggs the development of blastodisc is absent or development is arrested after a certain stage, and the central yolk-mass may show gradual disintegration. The fertilized eggs soon start swelling up and look like small pearl-like bodies and gradually get water hardened. By examining these eggs total number of good (where normal development is noticed) and bad eggs and percentage of fertilization, which is the main index of success of the induced fish breeding operation, can be determined.

On a few occasions the normal development of the eggs gets arrested due to a variety of reasons, and hence the stage of development when the embryo is seen by the naked eye as elongated and crescent shaped ('C' stage) is considered to be a safe stage and gives a reliable index of the expected yield of hatchlings from the fertilized eggs.

Stripping and Artificial Fertilization of Eggs

In major Chinese carps usually striping is done. The female is examined quite regularly three to four hours after the final injection. When the female releases eggs very freely in is ready for stripping. The eggs are striped inside an enamel tray or basin, about 500 ml in each container and later sufficient milt is poured over it. The eggs and milt are thoroughly mixed with a feather and by rotating the contents by hand to ensure proper fertilization of the eggs. This is also called artificial fertilization. After thorough mixing for about 5 minutes, gradually some water is poured into the tray and the mixing of the contents is continued. After a few minutes more water is poured in and gradually the excess of milt, the blood clots and other extraneous matter and completely removed through washing. By this time, the swelling of the eggs starts and the eggs gradually get water hardened.

In India stripping is mostly done using the 'dry method'. In this case the body of the fish is gently wiped with a Turkish towel and the eggs (in free flowing condition) are striped into a dray enamel basin over which the milt is poured. Care is taken that no extra waterfalls into the basin. In the USSR, where usually the size of brood fishes is quite large (6-9 kg), the fishes may be anaesthetized before stripping. In a few countries, 'wet method' of stripping is also practiced. Physiological salt solution is used to receive the milt first and later the eggs are stripped into the basin. Usually the sperms of Chinese carps remain active upto 2-3 minutes in 0.3 per cent saline. They die much earlier in ordinary water. The 'dry method' of stripping is much more convenient and effective than the 'wet method'.

The failure in spawning or readiness of the fish for stripping may be due to a variety of factors which are not fully understood. On several occasions the eggs even when naturally released remain unfertilized or the eggs swell up; even they show some initial cleavages, but after some time further development is altogether stopped and the eggs turn opaque. This may be due to forced release of slightly unripe or over-ripe eggs in pre or post-prime maturity condition which do not develop properly. This could also happen as a sort of parthenogenetic development of eggs as observed by several Russian workers. Such eggs cannot be fertilized by sperms. On a few occasions the presence of unfertilized eggs may also be due to non-synchronization of spawning behaviour of the female and male in natural spawning and poor and inadequate quantity and or quality of the milt. In majority of the cases where negative results were obtained, the possible factors responsible for it could be the females not being in prime stage of maturity and also adverse weather conditions such as rise in water temperature much above the optimum range.

Conditions Favourable for Induced Spawning

Optimum hydrological and climatic conditions are essential for the natural spawning of Indian and Chinese carps. But how far these factors influence these fishes when induced to breed by pituitary hormone injections is not known. Fish breeding experiments definitely indicate the temperature of water is one of the very important factors in successful induced spawning. The injected fishes successfully spawn only when the temperature is within in certain optimum range. The optimum range of temperature for Indian and exotic carps seems to be between 24°C and 31°C. The other important factors are sudden and considerable dilution of pond water by rain. Natural spawning of fish is mostly associated with the above factors. In induced breeding, too, successful spawning is largely associated with dilution by rains or rain water, although there are instances when fish breed even without any rain.

Observations made on the physio-chemical conditions of water and climatic conditions during spawning of fishes induced by pituitary hormones revealed the following facts:

1. Breeding takes place both in clear and turbid waters.

2. Spawning occurs when the temperature of water ranges from 24° to 31°C, the most suitable temperature being around 27°C, on the plains of India.
3. Cool, rainy days are more conducive to spawning.
4. That fishes breed at a fairly wide range of pH and dissolved oxygen content of water.
5. Hot, sultry days do not give satisfactory results.
6. Accumulated fresh rain water promotes successful spawning.
7. Breeding takes place both during day and night.
8. Flowing river water gives better results especially on warmer day, as the river water is cooler than pond water.

The workers on Chinese carps elsewhere, *viz.*, mainland China, Taiwan and USSR, also lay emphasis on the above described features of the environment to obtain better results in hypophysation. Among the ecological factors conducive for induced breeding, water temperature, cool and oxygenated water supply are considered important. Providing gently flowing water supply are considered important. Providing gently flowing water in the breeding environment, which may be a specially constructed canal, tank or tub, is also useful. The pituitary hormones (endo-and exogenous) in association with conductive ecological conditions help the selected brood fishes reach full maturity quickly and in spawning.

Spawning is an innate urge and is a state of absolute co-ordination reached between the external and internal environments of fish through its pituitary.

Induced Spawning and Hybridization

In the nature hybridization occurs among fishes. This usually takes place between the closely related species of the same genus and also among related genera. With the development of hypophysation technique an altogether new field has been opened up with tremendous potential and possibilities. Hybridization can now be done in a planned and controlled manner in order to evolve improved fish stocks with desirable qualities. The female and male injected fishes belonging to different species or genera can be stripped to obtain hybrid progenies. The intraspecific crosses between various varieties or strains are usually considered as selective breeding. Both fertile and sterile hybrids have been obtained, and this variation has been taken full advantage of in fish culture and general fisheries development. The sterile hybrids are very valuable for certain specific purposes in fish culture where fertility may not be a desirable feature as regards regulation of their population. The fertile hybrids, on the other hand, are valuable in obtaining new hybridogenic forms with improved qualities in sorting out, combining and selecting various desirable characteristics though successive generations and backcrossing, etc. Through hybridization and selective breeding fishes of improved qualities have been produced, particularly so in common crop, tilapia and sturgeons, but such studies in the case of difficult-to-breed major Indian and Chinese carps are comparatively recent and meager. In the present account greater emphasis has been placed on these major craps.

Hybridization Among Major Indian Carps

Outstanding success has been achieved in hybridizing the Indian craps, and interspecific and intergeneric hybrids have been produced (Chaudhuri, 1960, 1971).

Interspecific Hybrids

	Male Parent		*Female Parent*	*Hybrid*
(i)	Labeo rohita	X	Labeo calbasu	Rohu-calbasu
(ii)	Labeo calbasu	X	Labeo rohita	Calbasu-rohu
(iii)	Labeo bata	X	Labeo rohita	Bata-rohu
(iv)	Labeo bata	X	Labeo calbasu	Bata-calbasu
(v)	Labeo calbasu	X	Labeo gonius	Calbasu-gonius

In the first two cases 94 per cent fertilization was obtained. Hybrids were extremely viable, looked more like the male parent, and growth rates were faster than that of the mother species. In the remaining three crosses, the percentage of hatching was poor, and showed no distinct promise.

The first cross achieved maturity in two years and was induced bred, producing F_2 generation possessing a varying range of characters. The injected hybrids (F_1 generation) spawned naturally inside the breeding hapas, percentage of fertilization was high and no stripping was therefore necessary.

Intergeneric Hybrids

These have been produced by crossing 5 species belonging to 3 genera.

	Male Parent		*Female Parent*	*Hybrid*
(i)	Catla catla	X	Labeo rohita	Catla-rohu
(ii)	Catla catla	X	Labeo calbasu	Catla-calbasu
(iii)	Catla catla	X	Cirrhinus mrigala	Catla-mrigal
(iv)	Labeo rohita	X	Cirrhinus mrigala	Rohu-mrigal
(v)	Cirrhinus mrigala	X	Labeo rohita	Mrigal-rohu
(vi)	Cirrhinus reba		Labeo rohita	Reba-rohu
(vii)	Cirrhinus reba		Labeo calbasu	Reba-calbasu
(viii)	Cirrhinus mrigala		Labeo calbasu	Mrigal-calbasu

Catla-rohu Hybrid

About 60 per cent of the eggs developed and hatching was completed within 16 hours, but mortality of hatchlings was high. Growth of the fry of the hybrid was faster than that of rohu from the same mother. The hybrid characters were intermediate between the parents, with a distinct advantage that the head was smaller than of catla (a disadvantage in pure catla) and broader body than of rohu, thereby providing greater flesh for known weight of either of the parent species. However, growth in hybrid was not as fast as in catla, but this was compensated by the greater full maturity in 3 years. Although ill formed testes were observed in a few specimens, and an interesting case of a hermaphrodite specimen with one ovary and one testis was recorded, yet a good number of males and females were found to be normal and fully mature. The fecundity, however, seemed to be lower than of the parental species.

The crosses (*ii*), (*iii*), (*iv*) and (*vii*) did not appear viable and promising. However, rohu mrigal and mrigal rohu hybrids recorded over 90 per cent fertilization of eggs, the body characteristics were intermediate between the parents and both the hybrids matured in 2 years.

In mrigal calbasu hybrid 60 per cent of the eggs hatched out. The hybrid possessed slightly fringed lips, 2 pairs of black barbells resembling that of calbasu (mother) and body colour intermediate between the parents. The hybrids attained ınaturity in 2-3 years. One 8-year old female was bred by hormone injection.

Out of the above crosses, the hybrid catla-rohu appears quite promising. Four-year-old hybrids were used to obtain the F_2 generation but the maturity of the F_2 generation had not been achieved.

Out Crossing of Intergeneric Hybrids

By out crossing the hybrids, as detailed below, the following hybrids were produced.

	Male Parent		*Female Parent*	*Hybrid*
(I)	Catla catla	X	Mrigal–calbasu	Catla-mrigal-calbasu
(ii)	Labeo calbasu	X	Mrigal–calbasu	Calbasu-mrigal-calbasu
(iii)	Cirrhinus mrigala	X	Mrigal–calbasu	Mrigal-mrigal-calbasu

Hybridization of Indian Carps with Common Carp

The cross between *Labeo rohita* (male) and *Cyprinus carpio* (female) was attempted. Considerable mortality of fertilized eggs during hatching and soon after hatching occurred. Some adults could be grown, but these did not mature.

An outcross between male *Cyprinus carpio* and female rohu-calbasu hybrid was carried out successfully.

Hybridization of Chinese Carps

Intergeneric hybrids between the silver carp, big head and grass carp were attempted in India, the USSR and Taiwan. Most of the crosses were not promising, and in most cases the hybrids died in early stages. Andriasheva (1968) reported that in the USSR, Silver carp (O_+) and big head (O) cross (SB) was found to be promising and had been released for fish culture. The cross between silver carp (O) and grass carp (O) could not be successfully done in the USSR, though, in India, Sinha and Chakraborti (1970) succeeded in growing this hybrid in ponds for over I year. However, this cross did not appear to be promising.

Hybridization Between Indian and Chinese Carps

In India, attempts were made to produce the following inter generic hybrids:

	Male Parent		Female Parent
(i)	Silver carp	X	Rohu
(ii)	Silver carp	X	Catla
(iii)	Grass carp	X	Rohu
(iv)	Grass carp	X	Mrigal
(v)	Catla	X	Grass carp
(vi)	Rohu	X	Grass carp
(vii)	Catla	X	Silver carp
(viii)	Catla	X	Big head
(ix)	Rohu	X	Big head

In all the above cases, the hatchlings did not survive beyond 1 week.

Hybridization Among Other Fishes

In Japan several varieties of gold fish (*Carassius auratus*) and common carp (*Cyprinus carpio*) were produced through selective breeding. Most of the work is being done on Cyprinid and Cobitid fishes. Selective breeding and hybridization work with sturgeons and common carp in particular in the USSR is of great significance. Also, the hybridization work done on Tilapia in Malaysia by Hickling (1960, 1968), which resulted in producing about 100 per cent male progenies, is notable.

Induced Fish Breeding and its Potentialities

In recent years, the use of fish pituitary hormones in inducing spawning in the difficult to spawn cultivated species of fish has revolutionized fish culture all over the world. Fish farming in many countries where the major Indian and Chinese carps were introduced was greatly handicapped because these cultivated food fishes did not spawn in captivity and their seed for culture had to be imported from the natural native resources. This problem has been successfully solved by the development of the hypophysation technique. In addition to carps, grey mullet, milk fish and a few other important salt water species are cultivated in ponds in several countries. Fish production in brackish waters has been reported to have increased manifold with the introduction of these selected salt water species. However, these fishes also do not spawn in ponds and the need for developing a method to induce these to breed was keenly felt, and encouraging results have been obtained in this direction in recent years. In Russia, hypophysation technique is widely emphasized for the conservation of sturgeons in rivers, a problem caused by building dams across the rivers.

The major problem confronting the development of fish culture in India is the acute shortage of fish seed for stocking the available culturable waters. The success in breeding Indian carps under artificial conditions by injecting fish pituitary hormones is an important landmark in the history of fish culture in India. It has provided a new tool to supplement and even partly replaced the age old practices and has opened up enormous possibilities for the development and expansion of pisciculture in India and even in other countries where fish culture is practiced on similar lines.

The fish seed requirement of an average farmer could easily be met if only one pair of each species is bred satisfactorily. An idea of the potential of production of seed in carps could be obtained from the following examples. In one experiment, a rohu weighing 2.72 kg had laid 0.504 million of eggs and 92 per cent of these eggs had hatched out and 0.47million of young fry were obtained for stocking. In another experiment, 0.63 million of young fry were obtained from a female rohu weighing 3.85 kg. The expected yield of fry after rearing for 2-3 weeks in a well managed nursery pond at 50 per cent survival is expected to be about 0.315 million from this fish. Considering the minimum selling rate at Rs. 10 per 1,000 of pure stock of major carp fry, the price of 0.315 million amounts to Rs. 3,150. From the above example, it is evident that by breeding successfully a few pairs only, a fish farmer could attain not only self sufficiency but also could earn some money by selling the excess.

The technique is simple and can be learnt by the fish culturist quite easily. In future, 'pituitary banks' could supply the hormone in ampoules fcr fish breeding purposes. For adequate popularization of the technique a large number of demonstration centers have to be established all over the country. The village level workers could be trained by the Fisheries Officers at the Community Development and National Extension levels.

With the tremendous increase in demand and popularity of fish culture in the country the demand for the stocking material has also considerably increased. The fish seed that is available from natural waters would hardly cope up with the growing demand. On the contrary, there is every possibility of scarcity of fish seed in rivers in future.

Discharge of considerable quantities of factory wastes into the rivers pollute the rivers and adversely affect the riverine fisheries. Besides, large scale mortality of fish in rivers is apprehended from the highly toxic insecticides, widely used in agriculture. Majority of the insecticides used are also highly toxic to fish and fish fry.

Construction of dams across many rivers in India has restricted the spawning migration of fishes considerably, and has brought about changes in the environment had natural breeding grounds of many fishes, thus depleting the fishery in these waters. With the possibility of construction of more dams in future the conditions will be still worse.

Lastly, the natural calamities brought about by flood almost every year in most of our rivers are affecting fish seed collection from rivers to a great extent. Besides, embankments constructed along many rivers to prevent floods have also destroyed many natural spawning grounds of carps.

Considering the above factors, it is apprehended that in future the riverine fisheries will be gradually depleted if necessary conservation measures are not taken, and consequently the chances of procuring sufficient quantity of fish seed from the riverines resource will be remote. The need for depending on seed production by induced breeding of fishes thus appears to be all the more important.

An additional advantage of induced fish breeding is that the method provides self sufficiency to the fish farmer. He can produce the stocking material for his farm and need not depend on other sources. The induced fish breeding techniques has shown considerable promise in breeding not only the difficult to spawn major carps but also many estuarine fishes, particularly the mullets. The significant role of phypophysation in improvement of fish stocks and evolution of new strains through selective breeding and hybridization has been already discussed in short.

Economics of Fish Breeding by Hypophysation

The technique of hypophysation is highly viable economically. However, it may be added here that working out any standard economics of various aquacultural practices including seed production is a difficult job, because is it liable to vary considerably from year to year, place to place and region to region under different socioeconomic conditions. This point is convincingly illustrated by Bradach *et al.* (1972). "It is clear from these few examples that the economic viability of a given aquacultural practice is few examples that the economic viability of a given aquacultural practice is dependent upon a large number of complex and interacting factors, often more sociological than technical that are peculiar to the region if not unique to the individual enterprise. Broad generalizations and principles concerning the economics of aquaculture thus have little meaning. The blueprint of a profitable operation in one locations is no guarantee of its successful application in another physical environment or in a different cultural or political setting".

Economic estimates of induced fish breeding though hypophysation have been made in the past. Based on the economics calculated earlier in 1964, the cost of production of 10 million of major carps spawn was estimated to be around Rs. 2,500. In this estimation the cost of equipments and labour were mostly taken into account. Later, another estimate was made in 1969, wherein the cost of rearing the brood fish was included. The figure arrived at was Rs. 3,860 for 10 million spawn, with the passage of time resulting in the changes in cost structure subsequent revision was made at Rs. 9,856 for the major Indian carps and Rs. 67,270 for the exotic silver and grass carps for 10 million spawn (Bhowmick, 1976); the later estimate also included expenses on staff salary, etc. for pond management. It is evident from the above account that varying estimates have been and would be arrived at in future by different workers in different places and years. The main points of difference lie in including or excluding the cost of inputs like purchase and maintenance of brood stock, staff salary, etc. In the opinion of the authors, the inputs in the form of purchasing brood fish, donor fish, preparation of ponds, fertilization and supplementary feeding, etc., need not be included in the evaluation of economics of induced fish breeding, since if forms a part of the fish culture operation and the expenses involved in rearing the brood and donor fish and also the technical supervision can be fully recovered through the sale of spawned brood fish, donor fish after gland removal, and spawn, fry and fingerlings.

In the present account the revised cost of essential items required for the cost of production of 10 million spawn though hypophysation has been estimated on the general lines of National Fish and Seed Committee Report (1966), and the details have been calculated. The cost of production of 10 million spawn of both the major Indian and exotic carps has been estimated at about Rs. 4,800.

The technical supervision during the work is to be covered under the extension activity of the State or Central Extension Units.

While the above details provide an idea of the cost of production of spawn through hypophysation, there is another essential information required for planning properly the induced fish breeding programme. This pertains to the estimate of number of breeding sets required for induced breeding to produce a particular quantity of spawn. The tentative details have been worked out to obtain 10 million of major carp spawn. These are described below.

Rough Estimate to Obtain 10 Million Spawn

Though invariably the success in induced spawning of carps is over 80 per cent on an average when done by experienced hands. Yet for the sake of calculation of the overall percentage of success in the country as a whole it has been kept on the lower side at 60 per cent. According to this estimate out of 70 sets, at least 40 sets should successfully spawn. The average size of each female is expected to be around 3.0 kg, and the average yield of eggs, for all the major carps (Indian and exotic) taken together, is calculated to be around 0.4 million on the conservative side. The percentage of fertilization is usually around 80. Thus each female of 3 kg would yield around 0.32 million of fertilized eggs and these at 80 per cent hatching success would in turn yield about 0.25 million of spawn. Hence, the 40 successful sets would yield a minimum of 10 million of spawn.

Suitable composite fish seed farms should be established all over the country at government level (State and Central). To augment seed production through induced breeding, the size of the composite fish seed farm to raise the healthy brood stock, donor fish, fry and fingerlings may be about 6 ha to begin with. Also, another 2 ha water area or land adjacent to the farm may be highly desirable for future expansion. The following set up at the farm has been suggested.

Suggestions for Future Work

The suggestions for future work are given under two heads, *viz.*, (*i*) the laying down of a national policy and its implementation to propagate the hypophysation technique through out the country; and (*ii*) the future line of work.

National Policy

The technique of hypophysation of carps is highly dependable to produce quality fish seed and has immense possibilities toward augmenting fish production from cultivable water bodies. The need for laying down a national policy in this regard is most essential and the technique made widely applicable throughout the country.

After a careful review of the present situation with regard to fish seed production through hypophysation, the following steps appear essential to achieve the desired results.

1. The cultivable waters should be surveyed in the country and a realistic assessment of the total fish seed requirements for stocking these should be carried out.
2. Composite fish seed farms should be established in each state to meet its own fish seed requirements. These farms should be located in areas where there is good and prolonged monsoon and the ecological conditions are conducive for proper maturity and breeding of major Indian and Chinese carps. The size of these farms would depend upon the seed requirement of the area, and should be highly economically viable. The activities of these centralization farms could proliferate and reach the smaller fish seed unit's in remote rural areas.
3. Brood and donor fish stocks should be maintained.
4. A hatchery at the composite fish seed farms to provide adequate hatching facilities is important so that optimum yield of spawn can be ensured. This would eliminate the

natural hazards and also make the technique more dependable and economical.

5. Planning and procurement of all the necessary items should be done in advance.
6. "Pituitary bank" at national and regional levels and also at each farm is highly desirable.
7. Short crash training programmes should be imparted for training in hypophysation technique for carps at various Central and State Training Institutes and Centres.
8. Effective extension activities particularly at state level are most vital for successful implementation of this developmental programme.
9. In order to achieve the above objectives, it is of paramount importance that adequate incentive is provided to good workers.

In many countries including China, hypophysation technique has largely replaced the conventional methods of fish seed collection from natural resources. In our country, however, there is justification to promote seed procurement both the hypophysation and from natural resources simultaneously till the hypophysation method becomes the main source of supply of quality carp seed. The dependence on the riverine collection would of course be reduced gradually in phased manner, though not fully. The only source of production of seeds of Chinese grass and silver carps, in India, is at present though hypophysation.

Double breeding of major Indian carps in one breeding season by hypophysation has been successfully demonstrated at the Cuttack Research Station. This has opened up new possibilities of obtaining increased seed production by induced breeding from a limited number of brood stocks at the seed farm. Further, the importance of favourable ecological factors in induced fish breeding has to be borne in mind to achieve the best results.

Future Line of Work

In the earlier account the conducive ecological factors for induced fish breeding and the use of crude fish pituitary extracts for the purpose have been discussed. Attempts to isolate or purify the fish pituitary hormones have also been made in order to understand the reproductive physiology of the fishes and standardization of the dosage, etc. Contributions of Yamazaki and Donaldson (1968) towards purification of fish pituitary extract using Sephadex G-100 technique opened up new possibilities of researches in the field of reproductive physiology. Sinha (1974) also found positive results from isolated fractions of pituitary but the total quantity of pituitary required for breeding Indian carp (mrigal) was the same as the of crude pituitary. Purified or isolated fraction of the fish gonadotropin seems to have not much of an advantage over the widely used crude pituitary. The bioassay experimentation serves to throw some light on the internal environment of the brood fish before and during spawning. The exact role of the exogenous pituitary in regulating ovulation and spawning, particularly in the difficult to spawn major carps, is still little understood. It is quite expected that this control of ovulation may differ from species to species. A review of the present status of knowledge on reproductive physiology of fishes, particularly about the carps, *viz., Carassius auratus,* and catfishes, *viz., Heteropneustes fossilis* and *Clarias batrachus,* is quite advanced and considerable amount of work has been carried out (Sundararaj and others).

Induced Breeding in Fishes and its Application to Aquaculture: A Short History

Editor's Note

This paper attempts to provide a brief history of the progress made in research on the induced breeding of fishes with emphasis on the application of the technology in aquaculture. The role of environment parameters, such as, temperature, light, salinity, rainfall and other factors on gonadal development and spawning in fishes, have been reviewed. The various other inducing agents investigated to induce breeding in fishes, include the fish pituitary homogenates, semi-purified fish pituitary gonadotropins. HCG and other hormones of mammalian origin, the neuro-hormones–LHRH (decapeptide and nonapeptide) analogues, and most recently developed mammalian and piscine gonadotropin-releasing hormone agonists combined with dopamine antagonists. Also, the efficacy of sex-steroids, prostaglandins, pheromones and thyroid hormones in addition to some extra-biological, non-hormonal substances like clomid, sexovid, tamoxifen and others, in spawning fishes, have been briefly described.

The results achieved on the gonadal development of brood stocks, induction of spawning in fishes and the progress made in the art of larval rearing, have shown great promise. It had been possible to breed most of the major groups of commercially important fishes, namely, in difficult to spawn cyprinid fishes (Asiatic carps), mullets, milkfish, catfishes, sturgeons, salmons and trouts, seabass and rabbit fishes. An account of the various hormones administered in inducing these fishes to breed and obtaining commercial production of their seeds, has been given. However, the techniques of breeding the catadromous eels and the marine fishes, are yet to be perfected, and hence large-scale production of their seed for culture has not been possible.

Although, highly significant progress has been made over the past couple of decades in hormonal research and efficient technology developed, yet hypophysation is still largely practiced by many fishfarmers to induce-breed their farmfishes. The made the new technology more effective and popular, it is suggested that the technology be further refined and simplified and the cost of the alternate inducing agents be lowered so that the average fishfarmers could easily adopt the technique at their levels.

Introduction

In the vast majority of the developing counties in Asia, Africa and Latin America, where malnutrition due to protein–deficiency is prevalent, fish is the main source of cheap protein. Aquaculture has been recognized in recent years, as one of the major strategies to increase would fish production since the marine catches alone are incapable of meeting the increasing demand of fish protein for the rapid development and expansion of aquaculture, one of the main constrains, is the dearth of adequate quantity of quality seed to the major cultured fish species.

The major aquaculture species like Asiatic carps, mullets, milkfish, some catfishes and several other groups of fishes become refractory when cultured in ponds. They either do not mature in ponds or are unable to reproduce in the culture–environment. The seeds, therefore, have to be collected from their natural breeding grounds, the rivers or seas. The wild fry, thus collected from nature are often usually of poor quality (a mixture of economical and non-economical species) and the quantity is very often insufficient. This constraint generally restricted expansion of culture of these important species to areas where seeds from the wild are not available.

As such, the only alternative is to find out ways and means to somehow induce these fishes to breed in ponds. To achieve sources in these difficult-to-spawn fishes, adequate knowledge of factors responsible for the controlled reproduction of these economical food-fishes is essential. The studies on the controlled reproduction in fishes for the purpose of seed production in aquaculture practices, therefore, boil down to two important categories namely:

1. Study of the environmental parameters responsible for the development of gonads, their final maturation, ovulation or spermiation and spawning. Once the environmental stimuli are pinpointed, then spawning could be manipulated; and,
2. Study on the effect of various hormones of both piscine and mammalian origin, synthetic hormones and steroids, and also certain chemicals which would accelerate maturation of sex-cells, ovulation/sperminate and finally spawning in commercially important food-fishes. Basic research in vitro and in vivo studies on the laboriartory–test fishes which ordinarily are not commercially important or on those which breed in captivity, also would contribute considerably in providing useful information for applied research.

Research on induced breeding in fishes by hypophysation, and other hormonal and or environmental manipulations and also their application for piscicultural purpose practically started in early 1930's. The first publication by Houssay (1931) of Argentina, followed by the Brazilian pisciculturists Cardoso (1934) and von Ihering (1937) reporting their success in inducing breeding of native food-fishes by injection of piscine pituitary extract, was the turning point towards applied research in the line. The Russians used the Hypophysation technique in breeding the anadroimous sturgeons (Gerbilskii, 1938; Kazanskii, 1939). Since then, stray attempts were made in several other counties (Khan, 1938; Haslere *et al.*, 1939), but it was only after mid-fifties that renewed interest in spawning food-fishes by hormonal treatment and environmental manipulations was initiated and success was achieved (Chaudhuri, 1956; Chaudhuri and Alikunhi, 1957; Ramaswamy and Sundararaj, 1957a, 1957b; Atz and Pickford, 1959). Simultaneously, the publication of the monumental book, "The physiology of the pituitary gland of fishes" by Pickford and Atz (1957) provided fresh impetus to the fish-culturists and research workers by disseminating voluminous information on the reproduction physiology and endocrine control on the reproduction in fishes. Also Hoar (1957) provided useful information on the physiology of the endocrine organs in fishes.

Since then, rapid development in hormonal work has contributed much to our knowledge on the gonadal maturation, ovulation and spawning of the important food-fishes all over the world. The exist numerous literature, and a number of reviews have been published on the progress in the development and improvement of the hypophysation and other related techniques and their role in aquaculture (Chaudhuri, 1968; Shehadeh, 1975; Donaldson 1975; Chaudhuri 1976; Fontaine, 1976; Chaudhuri and Juario, 1977; Harvey and Hoar, 1979; Lam, 1982; Abraham, 1988; Billard, 1989; and Little and Dawson, 1989).

When chronologically examined, it has been observed that 1950's brought the first success in induced spawning of Asiatic carps, followed by mullets in 1960's and lastly in 1970's success was achieved in spawning the important food-fish, the milkfish. However, during the next decade (1980's), all out efforts were made towards further improvement of the breeding technique by standardization of both exogenous and endogenous hormones, introduction of analogues of mammalian as well as piscine luteinizing hormone-releasing hormones as inducing agents and the use of dopamine antagonist during the last few decades and more and more fresh information are available on the application of endocrinology to the culture of important cultivated fish species including new commodities, especially the marine commercial fishes. This necessitated dissemination of the latest information on this highly important subject for the benefit of the research workers in piscine reproductive physiology as well as fish culturists.

The present review paper provides a brief history of the experimental work carried out to study the reproductive endocrinology of fishes and the effect of various inducing agents including environmental parameters and also hormonal and or non-hormonal substances in the induction of maturation, ovulation and spawning of the various fish species. The ultimate aim of this work is the utilization of this information for the successful spawning of the major cultured species of the world for obtaining commercial production of their seed for the development and progress of aquaculture.

Inducing Agents

Environmental Factors

It is well-known that environmental factors do influence the reproduction of many animals including fishes. The endocrine system appears to act as like between the fishes and the environmental. Harvey and Hoar (1979) have very nicely depicted diagrammatically the effect of environmental stimuli on the reproduction of teleost fishes in a physiological pathway. According to them, "the major links in the physiologic chain of events leading from the reception of environmental stimuli to the release of mature gametes" are largely mediated through interactions between brain (neural connections)–hypothalamus (releasing hormones)–hypophysis (gonadotropins)–gonads (sex hormones). Important environmental factors believed to be influencing maturation and spawning in fishes are temperature, light (photoperiod), salinity, pH, turbidity, meteorological factors like rainfall, flood, water current, lunar periodicity and others. The role of one or several of these environmental parameters have been reviewed by several authors (Chaudhuri, 1956; Pickford and Atz, 1957; Jhingran, 1969; Donaldson, 1975; Harvey and Hoar, 1979).

Temperature

Temperature is an important factor controlling the sexual maturation and spawning in many fishes. Craig-bennett (1931) observed that temperature control the sexual cycle of three-spined stickle back (*Gasterosteus aculeatus*). Burger (1939) found temperature regulating spermatogenesis is Fundulus. According to Merriman and Schedl (1941), temperature as well as light are probably essential for complete maturation of the four spined stickle back (*Apeltes quadracus*). Khan (1945) observed that the Indian major carps, stimulated by environmental manipulation, spawn only when the water temperature is within the optimum range. Atz (in Pickford and Atz, 1957) stated "it is well recognized that there are critical temperature above and below which fish will not reproduce". Chaudhuri (1956) reported that goldfish could be induced to spawn at least 3 months before the normal spawning season by hypophysation and artificially regulating the water temperature to optimum level of 65°F; spawning efficiency was, however, minimized at temperature 56°-58°F, while no Chaudhuri (1960-1968) hormone-treated Indian major carps generally bred at water temperature between 24-31°C. Tripathi and Khan (1990) reported successful spawning of Indian major carps in West Bengal, India ahead of the breeding season by manipulation of water temperature either artificially or by lowering down temperature through use of cooler deep tube-well water.

Light (Photoperiod)

Light as an environmental factor influencing reproducing in higher animals is well documented. In case of fishes, however, conflicting results have been obtained and different conclusions reached even on the same species. Atz (in Pickford and Atz, 1957) has provided quite a long list of experiments, carried out mostly on test fishes in the laboratory, demonstrating the effects of increase or decrease of light of the reproduction of fishes.

Hoover (1937) induced spawning in brook routs ahead of the normal breeding season by a gradual increase in daily photoperiod followed by a gradual decrease. Harrington (1950) was able to induce pre-seasoned spawning of *Notropis bifrenaus* by subjecting the fish to 17 hrs. of light per day. According to Allison (1951), artificial lighting on the other hand delayed spawning in eastern brook trouts. Shiriashi and Fukida (1966) obtained delayed maturity in brook trout, rainbow trout and in *Oncorhynchus rhodurus* subjecting them artificially to long light periods while early maturation under short light period had been observed for these species. Verghese (1968) experimented with the Indian carp *Cirrhinus reba* and obtained early maturity when exposed to photoperiod longer than the natural daylight period. Lam and Sob (1975) found retardation in gonadal maturationin the rabbitfish, *Siganus canaliculatus* by exposing it to a long photoperiod of 18 hrs. Kuo *et al.* (1974a) reported success in obtaining out-of-season maturation in *Mugil cephalus* by manipulation of photoperiod and temperature. By controlling photoperiod Fores *et al.* (1988) were able to induce spawning in turbot Schophthalmus maximus. Red drum, *Sciaenops ocellatus* was induced to spawn in captivity by proper manipulation of photoperiod and temperature (Lowson *et al.,* 1989). Prentice *et al.* (1989) obtained maturation and spawning in *Cynoscion xanthulus* using abrupt temperature fluctuations.

From the various experimental results, as stated above and others, on the effect of the increase and decrease in photoperiod on the maturation and reproduction in these fishes, one general trend could be observed that the fall and winter spawning fishes, such as, salmons, trouts and Ayu, mature ahead of the normal breeding season if light hours are reduced (see Shiriashi and Fukuda, 1966; Shiriashi and Takeda, 1961). On the other hand, in case of spring and summer spawning fishes, such as Asiatic carps and others, it appears that increasing light hours before the breeding season results in early maturation, much ahead of the regular breeding season. However, this needs further experimentation for confirmation of the effect of increase or decrease of photoperiod in these two groups of fishes.

Salinity

Salinity of ambient water is very important for the gonadal maturation, ovulation and spawning of many commercial species of fishes, especially, the anadromous and catadromous fishes including several euryhaline species. The anadromous Pacific and Altantic salmons migrate from the marine environment of freshwaters to complete their reproduction cycles, while the catadromous eels leave freshwaters and migrate to deep sea for spawning. According to Donaldson (1975) this spawning migration may probably be related to some extent to meet the osmotic requirements of the genmetes and the developing embryos. Also, migration of the maturing mullets and milkfish from fresh and brackish waters to the sea may be due to the low salinity of the former environment since low salinity has been found to inhibit sexual maturity. Abraham *et al.* (1966) found that in freshwaters, oocytes of Mugil cephalus do not progress beyond the previtellogenic phase. Maturing mullets from Chilka Lake in India are observed to migrate to the sea for final maturation and spawning. In 1961, Chaudhuri and co-workers (Anon., 1961, 1963; Chaudhuri *et al.* 1977a) induced spawning in *M. cephalus* caught from the lake mouth in lake waters. Although repeated ovulations occurred, the eggs were not fertilized. It was observed that the sperms did not survive when came in contact with every low salinity of lake water. However, *M. cephalus* eggs could only be fertilized when stepped in sea water. Milkfish do not ordinarily attain full maturity when reared in fresh or brackish water ponds. Lin (1984) reported full maturation of milkfish when reared in seawater ponds. These could be induced to breed. Natural spawning was also observed in these ponds in Taiwan. Chaudhuri *et al.* (1977b, 1978) succeeded in artificially fertilizing milkfish eggs which hatched out in the laboratory in seawater having salinity 30-34 per cent.

Rainfall and Other Factors

Rainfall appears to be the most important of all external factors triggering final maturation, ovulation and spawning in most of the warm water fishes. Amorso and Marshall (1960) discussed the possible role of rainfall in regulating reproduction among the vertebrates. The Asiatic carps generally breed once a year during monsoon months in rivers, their natural habitat, when the rivers are in flooded conditions after heavy rainfall. Very often delay of rains also observed to have delayed their spawning. Hora (1945) reviewed the available information on the factors responsible for the spawning of Indian carps in rivers or in special type of ponds which are called 'bundhs' in West Bengal India. In these 'bundhs', environmental conditions similar to those at rivers are stimulated. Of the various factors probably responsible for the spawning of carps. Khan (1945) advocated rainfall accompanied by flood, besides temperature as the most important one. Mookerjee (1945) stressed high dissolved oxygen content of fresh rainwater, while Das and Das Gupta (1945) found an increase in pH of 'bundh' water at the time of spawning. Husain (1945) emphasized monsoon floor as the main factor which is supplemented by temperature and other factors. According to Dubey and Tuli (1961), spawning of carps in 'bundhs' mostly occurred after heavy rains, having water pH between 7.2–8.2 and water temperature between 26^0-33^{0}C. Bear and Thomas (1964) suggested the possibility of a substance called 'petrichor' present in the flood waters being the inducing agent (cited by Fontaine, 1976). This substance, present in the clay is released when the flooded water fills lup a dried-up basin. This situation is comparable to the dry-type 'bundhs' is India.

According to Swingle (1956), fishes, especially, mature fishes when kept in crowded condition in confined waters, fail to spawn because of the secretion or excretion of a hormone-like inhibitory factor called by him as 'Repressive factor'. When such factors in waters are eliminated or sufficiently diluted, these fishes might spawn. The in-rush of sufficient floodwaters replacing the old water in a 'bundh' probably eliminate the 'repressive factor' thereby inducing the carps to spawn. In order to study the effect of the so-called 'repressive factor', Chaudhuri (1956) in his experiments conducted during 1954 with the catostomids bigmouth buffalo (*Ictiobus cyprinellus*) and smallmouth buffalo (*Ictiobus bubalus*) and also with common carp and gold fish, obtained reated successful spawnings by transferring them from a crowded condition to freshwaters at suitable temperatures. That rainfall and fresh rainwater are conducive to spawning was repeatedly observed by Chaudhuri (1960) when he obtained highly successful spawning with high percentage of fertilization whenever injected carps were released in accumulated fresh rain waters and also on cool rainy days. Peterson (1972) found a significant correlation between a decrease in barometer pressure and movement of mature trout for holding tanks into the spawning trap. Of other environmental factors effecting the spawning of fishes lunar spawning rhythm has been observed by several workers. May (1929) mentions that the threadfin, Polydactylus sexfilis spawns with a well defined lunar rhythm which always occurred in proximity to the last quarter phase of the moon. Podder *et al.* (1976) reported lunar spawning rhythm in *S. vermiculatus* which spawns about I week after the new moon, that is, on or around the 1st quarter phase of the moon.

Piscine Pituitary Gonadotropins

Fish pituitary gland secrets gonadotropic hormone(s) which play an important part in the maturation of gonads and reproduction in fishes. In higher animals two different gonadorophis (*a*) the follicle stimulating hormone (FSH) and (*b*) luteinizing hormone (LH) have been identified. In fishes, however, there is some controversy as to whether they possess one or two gonadotrophic hormones. Two types of Gonadotrophic cells were observed by Otsuka (1956) in salmon; Olivereau and Herlant (1960) I eel and Olivereau (1962) in goldfish. On the other hand, only one type of gonadotropin has been observed by Sundararaj (1959) in catfish. Fontaine and Gerard (1963) in common crop; Yamazaki (1969) in goldfish; Yamazaki and Dopnaldson (1968) in salmon; Burzawa–Gerard (1971) in common carp and several other researchers. Dodd (1960) is of opinion that there is very strong evidence of LH like GTH is fish but there is little evidence of having a separate FSH like type hormone. According to several supporters of one GTH in fish, the single GTH is physiologically identical with the mammalian LH. Hoar (1969) also suggested the presence of one gonadotropin in piscine pituitaries and remarked "Although the nature of the fish gonadotropin is still not resolved, the data for a single factor seem to be accumulating steadily". Fontaine (1976) reviewed the various studies made on this subject, and is of opinion that a single GTH is present in fish and that it differs from both LH and FSH of higher vertebrates.

Fish pituitary hormones have been observed to be highly effective in including spawning in fishes. Since the very first demonstration of their effectiveness (Houssay, 1931), the fish pituitary extract (FPE) has been widely used by fish culturists in obtaining seeds of the commercially important cultivated species in many parts of the world. The Brazilian pisciculturists were the pioneer and they developed the technique of routine hypophysation of Curimata (*Prochlodus* sp.) and other local food-fishes (von Ihering, 1937). Fontenele (1955) described the technique providing information on the collection of pituitary glands, their preservation, and extraction. The FPE is then administered either intramuscularly or intraperitoneally, usually in four gradually increasing doses at six-hour intervals. Parts or whole glands were injected.

Russians also applied the technique of hypophysation in obtaining seeds from the anadromous sturgeons getting standard below the riverine dams during their spawning migration upstream. Initially, they were administering intra-cranial injection of FPE preferably using pituitary glands from the same or closely related species (Gerbilskii, 1983); Kasanskii, 1939).

Subsequent to these successful application of hypophysation technique, stray experiments were conducted in USA and a few other countries with varying success. Hasler *et al.* (1940) succeeded in inducing spawning in muskellunge and trout by injection of carp pituitary suspension. Ball and Bacon (1954) obtained viable eggs in creekchub by injecting carp pituitary extract (CPE). Palmer *et al.* (1954) observed ripening of male salmon when injected with salmon pituitary extracts; but in females, the spawning period was only slightly advanced. Chaudhuri (1956) succeeded spawning goldfish 2-3 months ahead of normal spawning season by injection of common carp, goldfish and carpsucker pituitary extracts and accelerated development of secondary sexual characters in male fathead minnow (*Pimephales promelas*).

In India, from mid-fifties concerted efforts were made to use the hypophysation technique in breeding in indigenous fish species; Chaudhuri succeeded in inducing spawning of the carp-minnow Espmus danricus in 1955, a riverine catfish Pseudotropius atherinoides in 1956 and the first success in spawning an Indian carp *Cirrhina reba* was achieved on 10 July 1957 by him. In all these, Indian major carp pituitaries were used for the hypophysation (Chaudhuri, 1960). Following these initial success, large-scale breeding of important species of Indian major carps were obtained (Chauduri and Alikunhi, 1957). This had been the first published report of successful induction of spawning of any Asiatic carp in the history of Asian aquaculture. Also, in India, Ramaswamy and Sundararaj (1957a, 1957b) experimented on the pond catfishes *Clarias batrachus* and *Hetropneustes fossilis* to obtain spawning by pituitary hormone injections.

Since 1957, the technique of hypophysation was further improved and extended to all over India where the technique is being used for commercial production of seed of these important food-fishes. The technique was also disseminated to several other countries through training of their fishery officers (Chaudhuri, 1963; 1969; 1972). In 1969, FAO/UNDP arranged training seminars on induced breeding of cultivated fishes that were held at Kolkata/ Cuttack/Mumbai in India, where special training in both theoretical and practical aspects of the hypophysation technique as developed at the Cuttack Station of the Central Inland fisheries Research Institute, was imparted to trainees from various countries in Asia (Anon, 1969).

The technique was further extended to breed the exotic Chinese grass and silver carp in India in 1962 (Alikunki *et al.*, 1963). Subsequently, commercial production of seed of these Chinese carps was obtained as a part of routing fish cultural procedure (Chaudhuri *et al.*, 1966). In Taiwan, Tang (1965) induced-bred Chinese carps by injecting carp pituitary extracts. Wu and Chung (1964), as cited by

Singh (1969), gave an account of success in induced spawning in the mainland China, where Chinese carps were bred by injection of common carp pituitary extract. In USSR, Aliev (1961) first reported success in breeding grass carp and silver carp by hypophysation. In Malaysia, Chen *et al.* (1969) bred Chinese grass carp, bighead and silver carp by injecting 'Synaworin' combined with a threat hold dose of common carp pituitary glands, but they also attained success by hypophysation alone.

The grey mullet Mugil cephalus was successfully spawned by Chaudhuri and co-workers of Central Inland Fisheries Research Institute. India in 1961, by injection of homoplastic pituitary extract (Anon, 1962, 1963; Chaudhuri *et al.*, 1977a). Tang (1964) and Liao *et al.* (1969) obtained successful spawning in grey mullet by injection of mullet pituitary glands in combination with synahorin. In Islrael, *M. cephalus* was bred by injection of extract of common carp hypophysis (Yashouv, 1969). In Hawaii, Shehadeh and Ellis (1970) used salmon pituitary extract along with synaborin. Subsequently, Shehadeh *et al.* (1973) reported spawning of grey mullet with partially purified salmon gonadotropin alone. Mullet pituitary was also used in India to induce spawning in *M. macrolepis* (Sebastian and Nair, 1975).

In breeding catfishes, generally homoplastic or common carp pituitary glands are used. Ramaswamy and Sundararaj (1975a, 1957b) used homoplastic pituitary extracts in breeding *C. batrachus* and *H. fassilis*. Tongsanga *et al.* (1963) bred *C. macrocephalus* regularly by administering homoplastic pituitaries. The channel catfish *Ictalurus punctatus* are generally breed by FPE or HCH (Bardach *et al.*, 1972). The Thai catfish *Pangasius sutchi* and *P. pangasius* are also regularly bred by administration of FPE (Potaros and Sitasit, 1976; Boonbrah *et al.*, 1968). Britz (1991) reported the use of homoplastic pituitary gland for spawning *Clarias ghariepinus* in commercial aquaculture in Africa.

Partially purified gonadotropins from salmon (Chinook and chum), common carp and tawes also have successful results in breeding several food-fishes (Burzawa-Gerard, 1971; Sinha, 1971; Donaldson *et al.*, 1972; Idler *et al.*, 1957). Chaudhuri *et al.* (1977c) compared the efficiency of crude pituitary extract with partially purified gonadotropin and observed that there is not much difference in the effectiveness, but the latter is exorbitantly costlier than the former. In the Philippines, for spawning milkfish, initially SG-G100 (partially purified salmon pituitary gland which gave successful results when administered in combination with HCG (Vanstone *et al.*, 1977).

While analyzing the results of experiments on the induction of spawning in the above mentioned fishes, it is observed that in general, hypophysation technique was widely used, especially, in routine seed production procedures. However, in several cases the investigators used other hormones and even then in vast majority of the cases, they preferred to give a threshold dose of FPE. Therefore till recently, either absolute alcohol-preserved or acetone dried homoplastic or common carp crude pituitary glands were being used for production of quality seed in aqucultural practices. According to Atz and Pickford (1959), the synergic cactions among different pituitary hormones perhaps account for the better results obtained by injecting crude FPE rather than purified or partially purified gonadotropic preparations.

The piscine gonadotropins, however, appear to process of partial zoological specificity as observed by Fontaine *et al.* (1972) who made a comparative study on the effectiveness of two purified GTH from carp and Pacific salmon on the adenylcyclase activity (AC) in homogenates of ovaries in goldfish. The former was found to be 36 times more active than the latter. The species specificity, according to Creaser and Gorbman (1939), does exist in GTH and when the variation is great between the donor and recipient species, the hormones become in defective. Atz and Pickford (1959) mention the phenomenon of species specificity "as the tendency of one species to respond poorly (or not at all) to pituitary hormones from another". This probably explains of the poor response in fishes when injected with mammalian pituitary hormones. Chaudhuri (1956) obtained best result from homoplastic FPE as compared to heteroplastic FPE. Absence of any sex-specificity among the Indian carps, however, have been demonstrated by Ibrahim and Chaudhuri (1970).

Pituitary and Other Hormones of Mammalian Origin

Hormones of mammalian origin, both natural and synthetic with various modifications, are progressively being used in recent years with a view to induce maturation and spawning in teleost fishes. These are administered singly or in combinations, very often with a threshold dose of piscine pituitary hormones. In the earlier attempts, however, Hormones of mammalian origin did not give very satisfactory results when administered in fishes. Fontaine (1976) stated that there is much more uncertainty in the comparative action of mammalian gonadotropins than that of fish gonadotropin.

The following investigations obtained mostly, negative results; von Ihering and de Azevedo (1937) on Prochildus and Astyanax, with Preloban and Prolan; Khan (1938) on Indian carps with anterior pituitary lobe of sheep; Vivien (1939) on a goby with horse hypophysis; Hasler *et al.* (1939) on rainbow trout with PMS and FSH preparation of sheep pituitary gland; Jhonson and Riddle (1937) on trouts with bovine FSH, PMS and HCG (Prolan); Landgrebe (1941) on eel with cattle anterior pituitary lobe extract; Kwamoto (1950) on grass carp with dried mammalian anterior pituitary lobe; Palmer *et al.* (1954) on adult salmon with mammalian GTH. However, more recent work gave better results. Sundararaj and Goswami (1966a) studied relative potencies of several mammalian hypophyseal hormones in inducing ovulation in hypophysectomized gravid catfish *H. fossilis* and obtained best results with LH. Ishida *et al.* (1970) was able to induce spawning in Ayu (Plecoglossus altivelis) by injection of mammalian GTH.

Human Chorionic Gonodotropin (HCG)

Pickford and Atz (1957) reviewed exhaustive literature involving the use of mammalian GTH, PMS, HCH and other on fishes and considered HCG to be more effective than others in the ovulation and spawning in teleosts since it has been largely experimented in inducing gonadal maturation and spawning in fishes. HCG has one advantage over several other hormones including piscine hypophyseal hormones, that its potency can be standardized in I.U. and the results of different investigations can be compared. Besides, HCG is comparatively cheaper than other hormones. HCG or CG is being largely used in recent years in the routine breeding of cultivated species either alone or in combination with a threshold dose of FPE.

Lam (1982) has provided quite a long list of positive results obtained by administration of HCG or various fish species; Successful spawning has been obtained with HCG in channel catfish (Sneed and Clemens, 1959); in carp and goldfish (Clemens and Grant, 1964); in the striped bass (Stevens, 1966); in *Marble goby* (Tan and

Lam, 1973); in striped mullet (Kuo *et al.* 1973); in *Clarias macrocephalus* (Carreon *et al.*, 1976); in Ayu (Hirose *et al.*, 1977); in seabread *Sparus aurata* (Gordin and Zohan, 1982); in yellowtail (Soletchnik *et al.*, 1988) and in *Clarias fuscus* (Young *et al.*, 1989).

HCG in combination with fish pituitary extract (FPE) has very often given better results and many fish culturists use the combination in their routine seed production programmes. The following gives a fairly long list of fishes, successfully induced to spawn with HCH+FPE or FPE+HCG or synaborin+FPE as in Chinese carps (Tang, 1965; Chen *et al.*, 1969), in grass carp (Bailey and Boyd, 1970), in milkfish (Vanstone *et al.*, 1977; Liao *et al.*, 1979), in grey mullet (Tang, 1964; Liao *et al.*, 1977), in grouper (Chen *et al.*, 1977) and in *Pangasius sutchi* (Varikul and Sritongsook, 1981).

However, in several cases HCG alone has been found to be ineffective as in Prochilodus (von Ihering and de Azevedo, 1937), and in grouper (Chen *et al.*, 1977). According to Kuo *et al.* (1973), a dose as high as 60 µg/g of HCG is needed to breed striped mullets. Also, prolonged use of HCG in breeding Chinese carps for several years was reported to have developed resistance to HCG in these fishes (Anon, 1977b).

Gonadotropin Releasing Hormones

From early 1970's the trend in hormonal research was shifted to an altogether new line of approach. Attempts were made to intervene at the hypothalamic-pituitary interface in the endocrinal chain than intervene directly on the hypophysis itself (Harvey and Hoar, 1979). Earlier work on this approach by Peter (1970) and Breton *et al.* (1971) demonstrated the hypothalamic control of gonadal activity in goldfish and in vitro gonadotropin secretion in common carp respectively. Breton and Weil (1973) were successful is elevating plasma level of GTH in common carp by injecting synthetic decapeptide LH-RH. Also, high doses of synthetic LH-RH induced ovulation in Ayu (Hirose and Ishida, 1974); in goldfish (Lam *et al.*, 1976) and in common carp (Sokolowska, 1978). A host of other workers obtained positive results with synthetic LH-RH (Lam, 1982).

Further, following elucidation of the chemical structure of LH-RH, many investigators successfully synthesized a number of analogues which are more potent than LH-RH in releasing LH and FSH and in ovulation in animals. Banik and Givner (1975) experimented with pro-oestrous rats, earlier treated with fluphenazine dihydrochloride to block ovulation, and compared the ovulation–inducing efficiency of synthetic LR-RH with one of its analogues [D-Ala^6, des-Gly-NH_2^{10}]–LH-RH ethylamine (Ayerst-25, 205). They found the analogues as 36 times more potent than LH-RH. The other LH-RH analogue [Ayerst-25, 650] was also observed to be much superior to LH-RH (Banik and Givner, 1977). Lam and co-workers (cited by Harvey and Hoar, 1979) obtained encouraging result with the nonapedtide LH-RH analogue (Ayerst 25, 205) than with simple LH-RH in inducing ovulation in goldfish. Donaldson *et al.* (1979) found greater potency for the LH-Rh analogue [D–Ser–(bu^1)^6–des–Gly^{10}]–LH-RH–ethylamine (Hoechst 76), when a very low dose of 0.011 mg/kg induced spawning in coho salmon. However, the analogue was preceded by a priming dose of salmon GTH.

Carp breeders in China reported great success in inducing majority of the Chinese carps to spawn by using a nonapeptide LH-RH analogue (same as Ayerst 25, 205). This analogue termed as LRH-A was found to be more effective in the release of hormone than the decapeptide LH-RH in mammals (Anon, 1977a, 1977b). This finding has been of great significance not only for the Chinese fishculturists but also for other workers engaged in breeding. Asiatic carps and other commercial species. However, Peter *et al.* (1988a, 1988b) after detailed study of the carp breeding practices in China, found that LH-RH analogue (LRH-A) is currently not being used alone by the fish farmers and they generally combine this with fish pituitary extracts or HCG for better spawning success. LHRH-A has been used by many researchers to induce spawning in various fishes including seabass and rabbit fish (Harvey *et al.* 1985); Indian carp *Cirrhinus mrigala* (Kaul and Rishi, 1986); seabass (Garcia, 1989) and English sole (Sanborn and Misitano, 1991). However, Billard and Marcel (1980) failed to obtain any positive results using LRH or LRHa in inducing ovulation in pike. Lam (1982) in his review paper summarized the results of the experimental data on the use of LHRH-A and remarked that the LH-RH and its super active analogues are quite promising although the cost is prohibitive. However, he raised the following points for consideration of the workers in the line: (*i*) prolonged latency period in ovulatory response, (*ii*) likelihood of self-potentiating effect of LH-RH and its analogues, (*iii*) excessive dosage may be detrimental, (*iv*) responsiveness of pituitary to LH-RH may vary seasonally with the reproductive of the fish.

During 1980's, endocrinological research leading to ovulation and spawning of economical food-fishes were largely concentrated on the hypothalamic gonadotropic hormone (GnRH) or its superactive analogues. Further studies in the endocrinology of teleostean fishes showed that two neurohormones regulate their GTH secretion; of the two, GnRH is the gonadotropin releasing hormone while the other dopamine is the gonadotropin release–inhibiting factor (GR-IF). To counteract the latter, dopamine antagonists are usually injected along with the analogue (Lin *et al.*, 1986; Peter *et al.*, 1987). LHRH-A and GnRH are mostly used in combination with a dopamine antagonist, such as, pimozide, domperidone, eglonel, metoclopramide etc.

Sherwood *et al.* (1983) reported of another breakthrough when isolation of salmon gonadotropin-releasing hormone (sGnRH) was achieved in Canada. SGnRH analogue was found to be more effective than the mammalian LH-RH analogue (Ayerst-25, 50). The difference between the primary structures of the two analogues is in respect of the positions of two amino-acids. Sherwood *et al.* (1988) tried a new method of obtaining sustained release of GnRH-A when cholesterol baswed cellulose pellets containing the analogue are implanted into the body of the ripe female spawners. This inexpensive method was found very useful in getting final maturation and spawning of fish on several successive days. Marte *et al.* (1988) spawned *Chanos chanos* by injecting mGn RH-A (mammalian), sGNRH-A (Salmon) and HCG and found all the three equally effective. The milkfish implanted with mGnRH-A also responded. Chaudhuri and Singh successfully bred catla (*Catla catla*) at Vientiane, Laos with single does of 10 µg/kg of LHRH-A in combination with 10 mg/kg of pimozide. Also, success was achieved in ovulating a hybrid female (cross between Mekong carp *Cirrhinus microlepis* and Indian carp *Labeo rohita*) by similar injection and fertilizing the ovulated eggs with rohu milt (Anon, 1986).

Zohar (1988) reported on the efficacy of GnRHa in inducing spawning in all attempted experimental fishes and suggested further research to develop superactive fish GnRH analogues. Sahu and Biswal (1988) failed to breed Indian major carps by injecting LHRH-A alone, but succeeded when the analogue was combined with pimozide. According to Thomas and Boyd (1988), LHRH sub (a), a Superactive analogue of LHRH, was found effective in ovulation and spawning in three sciaenid fishes. Parameswaran *et al.* (1988a, 1988b) have developed a method (known as 'Linpe method') using a combination of a GnRH analogue (LHRH-A or sGnRH-A) and a dopamine antagonist (Primozide or domperidone) which has proved to be quite successful in inducing ovulation and spawning in several Chinese carp species. The Chinese carps generally do not response to GnRH-A alone. Manickam and Joy (1989) achieved success in ovulating *Clarias batrachus* by injecting LHRHa and pimozide but got negative results when LHRHa or pimozide alone was administered.

Nandeesha *et al.* (1990a 1990b) reported large scale success in inducing breeding in Indian major carps by using "Ovaprim"-a combination of a salmon gonadotropin-releasing hormone analogue (D-Arg^{0}, Pro^{9}-Net) and domperidone dissolved in Calibrated quantities in a non-toxic organic solvent. Ovaprim (according to the authors) is considered to be 17 times more potent than LHRH-A. Glubokov *et al.* (1991) induced ovulation in bream Abramis brama with LHRH-A (des Gly^{10} [D-Ala^{6}]–LHRH), a synthetic analogue in combination with dopamine antagonists, eglonil and metoclopramide. Experiments with LHRH-A and sGnRH-A in the presence or absence of pimozide (PIM) or domperidone (DOM) were tried in loach by Lin *et al.* (1991). The results demonstrated that sGnRH-A in the presence of PIM or DOM has a higher potency than LHRH-A in stimulating GtH release and inducing ovulation in loach. Fermin (1991) induced maturation in bighead Aristichthys nobilis by administering LHRH-A and DOM. In India, murrel gonadotropin–releasing hormone, cGnRH, purified from *Channa punctatus*, has been used successfully by combining with PIM and Ca^{2+} to spawn Indian carps rohu and Mrigal (Halder *et al.* 1991).

The above list of investigations and host of other data show the trend of research on fish endocrinology and its practical application in aquaculture practices in eighties (1980's) and early nineties (1990's). The most recent introduction of fish gonadotropin-releasing hormone agonists and dopamine antagonists has definitely demonstrated further efficiency in obtaining spawning in cultivated fishes.

Steroidal Hormones

Studies on the effect of sex-steroids in inducing oocyte maturation and spawning in fishes are still largely in the experimental stage and have so far met with limited success. It is well known that gonadotropin stimulates the sex-steroids which in turn induce maturation and ovulation in fishes. Experiments on gonadal steroids have been not been very encouraging so far. Estorgen influences only in the early stages of oocyte development but does not appear to take part in the final maturation and ovulation. Sex-steroids in general, bring about regressive changes in the ovary because of the negative feed back which decreases the GTH secretion.

Khoo (1974) induced ovulation in goldfish by treatment with progesterone and found androgen, estrogen and pregninolone ineffective. Sundararaj and Goswami (1966a) induced spawning in the hypophysectomized cat fish *H. fossilis* by injecting hydrocortisone and cortisone. But failed with estradiol benzoate, testosterone propionate and progesterone. However, the results of experiments conducted by Tamaru *et al.* (1991) to study the effect of several steroids in inducing final maturation in vitro in *M. cephalus* oocytes obtained from a female, primed with carp pituitary homogenate, have shown that the steroid 17 a-20 ß-diOH-P was the most potent steroid to induce GVBD. Sundararaj and Goswami (1966b) pretreated hypophywsectomized *H. fossils* for several days with the drug metopirone (chemical interrenalectomy) and then injected with desoxycorticosterone acetate (DOCA) and obtained ovulation in the catfish. In their subsequent experiments the authors found 11-desoxycorticosterol, 11-desoxycorticosterone and 21-desoxycortisol to be the most effective among the steroids used (Goswami and Sundararaj, 1974). Jalabert *et al.* (1977) successfully spawned common carp at low temperatures (13-15°C) by the treatment with 17a-20ß P after priming with a low dose of CPE. Also, Jalabert *et al.* (1978) induced maturation in rainbow trout which are close to maturity by injecting 17a-20ß P either singly or after priming with CPE. Adachi *et al.* (1988) using in vitro incubation techniques induced GVBD in follicle-enclosed oocytes of Pagrus major by sex-steroids. Burlakov *et al.* (1988) reported successful induction of ovulation in grass carp with the steroid cortexolone. The best result, however, was obtained after giving a preliminary injection of carp pituitary.

Lam (1982) made a review of the work on steroids by different workers, and concluded that the inconsistent results obtained using corticosteroids and progestogens may be due to the fact that follicular ruptures may not always follow steroid–induced oocyte maturation since it largely depends on the secretion of endogenous ovulation mediator which again is dependent on the gonadotropic action.

Although the use of sex-steroids in induction of spawning in fishes is still very limited, their use in the sex-reversal of some cultured species, such as, tilapias, is quite common. However, their use in fishes need be restricted, especially, on fishes which might reach the consumers as table-fish.

Miscellaneous Inducing Agents

Prostaglandins

Prostaglandin, as defined in an active Doralds medical Dictionary, "is an active principle obtained from the prostate gland and seminal vesicles in man; it is a pressor, vasodilator and a stimulant to uterine muscles." Prostaglanding regulate various reproduction functions in higher vertebrates.

In the teleosts, prostaglandins have been observed to stimulate gondotropic secretion in the catfish *H. fossilis* (Singh and Singh, 1976); inducing ovulation in rainbow trout (Jalabert, 1976; Jalabert *et al.*, 1978) and also to regulate spawning behaviour in gold fish (Stacey, 1976). Stacey and Pandey (1975) were above to induce ovulation in goldfish by single injection of PGE_1, PGE_{2a}, is capable of inducing maturation and regulating spawning behaviour leading to successful spawning.

According to Jalabert (1976), prostaglandins as well as catecholamines (epinephrine) are believed to be mediators in oocyte maturation and expulsion of oocytes from the follicles in rainbow trout. The above information and other evidence supporting the role of prostagllandins in the ovlulation of fishes and in the regulation of spawning behaviour in fish throw a new light on its application in aquaculture practices.

Pheromones

Pheromones are substances secreted by an individual which may provide specific reaction in the opposite sex of the same species. They are believed to be sex-steroid in nature and their secretions appear to be under the control of hormones either directly or indirectly.

Todd *et al.* (1976) observed that pheromone secreted by a ripe female catfish (*Ictalurus* sp.) exerted highly attractive power over the males of the same species. Spermiating males of some species on the other hand has been observed to secreted pheromone which may induce spawning behaviour of the females. In India, Moitra and Sarkar (1975) reported mass spawning of mature carps in the breeding 'bundh's, when a few pairs of CPE-injected carp spawners were released in the same water. This phenomenon appears to be due to the exudation of pheromones from the injected spawners. While reviewing the importance of pheromones in fish reproduction, Fontaine (1976) has emphasized their important role in fish culture.

Other Hormones

The role of thyroid in fish reproduction is not yet very clear. According to Sage and Bern (1971), thyroxine has a direct negative feedback on TSH and GTH cells, while several other fish physiologists are of opinion that thyroid hormones play an important part in the regulation of sexual cycles in fishes. Sneed and Dupree (1961) observed that TSH enhanced the effect of HCG and FPE to induce ovulation in goldfish but failed to precipitate ovulation by itself. Harvey and Hoar (1979) suggested that the presence of thyrotropin in the crude pituitary extract might stimulate gonadal development thereby contributing to the success by hypophysation. They also hinted the probable role of growth promoting hormones in the gonadal development in fishes.

However, application of either thyroid hormones or growth hormones in piscicultural practices for obtaining maturation of oocytes and induction of ovulation are still at the experimental stages and needs further investigations.

Non-hormonal Extra-biological Substances

Several extra-biological, non-hormonal substances, such as, Clomid (clomiphene citrate), Sexovid (cyclofenil) and Tamoxifen have been tried as ovulating agents in some fishes. Fontaine (1976) mentioned about some extra-biological substances which are capable of speeding up some stages of reproductive function in fishes. Clomid seems to act via hypothalamus hypophynis gonadaxis. According to Harvey and Hoar (1979), Clomiphene citrate (clomid) is capable of competing with gonadal steroids for binding sites in the hypophysis and hypothalamus, thereby releasing gonadotropin and raising the level of gonadal steroids. Clomiphene citrate was first reported bandy and Hoar (1972) as an ovulating agent inducing ovulation in goldfish. Singh and Singh (1976) also succeeded induction of ovulation in the air breathing catfish *H. fossilis* with clomid and sexovid. These substances appear to influence thyroid physiology through hypothalamus and hypophysis by increasing TSH secretion. Tamoxifen accelerated ovulation in coho salmon when combined with SG-G100 (Donaldson *et al.*, 1978). However, Clomiphene citrate gave negative result in rohu (Bhowmick *et al.*, 1986) and in carp (Bieniarz *et al.*, 1979).

Determination of optimum dose of these antiestrogens is still a problem since excessive doses are found to inhibit GTH secretion. Donaldson *et al.* (1978), therefore, have expressed doubt regarding the practical use of these antiestrogens and suggest further research on these compounds.

Use of Hormones in Aquaculture Practices

Present Status

In the foregoing section, various inducing agents and their role on the maturation, ovulation and spawning in various fish species have been largely discussed. While preliminary experimental success of failures have been reported in several species, in others the results have been highly successful when applied in the field. In this section, attempts have been made to disseminate available information on the present status of hormonal research on the major cultivated fishes of the world leading to their practical use in aquaculture practices by ensuring a ready supply of quality seed for stocking purposes.

The major groups of cultured species and the use of hormones in their culture will be dealt with separately as described hereunder.

Asiatic Carps

The Asiatic carps comprise of the Indian major and minor carps [catla (*Catla catla*), rohu (*Labeo rohita*), calbasu (*L. calbasu*), bata (*L. bata*), marigal (*Cirrhinus mrigala*), reba (*C. reba*) and others; the Chinese carps [silver carp (*Hypophthalmicluhys molitrix*), bighead (*Aristichthys nobilis*), grass carp (*Ctenopharyngodon idella*), mud carp (*Cirrhinus *molitorella*), black carp (*Mylopharyngodon piceus*) and others]; Indonesia carps [tawes (*Puntius gonioonotus* and others) and the common carp (*Cyprinus carpio*)]. Of these, the Indian and Chinese carps are the difficult to spawn fishes which become refractory under captivity. They develop gonads but do not spawn in ponds or in other confined waters.

Although the culture of Chinese and Indian carps is an age old practice, especially in China and India, for centuries the only source of their seed was the big rivers where these carps breed during monsoon months, from where seeds were collected and sold to fish culturists in seed markets. Non-availability of seeds of these carps, therefore, restricted their culture in areas located far away from the natural breeding grounds of Asiatic carps.

The tawes and other Indonesian carps as well as the common carp bred in ponds (Chaudhuri and Tripathi, 1979). Hormone induction is not essential for their breeding, but for synchronization of large scale spawning, hormones have been used by some fish culturists.

With the first success reported in 1957, in inducing spawning of Indian carps by injection of carp pituitary hormones (Chaudhuri and Alikunhi, 1957; Chaudhuri, 1960), a technique has been developed to breed them in ponds and obtain commercial production of their seed. Subsequently, hypophysation technique was also extended for inducing breeding of Chinese carps (Alikunhi *et al.*, 1963; Wu and Chung, 1964; Tang, 1965; Chaudhuri *et al.*, 1966; Konradt, 1969 and Chen *et al.*, 1969). The detailed techniques of induced breeding of Indian and Chinese carps by hypophysation have been described by several workers (Chaudhuri, 1963, 1968, 1969, 1976; Tang, 1965; Singh, 1969; Harvey and Hoar, 1979; lam, 1982). More recently, Tripathi and Khan (1990) have reviewed the work on 'Carp seed production technology'.

Henceforth, carp pituitary extracts (particularly those of common carp) are being largely used by fish farmers of Indian subcontinent, China, Taiwan and throughout South-east Asia for obtaining seed of major cultured carps. However, many fish farmers in China, Taiwan and South-east Asia in addition, were injecting HCG or Synahorin followed by a threshold dose of carp pituitary extract (CPE). During 1970's Chinese fish farmers were using HCG alone in their regular fish breeding programme (Tapiador *et al.*, 1977).

Subsequently, with the breakthrough achieved in the use of LH-RH analogues, fish farmers in China started using the drug for the commercial production of seed of the cultivated carps (Anon, 1977b). However, as mentioned earlier, currently, the Chinese fish breeders are not using LH-RH-A alone for inducing spawning and when use, it is combined with CPE or HCH (Peter *et al.*, 1988a).

Further research in the line during 1980's led to the development of GnRH or its supper active analogues and these in combination with dopamine antagonists were successfully used in Chinese carps. The 'Lanpe method' (described earlier under 'Inducing Agents') is rapidly gaining acceptance with the fish farmers in China (Peter *et al.*, 1988a, 1988b).

In India, spawning success has been reported with a new preparation "Ovaprim", containing sGnRH combined with domperidone which has recently been tried on an experimental basis in many parts of India (Nandeesha *et al.*, 1990b), However, application of this technique by fish farmers in their routine aquaculture practices have not been reported so far.

The sum up, although hypophysation technique has several drawbacks such as, lack of proper standardization of the dosage or quantification of the different hormones administered or correct assessment of the gonadal stage particularly of female spawners, till to date, the technique is being largely followed by individual carp breeder in India and many other developing countries of Asia since the fish farmers find it quite convenient, very useful and more economical. In spite of a few failures, the hypophysation technique can produce the required quantity of quality seed with much less expenditure as compared to other natural and synthetic hormones. The cost of the semi-purified gonadotropins of the mammalian or piscine gonadotropins of the mammalian or piscine gonadotropin releasing hormones are beyond the means of majority of the fish farmers.

However, it is hoped that in future, the carp seed production technology will be further improved and a cheaper and readily available much better inducing agent will be developed for the benefit of the end users.

Mullets

Mullets are popular food-fishes, both in the Orient as well as in the West. Of this group Mugil cephalus, the grey or striped mullet is the most important cultured species. It has a worldwide distribution and being a euryhaline and eurythermal fish, it thrives well in marine, brackish as well as in freshwaters. The grey mullets, however, attain final maturation and breed only in the sea (see under 'Salinity'). The young ones migrate from the sea to the estuaries and also to freshwaters from where seeds are collected for culture.

For induced breeding experiments, initially, the wild spawners were collected from the sea during their spawning migration. Spawning success was first achieved in India in 1961 (Anon, 1962; Anon, 1963; Chaudhuri *et al.* 1977a) by injection of homoplastic pituitary homogenates and then in Taiwan in 1963 (Tang, 1964) with mullet pituitary extract in combination with synahorin. The hatchlings, however, did not survive. Subsequently, grey mullet breeding was successfully carried out at the Tungkang Marine Laboratory in Taiwan and at the Oceanic Research Institute, Hawaii. Liao *et al.* (1969) used mullet pituitary homogenate with synahorin, Schehadeh *et al.* (1973) with SG-G100 and Kau *et al.* (1973) with HCG alone. In Israel, Yashouv (1969) reported success in spawning two female specimens of *Mugil cephalus* reared in freshwater ponds, by injecting CPE.

The techniques developed in the above laboratories have demonstrated that the breeding of grey mullet posed no major problems, but survival of the hatchlings beyond 6 weeks or so had been a big constraint. Heavy mortality occurs during the first critical period of yolk absorption (2.5 days after hatching) and the second at 8 days after hatching. At the Tungkang Marine laboratory, large scale investigations were carried out to improve the induced breeding technique, to increase survival rate of the larvae by developing live larval feeds and also onwards inducing maturation in pond grown mullets (Liao, 1975). Liao (1977) reported induced breeding of two pond reared 3 year old females, grown from induced bred fingerlings, thus completing a generation cycle in captivity. Oceanic Research Institute also succeeded in breeding pond reared grey mullet which spawned naturally (without stripping) in the pond after injection (Kuo *et al.*, 1973). The same researchers also developed a very reliable method of ovarian biopsy and found that female spawners have oocytes of mean diameter around 0.65 mm response easily to hormone treatment (Kuo *et al.*, 1974b). More recently, Lee *et al.* (1988) of the same Institute tried two injections (a primer and a resolver) with any two of the hormones CPH, HCG and LHRH-A on grey mullet and observed that CPH/LHRH-A combination was the most effective and least costly. Lee and Tamaru (1988) while discussing various mullet breeding techniques currently available, recommended carp pituitary extracts plus LHRH-A as the most promising.

Breeding experiments with a few other mullet species also have met with success but large-scale breeding had not been pursued as yet.

Results so far obtained in induced breeding of *M. cephalus* have been highly encouraging. It has been possible to breed the mullet at will and success achieved in inducing maturation in ponds, thereby ensuring ready availability of spanners. Also, there had been much improvement in the larval rearing technique. Still much more needs to be done to standardize the entire artificial propagation processes for transferring the technology to the end-users.

Milkfish

Like the grey mullet, the milkfish *Chanos chanos* (Forsskal) is another euryhaline fish having a very wide range of distribution from the red sea to the west coast of N. America. Milkfish is a very popular food fish in the Philippines. Taiwan and Indonesia and cultured mostly in brackish water ponds and also in freshwaters. Since the species does not breed or even fully mature in culture environment, till late seventies, the only source of its seed was the coastal areas. Practically, nothing was known about the life-history, breeding habits and reproductive physiology of the species.

Chaudhuri *et al.* (1977b; 1978) described the nature of fertilized eggs, the embryonic and larval stages of milkfish by artificially fertilizing stripped eggs of wild spawners. Vanstone *et al.* (1977) were about to successfully induce spawning in milkfish collected from the sea, by injection of acetone-dried salmon pituitary extract combined with CH. Subsequently, induced breeding of the wild spawners were repeated and the technique of rearing of the larvae was developed at the Aquaculture Department of SEAFDEC, Iloilo, Philippines (Liao *et al.*, 1979). Later on, Juario and Duray (1983) described the technique of milkfish breeding and successful rearing of their seed.

Further, at the SEAFDEC Aquaculture Department, the progeny of the first-bred milkfish were reared in floating cages in the marine environment where they matured in three years (Lacanilao and Marte, 1980; Marte *et al.*, 1984). Juario and Natividad (1980) bred the cagereared captive milkfish by hormone treatment. Also at the Igang Station of SEAFDEC AQD, milkfish (3.5 to 5.5 yrs old), reared in the floating cages and fed with commercial pellets (42 per cent protein), attained maturity. The first observation of spontaneous spawning of these milkfish was made in 1980 having 55 to 65 per cent fertilization (Anon, 1980). Since then, natural spawning (without any hormone treatment) were observed every year in the cove-reared or cage reared milkfish. The fertilized eggs were then collected from the cages for rearing in the laboratory.

During 1980's further attempts were made to hasten maturation of milkfish in captivity and also other inducing agents were tried to breed them. Marte *et al.* 91987) administered mGnRH-A and sGnRH-A through cellulose-cholesterol pellets either by interaperitoneal implants or though intramuscular injection and induced ovulation and spawning in 75 per cent of the experimented female milkfish.

At Taiwan, the first artificial propagation of pond-reared milkfish was reported by Tseng and Hsiao (1979). Lin (1982, 1984) reared milkfish in seawater ponds and by injecting HCG or HCG + FPE annually from the age of 3 years, could breed those from 7th year onwards. Natural spawning also occurred in 1984. Liao and Chen (1984) reviewed the milkfish breeding work in Taiwan and concluded that maturation and spontaneous spawning of milkfish reared in ponds depend on the combination of several factors such as environmental, nutritional, hormone and social/behavioral manipulations. Lam (1984) reviewed the progress made in milkfish breeding and the problems thereof, and suggested further work at standardize routine milkfish propagation.

To sum up, the achievements so far made in the artificial propagation of milkfish are: (*a*) wild breeders could be bred by various hormone injections; (*b*) spawners could be raised in captivity and bred; (*c*) occurrence of spontaneous spawning of pond-reared/cage-reared milkfish and (*d*) a larval rearing technique have been developed. Now what is needed is further research for the standardization and perfection of the methods so far developed.

Consumers generally prefer them for the high quality of flesh and absence of floating bones in the flesh. The most popular cultured catfishes belong to the genera: *Clarias, Heteropneustes, Pangasius, Ictalurus* and *Silurus*. Of these, the *Clarias* and *Heteropneustes* species are very hardy and can tolerate adverse environmental conditions including very low oxygen content of water because of their accessory respiratory organs.

Clarias batrachus and *Heteropneustes fossilis* (native of Indian subcontinent) breed in ponds during rainy season and can be easily bred by merely transferring the gravid spawners to freshly collected rain water although some researchers prefer to breed them with hormone injections. *C. macrocephalus*, however, is more difficult to breed and either CPE or homoplastic pituitary (Tongsanga *et al.*, 1963) or HCG (Carreon *et al.*, 1976) are used for spawning. *C. gariepinus*, a much larger-size species, a native of Africa, and introduced in many countries, are bred by injection of acetone-dried common carp pituitary extract or by HCG. Britz (1991) after comparing various alternative inducing agents advocated the use of homoplastic pituitary homogenate in *C. gariepinus* because of low-cost and easy availability to the catfish farmers. *C. fuscus*, a Chinese species, is bred by CPE + synaborin (Chen, 1976). At Hawaii, Young *et al.* (1989) spawned *C. fuscus* with HCG alone.

Large-size commercial catfish *Pangasius pangasius* and *P. sutchi* are also induced to spawn for obtaining their seed. The Thai researchers, Boonbrahm *et al.* (1968) and Potaros and Sitasit (1976) succeeded spawning *P. pangasius* and *P. sutchi* respectively by injecting pituitary extract of *C. batrachus*. Subsequently, CPE + HCG was largely used in commercial production of seed in aquaculture practices in Thailand (Varikul and Sritongsook, 1981). Fijan (1975) used CPE to spawn the European catfish *Silurus glanis*. Epler and Bieniarz (1989) could not induce spawning in *S. glanis* by injecting CPE or LHRH-A, but succeeded with LHRH-A combined with pimozide.

Catfish farming in USA became popular in mid-fifties. The channel catfish (*Ictalurus punctatus*) has consumers' preference as a food-fish because of high quality flesh and other cultural qualities and also as a sport fish. The blue catfish (*I. furcatus*), white catfish (*I. catus*), flathead catfish (*Pylodictis olivaris*) and a few other bull heads are also cultured to certain extent. The channel catfish industry has been developed in many states in USA especially in the South.

Ordinarily, the channel catfish is bred in special ponds provided with pens or spawning containers. But with the increasing demand of seed, hormones are being used in commercial production of seed. Clemens and Sneed (1957) first reported of success by administering CPE to female spawners. The present day practice, however, as reported by Huner and Dupree (1984), is that the farmers administer CPE or HCG intraperitoneally to induce spawning in channel catfish.

Thus it is observed that compared to other major groups, the catfishes do not pose much problem in breeding them. They are either bred by manipulation of environmental parameters or are induced to spawn by injecting hormones of low cost. However, the majority of the catfish farmers are facing the problem of very low survival of the larvae because of the cannibalistic habit of the fry. Improved methods of larval rearing to reduce high mortality in the larvae stage should be developed to reduce the cost of production of stocking materials for their culture.

Salmonid Fishes

Salmons, trouts and Ayu are the main food-fishes cultured in coldwaters. For breeding majority of the salmonids, brood fishes are generally collected from the open waters. Pacific salmons (*Oncorhynchus* spp.) and Atlantic salmon (*Salmo salar*) and lake routs (*Salvelinus namaycush*) mostly come from open waters, whereas rainbow trouts (*Salmo gairdneri*) are generally raised in ponds, and the brown (*S. trutta*) and brook trouts (*Salvelinus fontinalis*) come from both. The usual practice is to strip ripe spawners following both 'dry' and 'wet' methods for artificial insemination. So, like the other groups of warm water fishes, induced breeding by hormone treatment is generally not needed.

However, scores of experiments have been conducted to hasten early or off season maturation of salmons and trouts by manipulation of environmental parameters and by hormone treatment. Work on the effect of photoperiod in accelerating or delaying the maturation of gonads have been discussed in earlier part of this review. The hormonal work on salmons and trouts have been mostly pertaining to in vitro investigations to study response on plasma gonatodotropin profile or hastening spermatogenesis/vitellogensis and induction of early maturation or ovulation, etc. Funk *et al.* (1973) were able to accelerate ovarian maturation in pink salmon with injection of Chinook salmon gonadotropin. Jalabert *et al.* (1978) obtained ovulation in rainbow trout by administering prostaglandins. Also Crim *et al.* (1988) could stimulate release of GTH (*in vivo* treatment) in testosterone- primed juvenile rainbow trout with mLHRHa.

In Japan, extensive studies have been made to induce maturation of Ayu (*Plecoglossus altivelis*) either by environmental manipulation (Shiraishi and Takeda, 1961) or by induction of ovulation by administration of various hormones, such as mammalian GTH (Ishida *et al.*, 1970), LH-RH (Hirose and Ishida, 1974) and HCG (Hirose *et al.*, 1977).

Sturgeons

Sturgeons are important food fishes in Russian and a few other countries. The eggs of sturgeons called 'caviar' is a delicacy. Sturgeons belong to the family Acipenseridae. These are anadromous fishes which migrate from the sea to freshwaters for breeding. In Russia, three important species of sturgeons which migrate from the Caspian sea up the rivers Volga, Ural, Kura and others, are (I) the beluga (*Huso huso*) (2) the sevruga (*Acipensor stelletus*) and the osetr (*A. guldenstadti*). Because of the construction of dams over the rivers, their spawning migration is obstructed. The Russians catch the stranded spawners below the dams and induce them to spawn by hormone injections. (Gerbilskii, 1938).

Pickford and Atz (1957) have provided a big list of hormonal work undertaken in Russia during the decades (mid-thirtees to mid-fifties) on the effect of FPE on the reproduction of sturgeons especially the two species. *A. stelletus* and *A. guldenstadti*. The first success in inducing breeding sturgeons by pituitary treatment in 1937 was reported by Gerbilskii (1938). Since then hypophysation technique was practiced in Russia. Gerbilskii (1965) described the induced breeding technique. In the earlier years, intracranial injections were given which were later on replaced by intramuscular ones. Fractional doses of acetone-dried fish pituitary extracts are generally administered for better results, although synthetic hormones are also used. To determine the exact time of ovulation, the eggs from injected sturgeons are examined by collecting a few through an incision made in the body. The Russians have developed a bioassay method known as 'Vy'un unit or the loach unit (Kazanskii, 1949). One such unit contains 0.2 mg of acetone-dried common carp pituitary gland powder which is capable of inducing ovulation in 50 per cent of the loach, Misgurnus fossilis.

For decades, this hypophysation technique was practiced which was mainly restricted to Russia. In recent years, hormone treatment is also practiced in and outside the former USSR Barannikova *et al.* (1989) stimulated maturation in *A. guldenstadti* by injection of luliberin and by pituitary preparations. Parauka *et al.* (1991) reported successful spawning in the Mexican sturgeon *A. oxyrhynchus* injected with CPE. Miura *et al.* (1991) obtained final maturation in male Japanese sturgeon Hucho perryi by injection of FPE. Also, ovulation was obtained in the Siberian sturgeon *A. baeri* after using carp hypophysis.

The developments of an efficient hypophysation technique in Russia to breed sturgeons has been a great achievement. The Russians are able to produce millions of finger lines (mainly for ranching) of these invaluable fishes whose very existence was being threatened because of getting stranded below the riverine dams/ locks during their spawning migration.

EELS

Eels are delicacies as an excellent table fish in many countries of the world, especially, in Japan. The freshwater eels belong to the genus Anguilla of which the European eel (*A. anguilla*), the American eel (*A. rostrata*) and the Japanese eel (*A. japonica*) are the popular ones. The eels are catadromous fishes and till a few decades back, nothing was known about their breeding habits or exactly where they spawn after migrating into the sea. The European and American eels have been reported to breed at a depth of 50m in the Saragasso sea in the Atlantic Ocean away from the east coast of USA. The leptocephalus larvae then slowly migrate to the coasts of N. America and Europe and the elvars get into the rivers for growing to adult eel. The usual cultural method is still restricted to the collection of elvars from the rivers and culture in tanks. They are fed heavily with artificial diets. Attempts are being made to mature the eels under artificial conditions and also by hormone treatment.

Studies on artificial propagation in European eels have been initiated since 1030's Fontaine (1936) was able to secure complete maturation in male *A. anguilla* by injecting prolan (HCG). Also Boetius and Boetius (1967) induced maturation in male European eel by injection of HCG under controlled temperature and salinity. Fontaine *et al.* (1994) obtained full gonadal maturation in female A. anguilla by treating with CPE. In the Japanese eel *A. japonica*, Yamamota and Yamauchi (1974) succeeded maturation in males with synahorin but the female needed FPE for maturation. Nose (1971), quoted by Donaldson (1975), could breed a single female of Japanese eel producing ova of average diameter. 1.06 mm. Epler and Bieniarz (1978) could get considerable advancement in the maturation of female European eel when injected with HCG + CPE, but no ovulation was observed. In a recent experiment, Lin *et al.* (1991b) reported induction of ovarin development to, or nearly to the prespawning stage in Japanese eel by schronic treatment with testosterone. But failed when LHRH-A or domperidone were combined with the former.

The studies conducted so far, have demonstrated that maturation of gonads, and in some cases induction of ovulation was possible but so far there is little information of any success in breeding the freshwater eels for production of their seed. Uptil now the farmers have to depend on the collection of elvars only from the wild for their culture. It is hoped that through the joint efforts by the reproductive physiologists, nutritionists and aquaculturists, very soon this stumbling block in the progress of eel culture will be removed.

Sea Bass

Sea bass is the popular name given to the cultured species. Lates calcarifer which is also known as gaint sea perch or cock-up. It is an economically important food-fish cultured in many countries in the tropical and subtropical regions of Western–Pacific and Indian Ocean. Sea bass is a euryhaline fish and is cultured both in fresh and brackish water ponds and also in floating cages in coastal waters. It is a piscivore and is considered to be an ideal predator to be stocked in tilapia ponds for controlling overpopulation of the forage species.

The first report of sea bass spawning in early 1970's came from Thailand. Sea bass was observed to spawn at the river mouth or lake outlet to the sea having salinity range 25-32 per cent. (Pongsuwana and Bhukasawan, 1977). However, artificial breeding of ripe sea bass was achieved during Aug-Sept. 1973 (Wongsumnok and Manevonk, 1973). Subsequently, spontaneous spawning of captive seabass under controlled environment was observed in 1975. The brookstock were raised from fry spawned in hatcheries (Tattanon and Maneewongsa, 1982). Manewongsa (1982) first reported success in induced spawning of seabass by injection of HCG. Puberogen and Pregnyl were also used. Following these findings, millions of sea bass seed were produced in Thailand either by environmental manipulation or by hormone treatment.

In the Philippines, at SEAFDEC AQD, Kungvankii *et al.* (1983) spawned seabass by injection of puberogen and HCG. The survival of fry was poor (6-12 per cent). In the same laboratory, Harvey *et al.* (1985) succeeded in sea bass spawning by implantation of cholesterol–based pellets of LHRH-A. Nacario and Sherwood (1986) used LHRH-A by implantation of pellets or by using osmotic pump and also by IM injection. The treated females spawned 2 days after pellet implantation of injection (once only) but those implanted with osmotic pump spawned consecutively for 4-5 days. A similar experiment with mGnRH-A and sGnRH-A conducted by almendras *et al.* (1988) resulted in multiple spawning in treated sea bass. The pellets were found more reliable, less costly and less stressful to the female spawners. Garcia (1989) observed a dose-related increase in the spawning rate of sea bass injected with LHRH-A. In the Western Papua New Guinea, sea bass were observed to spawn in their natural spawning grounds during October-February (Moore, 1982).

Successful spawning of *Lates calcarifer* have been reported from several countries but it is in Thailand alone that the technique development, has been largely utilized for the production of sufficient seed of the species to meet the need of the industry.

Rabbit Fishes (Siganids)

A good number of siganid species are available in the Indo-Pacific region. They are herbivorous and feed generally on sea weeds and other algae. The rabbit fishes are considered as good table fish and have a great potential for their culture in cages in coastal waters and also in brackish water ponds. Of the several siganid species, *Siganus canaliculatus* and *Siganus guttatus* are more common and there culture has become popular in a few countries.

The rabbit fishes breed in captivity spontaneously, but very often hormone is administered for synchronizing mass breeding in large scale seed production. Popper *et al.* (1973) induced spawning in *S. rivulatus* and May *et al.* (1974) in *S. canaliculatus*. Also, the latter species was breed by Soh and Lam (1973) by injecting HCG in *S. rivulatus* and *S. argenteus*. At SEAFDEC AQD. Iloilo, Juario and co-workers observed that spawning of *S. guttatus* occurred almost every month of the year without any hormone treatment. They also succeeded breeding of the species by injection of HCG (Juario *et al.* 1985). Ayson (1991) obtained spawning in *S. guttatus* by administration of HCG in two doses and observed that while the females having oocyte-size 0.46 mm spawn spontaneously, those having oocyte-size between 0.42-0.45 mm could be bred by hormone treatment. In an earlier experiment he demonstrated that for ensuring 100 per cent monthly spawning of captive *S. guttatus*, exogenous hormone is necessary (Ayson, 1989).

The above experiments demonstrated that in the development of mariculature of siganids, breeding of the commercial species would not be a problem. However, there is plenty of scope for further improvement of the larval rearing techniques to ensure higher survival of fry.

Miscellaneous Marine Commercial Species

The culture of truly marine fishes is still at the primary stage as compared to the fresh, brackish water and euryhaline commercial species. However, during the last few decades, culture of a few marine fishes in cages or pens in marine environment have been initiated in several countries, but excepting yellowtail and red sunbreams in Japan, most others are still in the experimental stages. More recently, attempts have been made to induce breeding in many marine fishes including groupers, flatfishes, pompanos and others for obtaining their seed for culture.

Scores of experiments on hormone treatment in marine fishes are being carried out all over the world. Quite a few of those have given encouraging results. Chen *et al.* (1977) reported for the first time the successful spawning of the grouper *Epinephelus lauvina* by

IM injection of acetone-dried chum salmon pituitary gland combined with HCG, Groupers are the protogynous group of hermaphrodite fishes and the success in induced spawning has opened the possibility of commercial production of seed for their culture. In Hong Kong. Tseng and Ho (1979) succeeded spawning the red grouper E. akaara with injection of puberogen. Subsequently, Kuo *et al.* (1988), obtained spawning in the grouper E. fario by multiple injections of HCG (Puberogen). Tucker (1972) spawned Nassau grouper E. striatus by HCG administration.

Several attempts have been made to spawn flat-fishes and initial success has been obtained in some. Smigielski (1975), obtained spawning in summer flounder Paralichthys denatus with frozen carp pituitary powder. Hirose *et al.* (1979) were able to strip the Japanese flounder Limanda yokohamoe after injection of puberogen and fertilize and stripped eggs. Harmin and Crim (1989) obtained spawning in the winter flounder (Pseudopleuronectes ameriucanus) by implantation of GnRH-A; also Sanborn and Misitano (1991) induced gonadal maturation in English sole parophrys vetulus, and prercocious ovulation in Salea salea (Ramos, 1986), both by using LHRH-A.

Preliminary attempts made in breeding propanos by hormone injections were not very successful. Finucane (1970) reported spawning of Florida pompano, Trachinotus carolinus by injection with CG but the eggs did not hatch. Suquet *et al.* (1988) also spawned *T. goodie* by hormone treatment but hatching did not occur.

Hormone treatment on several other miscellaneous marine fishes gave some positive response. Gordin and Zohar (1978) reported induction of ovulation and spawning in Sparus aurata with HCG. The spawning season was also extended by 5.5 months by hormone administration. Prentice and Thomas (1978) spawned orangemouth corvine (*Cynoscion xanthulus*) with LHRH and pimozide. Thomas and Boyd (1989) reported spawning of spotted seatrout *C. nebulosus* by oral administration of LHRH-A. Ovulation in the snapper *Pagrus auratus* was obtained by injection of HCG (Battaglene and Talbat, 1992); also Soletchnik *et al.* (1989) successfully spawned yellowtail snapper (*Ocyurus chrysurus*) in captivity.

From the above findings and scores of other studies made on hormonal induction of marine species, there is reason to be optimistic about the development of feasible techniques of breeding and larval rearing of commercially important marine fishes so as to ensure a steady supply of their seed for mariculture and also for sea-ranching.

Tilapias (Cichlids)

Tilapias are African species and are largely cultured in their native land. A few faster growing species, however, have been introduced into various countries of the world. Of these, the mouth–breeder oceochromis niloticus is the most popular as a table fish and is cultured extensively in many countries giving very high production.

Tilapia are prolific breeders and breed in ordinary ponds in the tropics almost throughout the year. Hence, special efforts are not needed for their seed production; on the contrary, preventing reproduction or controlling their overpopulation in cultured ponds are the major constraints faced by tilapia culturists. However, controlled spawning at will is sometimes necessary, specially, for experimental purposes, such as, selective breeding and hybridization, for genetical studies or induction of polyploidy and others. So far, hardly any research has been done on this particular aspect of tilapia breeding, more so with the nest building group, such as *O. niloticus* and others.

Initial attempts made on hormone induction did not meet with much success. Srisakkultiew and Wee (1988) tried several doses of injection on *O. niloticus* with FPE and HCG but failed to get any spawning excepting in two cases where partial success was achieved. However, Babiker and Ibrahim (1979) got encouraging results, when they could get ovulation in *Tilapia nilotica* by introperitoneal injection of contisol, corticosterone and estrodiol in a dose related manner and also with high dose of HCG. They failed to induce ovulation with estrone, estriol and progesterone. However, success in inducing spawning in tilapia has been reported by Rothbard and pruginin (1975) by manipulation of environmental parameters, such as temperature and light. Hyder (197C) also mentioned about temperature, light and rainfall as major external factors stimulating spawning activity of tilapia.

It would be interesting to study the effect of both environmental factors as well as hormone treatment in tilapia in order to induce the cichlids to spawn at will or induce sterility in them.

Conclusion

In the foregoing, an attempt has been made to present a brief history or research carried out to induce breeding of fishes either by manipulation of environmental parameters or by administration of various hormones as well as non-hormonal substances. While reviewing the results obtained through different inducing agents, no detailed discussions were made on the endocrinological aspect of the various investigations, but emphasis was mainly given to the application of the successful result for the production of seed for culture.

It has been observed that from 1930's till the end of 1960's, hypophysation was generally practiced as the most efficient inducing agent, especially, in case of the widely cultured cyprinid fishes, sturgeons and mullets. Initial experiments conducted with hormones of mammalian origin gave mostly negative results. However, within a few decades, HCG and a few other mammalian hormones were found to give positive response. A highly probable explanation for the initial failure, and success in later years could be that in earlier experiments fishes from the wild stock only, were subjected to hormone induction, whereas in later years in spawners obtained from the fish farms were either bred previously or were raised from the induced bred progeny. Hence, there is possibility of the farm raised fishes getting some sort of domestication. Pillay (Anon., 1979) reports that according to the Chinese fish farmers, fishes which have been bred a few times are more responsive and can be bred by HCG alone. This explains why the fishfarmers in China had switched on HCG replacing CPE or synahorin + CPE in their regular breeding programme. In case of milkfish, initially it was a Herculean task to breed the large size spawners caught from the wild. However, within a few years after the first success, the induced bred progeny reared in cages in the marine environment started spawning spontaneously without any hormone treatment.

Liley (1969), while discussing on the role of various hormones in the regulation of the reproductive behaviour in fishes, remarked, "It is impossible at this stage to draw any firm conclusion regarding the mechanism by which spawning is regulated in fish". However, during 1970's a new line of approach is endocrinological research was initiated to intervene at the hypothalamus hypophysis interface instead of hypophysis itself. This had been a definite advancement over the previous approach of administration of exogenous gonadotropins.

In 1982, Huisman, in his concluding remarks at the 'International symposium of Reproductive Physiology of Fish', 2-6 Aug. 1982, Wageningen, Netherlands, stated, "we are desperately in need of more accurate and dependable methods in order to guarantee a safe and continuous supply of young fish of different species for aquaculture purpose." However, the subsequent decade, 1982-92 has demonstrated spectacular on induced breeding of fishes. With the discovery of the mammalian and piscine gonadotropin-releasing hormone agonists and the use of dopamine antagonists, more consistent and predictive results have been achieved in breeding cultured fishes. This new approach is generally regarded by many workers as the "fourth generation hormonal induction system". Also, the preliminary trials made on the pulsatile release of GTH and implantation of long acting LHRH-A pellets in the body of the female spawners and the use of osmotic pump for the slow release of hormone had been the latest addition to the induced breeding technology.

Simultaneously, research on brood stock development has also progressed considerably. It is well known that spawning success largely depends on the gonadal conditions of the spawners and once the gonadal stage is in ideal condition, induction of spawning becomes easier. Such ideal conditions could be achieved by manipulation of environmental factors and proper nutrition of the brood stock, in addition to the endocrinological research to control gonadal development by induction of the process of vitellogenesis, oocyte maturation and ovulation in the females.

The preceding paragraphs reveal that the latest development in the induced breeding technique and its application to the field for intensification of commercial fishfarming has shown great promise. It has been observed that the hypophysation technique is being gradually replaced by the more sophisticated ones but still the former, is widely practiced by fish farmers mainly because of the lower cost, simpler technique and ready availability of the hormones as compared to the latter techniques. Efforts should therefore be made to remove these constraints so that the latest improved technique could be adopted by the end users.

From the pisciculturists' point of view, it is highly encouraging to note that practically all the major roughs of cultivated species could be induced to spawn and seeds made available for culture excepting the truly marine fishes. However, in recent years, increasing interest has been observed in the farming of marine fishes in cages in coastal waters. It is hoped that effective technology related to mariculture will soon be developed.

Finally, to sum up, the results achieve so far on induced breeding of cultured fishes to facilitate production of quality seed for aquaculture have been highly encouraging. However, lot more to be done towards standardization of the various procedures, right from brood stock development to further refinement of the advanced techniques of induction of spawning and improvement of larval rearing methods for obtaining higher survival of fry for stocking culture environment.

References

Abraham, M. 1988. Recent treads in research on induced spawning of fish in aquaculture J. Appl. Ichthyol. 4:49-46.

Arraham, M., Blane, N. and Yashouv, A. 1966. Oogenesis in five species of grey mullets (Teleostei, Mugilidae) form natural and landlocked habitats Isreal J. Zool., 15:155-172.

Alikunhi, K.H., Sukumaran, K.K. and Parameswaran, S. 1963. Induced spawning of the Chinese grass carp. *Ctenopharyngodon idellus* (C and V) and the silver carp. *Hypophthalmichthys molitrix* (C and V) in ponds at Cuttack, India. Proc. Indo-Pacif. Fish Coun., p. 181–204.

Bhowmick, R.M. 1979. Observation on the use of human chorionic Gonodrotropin prepared in the laboratory in induced spawning in the major carp Labeo rohita. Tech. Pop. 16. Presented at the symposium on Inland Aquaculture, Barrackpore, India.

Bhowmick, R.M. Kowtal. G. V. Jana, R. K. and Gupta, S. D. 1986. Observations on the use of various hormones and clomiphene citrate in hypophysation in India major carps. Veterinarski Archiv. 56: 285-292.

Chaudhuri, H. 1965. Effect of pituitary injections and other factors on pond fishes. Ala. Pol. Inst. Bull., 1965 (M. S.–Dissertation: Abstract published).

Chaudhuri, H. 1960. Experiments on induced spawning of Indian carps with pituitary injections. Indian J. Fish., 7: 20-48.

Chaudhuri, H. 1963. Induced spawning of Indian carps. Proc. nat. Inst. Sci. India (B), 29: 478-487.

Chaudhuri, H. 1968. Breeding and selection of cultivated warm water fishes in Asia and the far east–A review. FAO fish. Rep. (44) 4:30-66.

Chaudhuri, H. 1969. Techniques of hypophysation of Indian carps. FAO/UNDP Regional seminar on Induced Breeding of Cultivated Fishes, Fri/IBCF/II. P15 (mimeo).

Chaudhuri, H. 1972. Production of fish seed by induced breeding. Presented at the seminar on production of quality fish seed for fish culture, Barrackpore, India. Nov. 1-2, 1968. Proc. Published in Sept. 1972: 270-282.

Chaudhuri, H. 1979. Use of hormones is induced spawning of carps. J. Fish. Res. Bd. Can., 33: 940-947.

Chaudhuri, H. and Alikunhi, K. H. 1957. Observations on the spawning of Indian carps by hormone injections. Curr. Sci. 26: 381-382.

Chaudhuri, H. and Juario, J.V. 1977. Use of hormone in breeding cultivated warm-water fishes with special reference to milkfish Chanos chanos (Forsskal). Fish. Res. J. Philippines, 11: 1–6.

Chaudhuri, H., Tripathi, S. D. 1979. Problems of warmwater fish seed production. In Pillay, T. V. R. and Dill, Wm. A. (eds): Advances in Aquaculture, pp. 127-134. Fishing New Books Ltd.

Chaudhuri, H., Singh, S. B. and Sukumaran, K. K. 1966. Experiments on large-scale production of fish seed of the Chinese grass carp, *Ctenopharyngodon idellus* (C. and V) and the silver carp, *Hypophyhalmichthys militrix* (C and V) by induced breeding in ponds in India, Proc. Indian Acad. Sci., 63: 80-95.

Chaudhuri, H., Bhowmick, R. M., Kowtal, G. V., Bagchi, M. M., Jana, R. K. and Gupta, S. D. 1977a. Experiments on artificial propagation and larval development of Mugil cephalus Linnaeus in India. J. Inland Fish. Sco. India, 9: 30-41.

Chaudhuri, H., Juario, J. V., Primavera, J. H., Meteo, R., Samson. R., Cruz, E., Jorabejo, E. and Canto, J. Jr 1977b. Artificial fertilization of eggs and early development of the milkfish, Chanos chanos (Forsskal). SEAFDEC Aquaculture department, Tech. Rept. 3:21-38.

Chaudhuri, H., Bhowmick, R. M., Kowtal, G. V., Jana, R. K. and Gupta, S. D. 1977c. Efficacy of partially purified salmon pituitary gonadtropin (SD-G100) in inducing spawning in Indian major carps. J. Inland Fish Soc. India 9:162-164.

Chaudhuri, H., Juario, J. V., Primavera, J. H. Samson, R. and Mateo, T. 1978. Observations on artificial fertilization of eggs and the embryonic and larval development of milkfish, Chanos chanos (Forsskal). Aquaculture, 13: 95-113.

Fontained, M. 1976. Hormones and the control of reproduction in aquaculture. J. Fish. Res. Bd. Can., 33: 922–939.

Jhingran, V. G. 1969. Review of the present status of knowledge on induced breeding of fishes and problems for future research. FAO/UNDP Regional seminar on Induced Breeding of Cultivated Fishes. Fri/IBCF/27, p30. (mimeo).

Sinha, S. B. 1969. Techniques of hypophysation of Chinese carps. FAO/UNDP Regional Seminar on induced Breeding of Cultivated fishes. FRI/IBCF/12; 17p.

Singh, A. K. and Singh, T. P. 1976. Effects of clomid, sexovid and prostaglandins on induction of ovulation and gonadotropin secretion in a freshwater catfish, Heteropneustes fossilis (Bloch). Endokrinololgie, 68: 130-136.

Chapter 4

Use of Hormone in Aquaculture

Editor's Note

Aquaculture technology has progressed tremendously during the past several decades. The progressive use of hormones in the various stages of aquaculture management has contributed immensely to the development of fish culture as an industry. The failure in obtaining sufficient seed of the major groups of cultivated commercial species had been a stumbling block in the development of fish culture for centuries. The successful application of hormones in inducing spawning in the majority of the important cultivated fishes has removed this main obstacle and it has been possible to produce adequate quantities of quality seed of these species for culture. More recently, in addition to the induction of spawning, hormones are being used in several other important phases of fish cultural practices, namely, brood stock development, larval rearing and growing fish in grow-out ponds.

For induction of spawning, the various hormonal inducing agents used, are the fish pituitary extracts (FPE), partially purified gonadotropins, HCG alonge or in combination with FPE, various other hormones of mammalian origin, luteinizing hormone releasing hormone (LH-RH) and its superactive analogues, the mammalian (mGnRH) and piscine (sGnRH) gonadotropin–releasing hormone agonists combined with dopamine antagonists and others. The role of these, and other hormones, such as, cortisteroids, progestogens, prostaglandins, pheromones and thyroid hormones in the induction of ovulation spermination and spawning in fishes have been discussed.

For the development of brood stock, hormones are used to obtain proper development of gonads and also to effect transfer of some useful substances from mother to the eggs for production of improved quality of eggs. Another new trend in hormonal research in fish culture is the use of cortisone, thyroxin (T_4) and triiodothyronine (T_3) for enhancing the growth and survival of fish larvae. Experimental studies in the use of growth hormones have demonstrated accelerated growth of fishes in grow-out ponds. The technique of sex-reversal in some fishes by oral administration of sex-horhmones to facilitate monosex culture for increased fish production is also briefly mentioned.

Use of Hormones in Induced Spawning of Carps

Reproductive Physiology of Carps and Other Fishes

The major India and Chinese carps (*C. catla, L. rohita, C. mrigala, H. molitrix, A. nobilis, C.molitorella* and *C. idella*) do not breed in ponds but they almost always attain maturity in such an environment. Lal (1962 and 1963) described the eytological changes that characterize the developing oocytes of *C. mrigala* and recognized six well-defined stages through which the ova pass before being finally shed. Kurth and Chow (cited by Chen *et al.* 1969) identified seven developmental stages in grass carp oocytes in the tropics. They observed that the eighth stage (prematuration stage) is rarely reached, but once attained, the failure of the fish to spawn results in the breakdown and resorption of the eggs internally. Yamamoto and Yamazaki (1961), however, recognized eight stages in the development of oocytes of the goldfish (*Carassius auratus*). According to them, the ripe stage is characterized by the presence of the nucleus at the micropyle which could be seen through the zona radiate at the animal pole. The living eggs, at this stage, lose their opacity and turn translucent.

The mechanism of ovulation in goldfish has also been described by Yamamota and Yamazaki (1967). The blood capillaries in the follicular layer become remarkably expanded before the onset of ovulation. A narrow silt appears between the fully ripe oocytes and follicle cells. The junction of the oocytes with the follicle cells becomes loose and the follicle ruptures in the region facing the visceral cavity, irrespective of the location of the pole of the oocyte.

The mechanism of ovulation in the difficult to spawn carps has not been studied thoroughly. The final stage of maturation in these fishes is never attained in ponds. Slack (1962) opined that the failure to spawn naturally in ponds in a tropical country is not due to suppression of gonad development but to some environmental factors responsible for bringing about spawning behavior and ovulation. The major Indian carps, however, breed in certain specialized types of ponds where perhaps the external stimulus is provided by natural environmental conditions (Hora 1945) which trigger the endocrine mechanism of the fish resulting in its spawning (Khan 1945). This mechanism is otherwise triggered in a fully ripe fish by administration of fish pituitary extracts.

Fish Pituitary Gonadotropins

While two gonadotropins responsible for ovulation and spawning have been identified in *Anguilla* (Olivereau and Herlant 1960), *Carassius auratus* (Olivereau 1962), *Oncorhynchus* (Olivereau and Ridgeway 1962), and *Mugil capito* (Olivereau 1968), only one has been identified in *Heteropheustes fossilis* (Sundarraj 1959), *Cyprinus carpio* (Fontaine and Gerard 1963), *Poecillia latipinna* (Ball and Baker 1969), *Oncorhynchus nerka* (McBride and Van Overbeeke 1969), *Carassius auratus* (Yamazaki 1969), and *Anoptichthys jordani* (Mattheij 1970). Hora (1955) also suggested that possibly only one gonadotropin is secreted by the fish pituitary. The two gonadotropins in the teleostean hypophysis, which regulate the mechanism of reproduction, are believed to be FSH-and LH-like (Kazanskii 1951 and 1952; Ball 1960). Witschi (1955) has shown that the teleostean hypophysis contains an LH-like hormone which is similar to that found in mammals. The presence of the above two types of hormones in the fish pituitary led scientists to believe that their action is similar to that of mammals. Administration of the mammalian preparations, LH and FSH, induced ovulation in some species of fish, confirming their role in spawning (Pickford and Atz 1957). However, the possibility that the naturally occurring gonadotropin is a single, multipurpose hormone which is broken down by chemical fractionation cannot be excluded (Deminalti and Weill 1958, cited by Atz and Pickford 1959).

The pituitary hormones being protein like in nature, it is natural to expect that the chemical structure of the gonadotropins would be different in different species of fish (Pickford and Atz 1957). "Phylogenetic specificity," "species specificity," or "sex specificity" were considered to have a bearing on successful induced spawning. Kazanskii (1940) pointed out that very little specificity exists among genera but it is prominent among orders. No "species specificity" exists as far as the carps are concerned and it has been our experience that the major Indian carps can be bred by administration of pituitary gland extracts from each other or from Chinese carps and common carp and vice versa. Recently, pituitary glands from catfishes have also been used for induced breeding of carps (Barrackpore 1972). "Sex specificity" has no role in induced spawning of carps (Ibrahim and Chaudhuri 1970). However, detailed biochemical and biological assay is necessary to confirm the various types of "specificity" as the results obtained so far have been conflicting.

Gerbil' Skii (1940) noted seasonal variations in the gonadotropin content of the pituitary. Menezes (1945) found relatively little gonadotropin in the pituitary glands of immature fish. Glands for preparing extracts for induced spawning of fish are, therefore, collected only from maturing or fully mature fish.

Role of Mammalian and Other Hormones in Induced Spawning

While various synthetic and mammalian hormones have precipitated spawning or brought about ripening of eggs in a few species, these hormones have yielded positive in Indian and Chinese carps only when used in combination with fish pituitary hormones. Commercial oxytocin (extract of posterior lobe of bovine pituitary extract) when injected at 10-13 I.U. after a preparatory dose of 2 mg of pituitary extract/kg body weight of fish did not result in ovulation of major carps. Lutocycline or ovocycline alone did not have any effect (Barrackpore 1967).

Pregnancy urine (PU) and various extracts derived the reform *viz.,* human chorionic gonadotropin (HCG or CG), prolan, and Antuitrin-S have brought about ovulation and spawning in several species as well as release of immature eggs on many occasions in others (Pickford and Atz 1957). The similarity of HCG to LH and the presence of hormonal contaminants in UP and others tends to suggest again that it is the LH-like hormones which bring about ovulation and that their content in a particular extract is responsible for the positive or negative results.

HCG alone could not induce successful spawning in the major Indian carps but when injected in combination with a lower dose of carp pituitary extract (3-6 mg/kg) + 1 mg HCG, it effected spawning. Singly, Synaborin was equally ineffective but in combination with pituitary extract (25 RU of Synahorin + 4-6 mg of pituitary extract/ kg) it induced spawning in *L. rohita* (Barrackpore 1968). Hora (1968) has also observed that fish are relatively unresponsive to mammalian placental gonadotropins. Antuitrin-S (HCG) injected alone at 100-1500 TU/kg or along with pituitary extract (2-4 mg/kg) did not have any effect on the ovulation in *L. rohita* (Barrackpore 1970). Synahorin, in combination with fish pituitary extract, has been used in Taiwan for induced spawning in grass carp (Lin 1965) but Chen *et al.* (1969) have found that this species could be bred even without synahorin. Yamamoto and Yamazaki (1967) have reported that HCG, APE, PMS, and a mixture of HCG and APE can induce ovulation in intact or hypophysectomized gravid goldfish. It is well known that HCG resembles LH in its physiological action and APE contains a large quantity of FSH, a small quantity of LH, and some other hypophyseal hormones in traces.

PMS, which resembles FSH in its actions, has no LH activity at all. Either alone or in combination with a threshold dose of homoplastic pituitary gland it failed to induced spawning both in *Clarias* and *Heteropneustes* (Ramaswami and Lakshman 1959).

It has been found that adrenocorticotropic hormone (ACTH) has no effect on catfish ovary even in combination with a threshold dose of the pituitary gland. Similarly, oxytocin also appears to have no effect, even when combined with pituitary gonadotropins on the gonads of these catfishes (Ramaswami and Lakshman 1959). Sinha (1971) obtained no response either in the form of courtship behavior or ovulation when the posterior lobe hormone, ACTH, LH FSH, CG and Synahorin were infected into *P. gonionotus*.

Very few workers have used steroid hormones to induce spawning in fishes. Kirshenblat (1952, 1959) first demonstrated that progesterone and deoxycorticosterone induced ovulation in *Misgurnus* fossilis, whereas estrogens and androgens had no effect.

Ramaswami (1962) failed to induce spawning in *H. fossilis* by treatment with gonadal steroid hormones. Sundararan and Goswami (1966) also failed in their attempt to induce ripening of eggs by administration of oestradiol benzoate, testosterone propionate and progesterone. It has been observed that both male and female sex steroids, when administered over longer periods, bring about regressive changes in the ovaries by virtue of their feedback effect on the pituitary of the recipient (Sundararaj and Goswami 1968). A series of adrenocorticosterids (DOCA, hydrocortisone and their respective acetates) were found to induce ripening of eggs in intact *H. fossilis* (Ramaswami 1962). However, on a dose level basis, both hydrocortisone and cortisone were much less effective than luteinizing hormone. Khoo (1974) observed that treatment of gravid goldfish (*C. auratus*) with progresterone and corticosteroids induces ovulation.

Pandey *et al.* (1973) used clomiphene citrate to induce ovulation in *Carassius auratus*. However, both intramuscular and intraperitoneal injections of chlomiphere citrate at 8-28 mg/kg in two doses did not precipitate spawning in *L. rohita*.

Considerable work has been carried out to study the effect of various hormones on the reproductive physiology of several species of fish, but hardly any information is available on the regulation of hormones in the Indian and Chinese carps. Due to their uncertain and unsuccessful results, the synthetic and other hormones have not found much favor with pisciculturists. The crude fish pituitary extract is, however, being used almost all over the world for inducing spawning in cultivated carps (Makeeva 1963; Prikhod'ka and Nosal 1963; Tang *et al.* 1963; Wu and Chung 1964; Konradt 1968; Chen *et al.* 1969). Perhaps the synergic action of growth hormone, thyrotropic hormone and sex steroids accounts for the better results obtained by administration of crude rather than purified gonadotropic preparations (Atz and Pickford 1959). In India, the crude pituitary extract is prepared in distilled water and used extensively to breed both the Indian and Chinese carps on a commercial scale. Only recently, a miniature "Pituitary Bank" has been established at the Central Inland fisheries Research Substation, Cuttack, from where ampouled pituitary gland extract is supplied to fishes organizations and fish farmers for ready use.

Technique of Induced Spawning the Major Indian and Chinese Carps

The technique of hypophysation in India has been described in detail by Chaudhuri (1963).

Collection of Pituitary Gland

The pituitary gland is removed from ripe male or female carps during May and June. The common carp (*C. carpio*) is a good donor and glands from this species can be collected throughout most of the year. Fishes which have been induced a spawn through administration of pituitary extract are also used as pituitary donors within 6-8 h of their spawning. The glands are preserved in absolute alcohol and stored in a refrigerator or in a cool place.

Preparation of Pituitary Extract for Injection

A known quantity of the gland, depending upon the requirement, is placed in a tissue homogenizer and homogenized with a very small quantity of distilled water. The homogenate is thoroughly washed into a centrifuge tube and the extract centrifuged. The supernatant liquid, which contains the pituitary hormones, is drawn into a hypodermic syringe and made up to the required volume of injection. Recent experiments on preservation of fish pituitary extract in glycerine and ampouling have been quite successful. Twenty to 40 mg of pituitary glands per milliliter of water can be preserved in two parts of glycerine and ampouled for ready use of the fish farmer (Ibrahim and Chaudhuri 1966).

Selection of Breeders and Injection

The major Indian carps can be easily sexed during the monsoon months by feeling their pectoral fins, the dorsal surface of the males being rough while those of the females smooth (Chaudhuri 1959). Fully ripe males and females are selected for hypophysation. The female is given a priming dose of 2-3 mg/kg body weight and, after and interval of 4-6 h, a final dose of 5-8 mg/kg body weight. It is at the time of the second injection of the female that the male is also injected at 2-3 mg/kg. The Chinese carps are similarly treated but the doses are slightly higher. The initial dose for a female is normally 3-4 mg/kg while the final dose ranges from 8-11 mg/kg. The male is injected at 3-4 mg/kg. A set of breeders comprises two males and one female by numbers which are generally in a ratio of 1:1 by weight.

Ovulation and Spawning

The males and the female are kept together in the hapa (a rectangular cloth enclosure) where they normally breed within 4-6 h of the second injection of the female. Chinese carps are normally stripped. About 4 h after injection a close watch is kept on the female and as soon as it is ready for oulation, the fish is taken out and the eggs stripped in batches into enamel trays (Chaudhuri *et al.*, 1966). The eggs are fertilized by stripping the milk from the males. The dry method of stripping is followed as it has been found to give better results than the wet method. The fertilized eggs are washed several times with water and then kept in a hapa for water-hardening.

The number of water-hardened eggs of the Indian and Chinese carps are then estimated quantitatively and transferred to hatcheries. The incubation period varies between 15 and 18 h at water temperatures ranging from 26 to 31°C.

It needs to be noted that water temperature is an important factor for the success of induced spawning of fish. Complete spawning and a high percentage of fertilization is usually achieved within an optimum range of 24-31°C. Cool rainy days have been observed to be conducive to successful spawning.

Isolation and Purification of Fish Pituitary Gonadotropins

Several attempts have been made in recent years to isolate and purify teleostean gonadotropic hormones (GTH) from pituitary glands. Otsuka (1956) prepared 40 per cent ethanolic extracts and 20 per cent saline extracts of acetone dried salmon pituitaries with further fractionation by rivanol and acetone. A solution of acetone and acetic acid was used by Robertson and Rinfret (1957) to extract the gonadotropin from the pituitary glands of *Oncorhynchus ishawytscha*. Salmon gonadotropin was purified and an assay method based on the spermination response of hypophysectomised goldfish developed by Yamazaki and Donaldson (1968). Donaldson *et al.* (1972) prepared Chinook salmon gonadotropin (SG-G 100) from the pituitary glands of sexually mature fish. Fontaine and Gerard (1963), Clemens and Grant (1964a), Clemens *et al.* (1964) and Burzawa-Gerard (1971) purified and characterized the carp (*C. carpio*) gonadotropin. While Breton (1968) purified the gonadotropins of *Coregonus lavaretus* and *Gardonus* sp., Sinha (1969,

1971 and 1974) worked on the purification of gonadotorpins from the pituitaries of gonionotus, *A. nobilis, H. molitrix, C. idella,* and *C. catla.*

The carp gonadotropin was observed to have a molecular weight of 27,000 with a shoulder having a molecular weight of 15,000 (Burzawa-Gerard 1971). The molecular weight of in fact salmon gonadotropin ranged from 28,500 to 29,400 at pH 7.8 and compared closely with the values obtained for carp gonadotropin (Donaldson 1973). The behavior on ion exchange columns of both carp and salmon gonadotropin suggests that they are acidic and thus more similar to mammalian FSH than LH (Fontaine and Gerard 1963; Donaldson *et al.* 1972) Sinha (1969) obtained three different fractions by molecular sieving of the whole pituitary extracts on Sephadex G-100. The second fraction alone brought about courtship behavior and ovulation in the carps (Sinha 1971 and 1972) and in the catfish (Sundarraj *et al.* 1972), Suggesting thereby the presence of an active gonadotropin.

Clemens and Grant (1964a), while purifying the carp gonadotropin, used the goldfish testicular hydration assay. They referred to this gonadotropin as the gonadal hydration factor. When injected with carp pituitary extract the gonadal water content of female goldfish increases which appears to be associated with ovulation. The hormone regulating the hydration process appears to be a gonadotropin and the involvement of electrolytes and other endocrines is also suspected (Clemens and Grant 1964b). Sinha (1972) studied the gonadal hydration response of *P. gonnionotus* by using the second fraction of the purified carp pituitary extract and suggested that the gonadotropin in the second fraction is involved in osmoregulation of fish. An injection of an additional amount enhances the rate of ion water transport, especially to the gonad, resulting in its final maturation.

Bioassay studies in India have been limited to using glycoprotein solvents for extraction of gonadotropins (Ramaswami and Lakshman 19600; Bhowmick and Chaudhuri 1968). Spawning was achieved when pituitary glands immersed in 1.25 per cent TCA for 6 and 12 h were injected into *H. fossilis*. The acidic fluid in which the glands were immersed was found to be equally potent (Ramaswami and Lakshman 1960). However, neither the glands nor the fluid could induce spawning when the glands were immersed for 6 h in 2.5 per cent TCA, probably due to denaturation of the gonadotropins. Similar results were obtained with Indian carps when carp glands were immersed for 6 h but ovulation could be induced successfully even when the duration of immersion was reduced to 3 h. Extracts obtained after immersing the glands for 6 and 12 h in 1.5 per cent TCA could also induce spawning. It is, however, felt that the gonadotropins are not fully extracted in 1.5 per cent TCA and a much higher dose is required to precipitate spawning (Bhowmick and Chaudhuri 1968).

SG-G 100 has been recently used in India to precipitate spawning in the major Indian carps. However, it has been found to be effective when doses comparable to the crude carp pituitary extract are administered. Sinha (1974) also found that the effective doses of injection of the lyophilized second fraction to *C. mrigala* and *H. fossilis* were almost the same as the crude carp pituitary extracts. Since the crude carp pituitary extract is effective and the dosages required are almost the same as lyophilized fractions there seems to be no additional advantages in using, such preparations for commercial induced breeding programs.

A review of the present status of studies on the reproductive physiology of fishes, especially those belonging to the order Cyprinoidea in the forgoing pages clearly indicates that considerable information is available on the carps, *Carrassius auratus* and *Carassius carassius* and the catfishes, *Heteropneustes fossilis* and *Clarias batrachus*. However, a little is known about the difficulties to spawn major Indian and Chinese carps. It is essential that detailed histophysiological investigations of their maturity cycle, hormonal regulation, and the mechanism of ovulation and spawning be made. Efforts should also be made to evolve a suitable substitute, either a natural or synthetic hormone, a steroid, or a chemical for replacing the crude fish pituitary which is at present extensively used inbreeding these carps. Success has been achieved in isolating and purifying fish pituitary hormones but these do not yet have much advantage over the crude pituitary glands in the carp, recent studies on the endocrine control of oogenesis and vitellogenesis and induction of ovulation by steroid hormones in certain fishes have opened up new possibilities in the reproductive endocrinology of teleosts. Investigations to assess their effectiveness on the cultivable species of carps and the economics thereof may be made so as to utilize them for commercial production of fish seed in piscicultural practices all over the world.

References

Atz, J.W. and G.E. Pickford, 1959. The use of pituitary hormones in fish culture. Endeavour. 18: 125-129.

Barrackpore Central Inland Fisheries Research Institute. 1967. Annual Report 1967: 58 p.

Barrackpore Central Inland Fisheries Research Institute. 1967. Annual Report 1968: 93 p

Barrackpore Central Inland Fisheries Research Institute. 1967. Annual Report 1970: 117p

Barrackpore Central Inland Fisheries Research Institute. 1967. Annual Report 1972: 113p

[NB: Presented by Professor Chaudhuri in Pacific Service Congress 1975]

Bhowmick, R. M., and H. Chaudhuri, 1968. Preliminary observation on the use of trichloracetic acid for extraction of gonadotrophic hormones in induced spawning of cases, J.Zool. Soc.India 20:48-54.

Chaudhuri, H. 1959. Notes on external characters distinguishing sex of breeders of the common Indian carps. Sci. Cult. 25:258-59.

Chaudhuri, H. 1963. Induced spawning of Indian carps. proc. Natl. Inst. Sci. India Part B 29: 478-87.

Chaudhuri, H., S. B. Singh and K. K. Sukumaran 1966. Experiments on large scale production of fish sewed of the Chinese grass carp, *Ctenopharyngodon idellus* (C. and V) and the Silver carp. Hypophthalmichthys molitrix (C and V) by induced breeding in ponds in India Proc. Indian Acad. Sci. Sect.B 63: 80-95.

Donaldson, E. M. 1973. Reproductive endocrinology of fishes. Am. Zool. 13: 909-927.

Donaldson, E. M., F. Yamazaki, H. M. Dey, and W. W. Philleo, 1972, pituitary glands. Gen. Comp. Endocrinol. 18: 469-481.

Ramaswami L. S. and A. B. Lakshman. 1959. Action of mammalian hormones on the spawning of catfish. J. Sci. Ind. Res. 18C: 185-191.

Ramaswami, L. S. and A. B. Lakshman. 1960. Bioassay studies of the catfish and skipper frog pituitary gland after treatment with some glycol protein solvents. Proc. Natl. Inst. Sci. India Part B25: 68-79.

Sinha, V. R. P. 1969. Chromatograpy of fish pituitary extracts on Sephadex G-100. J. Chromotogr. 44:624-628.

Sinha, V. R. P. 1971. Induced spawning in carp with fractionated fish pituitary extract. J. Fish Biol. 3: 263-272.

Sinha, V. R. P. 1972. Gonadal hydration response and major electrolyte levels in serum and seminal fluid of *Puntius gonionotus* (Teleostei) injected with the fractionated fish pituitary gonadotropin. J. Inland Fish. Soc. India 6: 150-154.

Sundarraj, B. I. 1959. A study on the correlation between the structure of the pituitary gland of the Indian catfish heteropneustes and the seasonal changes in the ovary. Acta Anat. 3:47-80.

Chapter 5

Pituitary Gonadotrophin in Indian Major Carps

Efficacy of Partially Purified Salmon Pituitary Godadotrophin (SG-G100) in Inducing Spawning in Indian Major Carps

Editor's note

Ever since the success of induced breeding of fishes in many countries, the fish culturists do not extract the gonadotrophin from the pituitary but use the whole pituitary extract and achieve good success. In India too, since the first success of induced breeding of major carps, catla (*Calta catla*), rohu (*Labeo rohita*), calbasu (*L. calbasu*) and Mrigal (*Cirrhinus mrigala*) in confined waters, the fish farmers have used pituitary extract and obtained seed of these fishes on a commercial scale. In the meanwhile, attempts on the isolation of gonadotrophic hormone from the pituitary glands of food fishes, have made considerable headway.

Partially purified salmon pituitary gonadotrophin (SG-G100), received from Canada (IDRC), was tried on rohu to find out its effect in inducing spawning with suitable controls injected with carps pituitary extract. The results of the experiments carried out during 1975 at Cuttack are presented. While rohu was the test fish, SG-G100 was tried on Catla and Mrigal also.

Introduction

The technique of induced breeding described by Chaudhuri was followed with suitable modification. The doses were expressed in terms of mg of the substance used per kg body weight of the recipient. The female received two injections at an interval of 6 hours, the first dose being 2-3 mg/kg and the second dose varied from 2-13 mg/kg. The male received only one injection @ 2-3 mg/kg at time of second injection to the female.

Pituitary gland collected from pond reared and induced breed carps and as well as the glycripine preserved pituitary extract were used for the controls.

Experiments were conducted on 29 sets of rohu, 15 sets of catla and 4 sets of Mrigal (a set comprised of one female and two male fishes). The controls comprised of 13 sets of rohu. Water temperature during the experiments varied between 28 to 33^{0}C. A total of 4.891 million of spawn was produced.

Out of 29 sets of rohu injected with SG-G100, 25 sets responded recording 72.4 per cent positive results yielding 2.846 million viable spawn. It is evident from Table 1 that the minimum effective dose was 5 mg/kg body weight of the recipient females. However, a dose of 4mg/kg resulted impartial ovulation in one female. The maximum dose administered was 13 mg/kg body weight which also gave successful spawning. The males were injected at a dose of 2-3 mg/kg body weight.

The controls, comprising of 13 sets of rohu, injected at 5-13 mg/kg with carp pituitary extract resulted in normal spawning while partial spawning was noted at a dose of 4 mg/kg. A total of 1.145 million of spwan was produced from the controls recording 61.5 per cent success.

Fifteen sets of calta were injected with SG-G100 of which 11 gave positive results recording 73.3 per cent spawning success. The females were injected at 10-15 mg/kg in two spilt doses. The effective doses that produced complete spawning ranged from 13-15 mg/kg body weight of the female recipient and yield 0.61 million of spawn.

All the 4 sets of Mrigal injected with SG-G100 at 9-10 mg/kg body weight of females yielded positive results producing 0.29 million of viable spawn (Table 1).

Table 1: Effect of Purified Salmon Pituitary SG-G100 on the Spawning of Indian Major Carps

Species	*No. of Sets*	*Dosage Applied mg/kg*	*Percentage of Spawning Success*	*Total No. of Estimated Eggs (Millions)*	*Spawn Obtained (Millions)*	*Remarks*
Labeo rohita	2	*4	50.0	0.028		*Single dose.
(Experiment)	2	**4				** Two dose
	2	5	100.0	0.151	0.095	
	3	6	33.3	0.140	0.095	
	3	7	33.3	0.168	0.095	
	3	8	100.0	0.882	0.615	
	3	9	66.6	0.364	0.190	
	6	10	100.0	2.086	0.935	
	2	11	100.0	0.784	0.490	
	2	12	100.0	0.420	0.246	
	1	13	100.0	0.154	0.085	
(Control)	13	4-13	61.55	2.650	1.145	
C. catla	1	10	100.0	0.25		
(Experiment)	4	12	25.0			Ovarian eggs
	3	13	66.6	0.3125	0.06	
	1	14	100.0	0.450	0.60	
	6	15	100.0	1.315	0.49	
C. mrigala	1	9	100.0	0.18	0.06	
(Experiment)	3	10	100.0	0.67	0.23	

It is evident from the results that 5 mg of SG-G100 kg weight of the female recipient was the minimum effective dose which induced successful spawning in females yielding viable hatchlings. Since, controls also required a minimum of 5 mg of whole carp pituitary extract for successful spawning, SG-G100 does not offer any added advantage for inducing spawning in rohu. Atz and Pickford have also stated that crude pituitary extract yielded better results than purified preparation and they infer that this may perhaps be due to synergistic action of gonadotropins and other hormones present in a pituitary extract, Further, 5 mg of crude pituitary/kg body weight is the minimum dose recommended for inducing spawning in Indian major carps. However, a single dosage of 4 mg/kg appears to be inadequate since only partial ovulation was induced in both experimental and control fish.

[H. Chaudhuri, R. M. Bhowmick, G.V. Katwal, R.K. Jana, S.D. Gupta (1977) in I.J.E.S.]

References

Atz. J. W. and G. E. Pickford 1959. Endeavour, 18: 125-29.

Bhowmick, R. M. and G. V. Kowtal, 1973. J. Inland Fish. Soc. India. 5: 218-22.

Bhowmick, R. M. and H. Chaudhuri 1968. J. Zool Soc. India, 20: 48-54.

Chaudhuri, H. 1963. Proc. Nat. Inst. Sci. India (B), 29(4): 473-87.

Chaudhuri, H. 1976. J. Fish Res. Board. Con. 33: 940-947.

Chaudhuri, H and K. H. Alikunhi, 1957. Curr. Sci. 26: 381-82.

Donaldson, E. M. 1973. Am. Zool. 13:909-27.

Sinha, V. R. P. 1974. J. Inland. Fish. Soc. India, 6: 150-54.

Chapter 6

Trichloro Acetic Acid for Extraction of Gonadotrophic Hormone

Preliminary observations on the use of Trichloracetic acid for extraction of gonadotrophic hormones in induced spawning of carps.

Editor's Note

This paper gives an account of preliminary bioassay studies made during July-August 1965.

Introduction

Fish gonadotrophic hormones in recent years; have attracted the attention of the fish culturists throughout the world. While extensive literature exists on the mammalian gonadotrophic hormones, only scanty information is available on the exact nature and physiology of the important hormones secreted by the pituitary complex consisting of the follicle stimulating hormone (FSH) and the luteinising hormone (LH) or interstitial cell stimulating hormone (ICSH), was well known, but the first sepraration of the FSH and LH, fractions could be achieved only in 1931 (Fevold *et al*). More recently, the presence of FSH and LH gonadotrophins in fish pituitaries have been demonstrated by a number of workers (see Pickford and Atz 1957; Hoar, 1957; and Pickford, 1959). Witschi (1955) observed that the Teleostean pituitary in low in FSH but contains normal quantity of LH as compared to mammals. Otsuka

(1956) described extraction of FSH and LH from the Salmon pituitaries.

It has been established that the FSH stimulates development of the ovarian follicle while LH brings about ovulation and corpus luteum formation in mammalian females. In mammalian males, FSH stimulates spermatogenesis and ICSH stimulates development of interstitial cell and secretion of male sex hormone in the tests. In fish, the functions of these gonadotrophinare supposed to be comparable to those of the mammals (Pickford and Atz., 1957 and Ramaswami and Lakshman, 1960). The functional properties of the fish gonadotrophin have been profitably utilized by the fish culturists in using these hormones for inducing breeding in fishes. However, fish culturists almost all over the world, in their routine fish cultural practice, do not separate the gonadotrophins from other pituitary hormones, but achieve fairly good success by using the crude extract from the pituitary gland. Since the first success in India, in inducing Indian major carps to spawn by pituitary injections (Chaudhuri and Alikunhi 1957), much work has been done leading to standardization of dosages and commercial production of fish seed (Chaudhuri *et al.*, 1962; and Chaudhuri 1963). In all these experiments only crude fish pituitary extracts were used which yielded highly successful results and this could be probably because of synergic action of somatotrophic, thyrotrophic and other hormones present in the crude preparation. However, attempts are being made by many workers to separate the FSH and LH fractions of the fish pituitary glands. Since FSH and LH gonadotrophins are glycomprotein in nature, investigations have also been carried out to develop a bioassay method by treating the glands with suitable glycoprotein–solvents such as Trichloracetic acid (TCA). Pyridine etc. (Barrnett *et al.*, 1965; Ramaswami and Lakshman, 1960).

Observations

Three experiments were carried out to study the effect of fish gonadotrophins, extracted in 1.5 per cent TCA, in inducing breeding of carps.

In the first experiment, 9 sets (each set consists of two males and one female) of major carps, Catla, Rohu, Mrigal were injected with 1.5 per cent TCA fluid in which both homoplastic and heteroplastic pituitary glands were immerse for 6 hr. out of the 9 sets tried, 8 sets spawned successfully (Experiment I–Table 1).

Table 1: Effort of First Gondadotrophins Extracted by Trichloragetic Acid in Iduing Spawning of Major Carps

Expt. No.	*% TCA*	*Duration of Immersion*	*Recipient Female £ Dose (mg/kg)*					*Remarks*
			Species	*Wt. in kg.*	1st	2nd	*Result*	
I	1.5	6 hr.	Rohu	2.0	4.5	15.0	+	Injected with hormones extracted in TCA
	1.5	6 hr.	Rohu	2.0	6.0	13.5	+	-do-
	1.5	6 hr.	Rohu	1.5	6.0	14.0	+	-do-
	1.5	6 hr.	Rohu	0.5	6.0	12.0	+	-do-
	1.5	6 hr.	Rohu	1.5	6.0	16.0	+	-do-
	1.5	6 hr.	Rohu	1.75	6.0	14.0	+	-do-
	1.5	6 hr.	Rohu	2.0	6.0	16.0	+	-do-
	1.5	6 hr.	Rohu	1.75	8.0	18.0	+	-do-
	1.5	6 hr.	Rohu	1.25	7.0	21.0	+	-do-
II.	1.5	6 hr.	Rohu	2.25	6.0	15.0	+	-do-
	1.5	6 hr.	Rohu	1.25	7.0	16.0	+	-do-
	1.5	6 hr.	Rohu	0.05	6.0	18.0	+	-do-
	1.5	6 hr.	Rohu	1.75	4.0	8.0	-	Immersed gland alone used; 3rd inj. @12 gm/kg.. given
	1.5	6 hr.	Rohu	1.50	5.0	16.8	-	immersed gland alone used; 3rd inj. @ 16 mg/kg given
III	1.5	12 hr.	Rohu	2.25	2.00	6.0	-	Injected with hormones extracted in TCA
	1.5	12 hr.	Rohu	1.00	3.0	7.0	-	-do-
	1.5	12 hr.	Rohu	1.00	4.0	8.0	-	-do-
	1.5	12 hr.	Rohu	1.00	5.0	9.0	-	-do-
	1.5	12 hr.	Rohu	1.00	6.0	10.	-	-do-
	1.5	12 hr.	Rohu	1.00	6.0	12.0	+	-do-

Contd...

Table 8.1–Contd...

Expt. No.	*% TCA*	*Duration of Immer-sion*	*Recipient Female £ Dose (mg/kg)*					*Remarks*
			Species	*Wt. in kg.*	*1st*	*2nd*	*Result*	
IV	2.5	6 hr.	Rohu	1.0	6.0	16.0	-	-do-
	2.5	6 hr.	Rohu	1.0	6.0	18.0	-	-do-
	2.5	6 hr.	Rohu	1.0	6.0	20.0	-	-do-
V	2.5	3 hr.	Rohu	1.0	6.0	16.0	+	-do-
	2.5	3 hr.	Rohu	1.0	6.0	18.0	+	-do-
	2.5	3 hr.	Rohu	1.0	6.0	20.0	+	-do-

Three sets of control were also kept which received the usual dosages of 8-12 mg/kg body weight of the recipient females and 2 mg/kg body weight of the recipient males.

In the second experiment, 3 sets of Rohu were administered with 1.5 per cent TCA fluid in which homoplastic glands were immersed for 6 hr. Another two sets of Rohu were separately injected with the extract of the immersed pituitary gland. While the former 3 sets gave successful results no spawning was in the latter two sets. Controls were kept as in the first experiment which gave successful results (Expt. II- Table 1).

In the third experiment the duration of immersion of pituitary glands in 1.5 per cent TCA was increased to 12 hr. and 6 sets of Rohu were injected with the acid fluid. Three sets were kept as control. With a view to determine the minimum effective dose (MED), various dosages starting with 8 mg. to 18 mg/kg body weight of the recipient females were tried. The set which received the highest dose alone spawned successfully. (Experiment III Table 1). The controls as usual responded when injected with 8-12 mg/kg body weight in females.

Another set of experiments was conducted where hormones were extracted in 2.5 per cent TCA. In the first experiment Rohu glands were immersed in 2.5 per cent TCA for 6 hr. after which the acid fluid was injected to 3 sets Rohu but all the sets have negative results (Expt. IV- Table 1). However, when the duration of treatment was reduced to 3 hr. in the second experiments, positive results were obtained in 3 sets of Rohu (Expt. V. Table 1).

A solitary experiment was carried out with a view to study the potency of gonadotrophins extracted by TCA and stored under refrigeration for later use. Pituitary glands of Mrigal were immersed in 1.5 per cent TCA fluid for 12 hr. The fluid was then transferred to an air tight phial and kept inside the refrigerator for 10 days. The acid fluid was then injected to 3 sets of Rohu, while 2 sets were kept as control. Only one out of the 3 treated sets spawned.

Discussion

From the experiments described in the foregoing, it is observed that injection of gonadotrophins extracted by immersing pituitary glands for 6 hr. and 12 hr. in 1.5 per cent and 3 hr. in 2.5 per cent TCA precipitated successful spawning in Indian major carps, but injection of extracts from the immersed glands failed to do so. Li and Evans (1948) in their attempts to purify LH have observed that FSH was soluble in 2.5 per cent TCA whereas LH was completely precipitated in the same solution. Barrnett *et al.* (1956) experimented with male albino rats and showed that 2.5 per cent TCA extracted both FSH and TSH (Thyroid stimulating hormone) leaving behind a considerable quantity of LH in the gland. Ramaswami and Lakshman (1960) achieved positive results in the catfish, *Heteropneustes fossilis* when pituitary glands immersed in 1.25 per cent TCA for 6 hr. and 12 hr. as well as the acid fluid were injected on test animals. However, both of these failed to bring about ripening of eggs when the duration was extended to 24 hr. The same authors observed that neither the glands nor the extracted fluid gave positive results when glands were immersed for 6 hr. in 2.5 per cent TCA. This indicates that probably the action of TCA on pituitaries of higher vertebrates as described by Barrnett *et al.* (1956) on rats, is not similar to that of the lower vertebrates. Ramaswami and Lakshman (1960) conclude "glands immersed in 1.25 per cent TCA for 24 hr. or in higher concentration of TCA (2.5 per cent) the fluid probably not only extracts the glycoprotein but also denatures them as the test animals receiving it do not yield ripe eggs; the extracted pituitary gland also failed to ripen the eggs of the test animals as the gonadotrophin was completely extracted and inactivated.

Our experiment with 2.5 per cent TCA also gave negative results when glands were immersed in the fluid for 6 hr. but successful results were achieved when the duration of immersion was reduced to 3 hr. The results with 1.5 per cent TCA indicate that probably the gonadotrophins are not fully extracted by TCA since a higher dosage (18-28 mg/kg) of extracted gonadotrophins was required for successful spawning than in the control (8-12 mg/kg). It is not known whether the extra gonadotrophins were completely or partly denatured and partly left behind in the glands. However, the negative results obtained by injecting the extracts of the immersed glands in distilled water are not conclusive and need further confirmation.

[It was a joint paper by R. M. Bhowmick and H. Chaudhuri (1968) J. Z. Sic. India]

References

Barrnett, R. J. Ladman, A. J., McAllaster and E. R. Siperstein. 1956. The localization of glycoprotein hormones in the anterior pituitary glands of rats investigated by different protein solubilities histological stains and bio-assays. *Endocrinology,* 59 (4): 398-418

Chaudhuri, H. 1963. Induced spawning of Indian carps, Proc. Nat. Inst. Sci. India, 29 B($): 469-487

Chaudhuri, H. and K. H. Alikunhi, 1957. Observation on the spawning in Indian craps by hormone injection, Curr. Sci., 26 (12): 381–82.

Chaudhuri, H., Bhowmick, R. M., Pal, R. N. and A. B. Munshi. 1962. Observations on induced breeding of major Indian carps during 1960 at Cuttack. Proc. Indian Sci. Congr., Cuttack

Fevold, H. L., Hisaw, F. L. and S. L. Leonard. 1931. The Gonad Stimulating the the Luterinizing Hormones of the Anterior Lobe of the Hypophysis. Am. J. Physiol 97: 291-301.

Hoar, W. S. 1957. The Gonads and Reproduction in M. E. Brown (Ed) The Physiology of Fishes, John Willy, New York.

Li, C. H. and H. M. Evans. 1948. Chemistry of Anterior pituitary Hormones, In G. Pincus and K. v. Thimann (Ed,) the Hormones. Vol. I Academic Press Inc. New York.

Otsuka, S., 1956. Comparative studies on the pituitary hormones of vertebrates, IV. Effect of crude LH and FSH on the ovulation of hypophysectomized newt., Endocrinal. Japan, 3 (3) 181-185.

Pickford, G. E. and J. W. Atz 1957. The nature and physiology of the pituitary hormones of fishes, In A. Gorbman (Ed.) Comparative Endocrinology, John Wiley, New York.

Pickford, G. E. and J. W. Atz. 1957. The Physiology of the pituitary gland of fishes. New York Zoological Society, New York.

Ramaswami, L. S. and A. B. Lakshman, 1960. Bioassay studies of the catfish and skipper frog/pituitary glands after treatment with some glycoprotein solvents. Proc. Nat. Inst. Sci. India. 26 (B): 30.

Witschi, E. 1955. Vertebrate gonadotrophins, Mem. Soc. Endocrinol. 4: 149-165.

Chapter 7

Induced Breeding and Hatchery Management in Carps

Hatchery Management

Introduction

The Indian major carps, catla, rohu, mrigal, and calbasu are the popular economic species cultivated in ponds of centuries in India. Also, the exotic Chinese carps, namely, the grass carp, silver carp and bigheads have been introduced in India and are cultured along with the Indian major carps. The major problem in cultivating these economic food-fishes, is the non-availability of their seed for culture in adequate quantities, as these fishes do not ordinarily breed in ponds under artificial conditions, although they attain maturity in ponds. They usually breed in riverine habitats during monsoon months from where the seeds are collected and sold to fishfarmers in recognized markets. However, the stocking materials (spawn), thus collected from the flooded rivers, are generally a mixture of economic and uneconomic varieties and very often contain high percentage of miscellaneous unwanted species harmful for the growth and survival of the economic species. Since sorting out the desirable varieties from others at this stage is practically impossible, the fishfarmers have no other alternative but to stock the entire lot thereby, suffering loss and wasting valuable nursery space. In view of the constraints in the procurement of sufficient quantity of quality fish seed, the need to develop an alternative dependable source of seed was greatly felt.

In West Bengal, the fish-farmers have developed a method of breeding craps by simulating riverine conditions in certain special types of ponds called 'bundhs'. Mature carps breed in the bundh after in rushing of sufficient rainwater for extensive catchment area. However, such a situation is possible only in certain areas with a topography of undulating terrain. Attempts were made to study the environmental factors responsible for spawning of carps in their natural habitat, the rivers and also in the 'bundh' type of ponds in West Bengal. The preliminary studies indicated that the environmental factors, such as, rainwater, flood, optimum temperature, pH and dissolved oxygen content of water were the ones supposed to be responsible for spawning. Among the internal factors, it is well known that the gonad stimulating hormones of the pituitary gland are supposed to play an important role in the development and maturation of gonads and spawning in fishes.

Concerted efforts were made by Fisheries Scientists to decades to induce these difficult to spawn fishes by injection of various hormones. Initially, the mammalian pituitary and other hormones of mammalian origin were tried but these did not give satisfactory results. It was in the year 1957, the first success was achieved when Indian major carps were induced to breed by injection of fish pituitary hormones obtained from the major carps themselves. The success initiated the development of the technique very successfully used for the production of millions of quality seed of both Indian and Chinese carps under artificial conditions in ponds. The technique is not very complex or expensive and can be practiced by fish farmers with little efforts. Since these carps are very prolific breeders, one or two sets of average size spawners would give sufficient seeds to meet the requirements of an individual fish farmer.

The subsequent paragraphs provide a brief description of the technique of induced breeding of carps by hormone injection.

Pituitary Gland (Hypophysis) in Fishes

The pituitary gland in fish is located at the floor of the brain box ventral to the brain proper. It secretes most of the important hormones that regulate varieties of physiological processes in fishes, the most important ones for breeding being the gonad stimulating hormones or the gonadotropins. The gonadotropins present in the hypophysis of a mature fish stimulate the development of gonads (ovaries and testes). The gonads on being activated by gonadotropins, secrete the sex hormones (estrogen and androgen) which in turn stimulate maturation of sex-cells (ova and sperms) and the release of the gametes (ovulation and spermiation) or spawning in fishes.

Collection and Preservation of Pituitary Gland

Pituitary gland is usually collected from mature donors, freshly killed or well-preserved in ice. A portion of the scalp is at first removed and the brain is exposed. The gland is located ventral to the brain on the floor of the brain box. It is then carefully picked up by a tweezer and put in absolute alcohol. Absolute alcohol dehydrates and defttens the gland. The glands are then stored in fresh absolute alcohol and kept at room temperature or preferably under refrigeration if storage is required for a longer period. Acetone can also be used in place of absolute alcohol and acetone dried glands are stored under refrigeration.

Preparation of Gland Suspension for Injection

The preserved glands are weighed in a chemical balance and the required quantity is macerated in a tissue homogenizer with a little distilled water or 0.3 common salt solution (physiological solution). It is then diluted with the required quantity of the same liquid and centrifuged. The tissue particles get settled at the bottom and the supernatant liquid with the dissolved hormones is then taken in a hypodermic syringe for injection.

Dosage of Pituitary Extract

Determination of proper dosage of pituitary gland extract depends mainly on the weight and proper stage of maturity of the breeders to be injected. Generally, the female spawners are given two injections, a preliminary dose of 2-3 mg/kg body weight and a second dose of 5-7 mg/kg body weight, after 6 hrs. The males receive only a single dose of 2-3 mg/kg at the time of the second injection to the female. The volume of the fluid injected is generally around 0.2 ml per spawner. The injected spawners are then released together inside a breeding hapa.

Care of Brood Stock

The success of induced breeding depends to a great extent on the care taken to raise brood stocks. The breeders may either be grown in the farm or collected from outside and stocked in large-size stocking ponds. The selected breeders should be at least 2-3 years old. These are stocked at the rate of 1000-2000 kg/ha and fed daily with 1 per cent of their body weight. Periodic examination of breeders is necessary so as to note then condition and parasitic infection, if any. Utmost care should be taken to raise healthy brood stock with well-developed gonads and preferably of a good genetic stock.

Identification of Sex in Ripe Breeders

During the breeding season, the sexes of carp breeders can easily be distinguished by examination of the pectoral fins. The pectoral fins of males are slightly larger and thicker than that of females and the dorsal surface of the pectoral fins of males are rough at touch whereas very smooth in case of females. Besides, fully ripe males would ooze milt on slight pressure on the abdomen and fully gravid females would have bulging abdomen and swollen vents.

Breeding Technique

Selection of Breeders for Injection

With the advent of monsoon when the water temperature goes down, fish breeding experiments are then initiated; Breeders are netted out for selection for the breeding. A fully ripe male freely oozes milt when the abdomen is gently pressed. Choosing a suitable ripe female, however, is quite difficult. Fairly good results are obtained with female spawners having soft, bulging abodomen with swollen, reddish vents. Selection is more difficult with the Chinese carps. To be sure, it is advisable to suck in a few eggs by inserting a catheter through the vents of grass or silver carps and examine those eggs for final selection. Healthy breeders having size preferably around 2.5-5 kg are selected for injection. The number of eggs produced by this size of carps are quite satisfactory.

Method of Injection of Breeders

For injection, the selected breeders are kept individually in a separate handnets and weights are noted. The required dose is then calculated from the weight. The spawner inside the hand net is then gently placed on a soft cushion and intramuscular injection is given on the back or at the region of the caudal peduncle. The entire operation is finished as quickly as possible and then the breeder is transferred inside a breeding hapa fixed in a pond. The breeding environment with flowing water, however, is preferable circular cemented spawning pools/tanks with constant blow of water is definite improvement over stagnant ponds. One set of breeders consisting of one female and two males is ordinarily released inside one breeding hapa. The mouth of the hapa is then securedly closed to prevent escape of the injected fishes. Intraperitoneal injections have also been observed to give good results.

Spawning

In case of Indian carps, spawning ordinarily takes place around 3-6 hrs. after the release of the injected male and female spawners. Some sort of sex-play is very often observed before the female releases the ovulated eggs in water followed by shedding of milt by the males. Fertilization of eggs by sperms occurs in water. within an hour after fertilization, the eggs start swelling by absorbing water and attain the size of a pea or so. The fertilized eggs are demersal and slowly sink to the bottom in quiet water but are buoyed up to the surface of water with the slightest disturbance in the surrounding water.

The technique described above is usually followed in case of Indian major carps. But in case of Chinese carps, natural spawning does not ordinarily occur and the usual practice is to strip the eggs from the injected female and then artificially fertilizing those by stripped milt from injected males. Stripping is a laborious process. Right from about 3-4 hrs. after the second injection, the females are to be occasionally examined and as soon as they are observed to be freely oozing eggs, all the ovulating eggs are stripped in small basins and are then fertilized by stripping milt on the eggs. Ordinarily, the dry method of insemination is followed and the milt is thoroughly mixed with the eggs with a feather. The fertilized eggs are then washed with fresh water to remove the excess of milt and kept soaked in sufficient water for swelling.

Hatching Technique

About 6-8 hrs. after fertilization when the embryo inside each egg starts twitching movement, the eggs are collected; since by then the eggs yet properly water hardened and can withstand handling needed during collection, estimation of total quantity of eggs laid, and their transfer to hatcheries under ideal condition, the percentage of fertilization of eggs many be as high as over 98 per cent. The eggs laid by good size (3-4kg) female rohu may vary from 1.0–1.5 million.

Hatching Environment

The hatchery consists of 'double hatching hapa' fixed in clear pond water. It is made of round-meshed mosquito-net cloth and the outer one of close meshed cloth. The inner hatching hapa is fully stretched so that the bottom is kept perfectly flat forming a hanging tray. The eggs are uniformly spread all over the inner hatching hapa as a thin layer at the bottom. The eggs usually hatch out within 15-18 hrs. after fertilization at 27-31°C. After hatching, the hatchlings pass though the round meshed of the inner hapa and escape to the outer hapa. Finally, when all the eggs have hatched out, the inner hapa containing bad, unfertilized eggs and the egg cases, is removed. The hatchlings inside the outer hapa are left undisturbed till the third day after hatching, by that time the yolk gets almost absorbed, mouth formed and the young fry (spawn) starts feeding. The fry are then collected from the outer hapa, number estimated and stocked into prepared nursery ponds.

The hatchery, as described above, is the simplest type and is suitable for fish farmers who do not have any facilities for running water supply or having any pumps or a deep tube well so that underground water could be pumped to overhead tanks from where continuous supply of water could be maintained in the hatchery. More sophisticated hatchery systems are in use in developing countries and also in several developing countries. In Russia and in some hatcheries in India, hatching of carp eggs is done in cylindrical glass jars with continuous circulation of oxygenated water under controlled temperature. The eggs are under constant movement. The flow of water in the jar is so regulated that the eggs do not flow out. When the eggs hatch out, the newly hatched larvae have a natural tendency to swim up and finally they escape out of the jars along with the flow of water and are collected in a hatchery container called the receptacle. Indoor hatcheries have several advantages over the outdoor ones. In the latter, very often heavy mortality of fry/hatchlings occur due to natural hazards including high fluctuation of water temperature, oxygen deficiency and loss of eggs hatchlings due to sudden floor, storm and others.

In addition to glass jar hatcheries, there are various other types of hatcheries used in different countries. In China, floating 'net boxes' and cemented circular thanks with continuous flow of water in used. In several Chinese carp hatcheries, single or multiple chambered hatching pools with diagonally installed water inlet pipes are fitted at the bottom for maintaining a constant flow of water during the incubation period. In India, hatchery equipments (portable hatcheries and hatchery containers) made of low density polyethylene material are used in many hatcheries.

The induced breeding technique as described in the foregoing, is the one that was developed initially after the first breakthrough was achieved in the artificial propagation of Asiatic carps. The technique, however, has evolved further in the course of the last four decades. Hormones other than fish pituitary hormones were used. HCG (Human Chorionic Gonadotropin) was largely used in China, very often with a threshold dose of fish pituitary extract. In 1970's LH-RH Analogue, both decapeptide and nonapeptide were used by Chinese farmers. However, dopamine, a neuro-hormone was found to inhibit the release of gonadotropic releasing hormone. More recently, salmon gonadotropin releasing hormone Analogue (sGnRH-A) in combination with domperidone (a dopamine antagonist) is being largely used in China. This method is known as the 'Linpe Method' and the hormones used called 'Ovaprim' (the trade name: prepared by Syndel Lab., Canada) is being used for breeding Asiatic carps in several countries of Asia and the pacific. However, still the technique of using fish pituitary extract as described at the beginning is largely used by the poorer fish farmers of the developing countries because of the simpler technique, cheaper and ready availability of the hormones as compared to the above mentioned techniques. Still, a more simplified technique is suggested for the use of individual fish farmers who are unable to get the opportunity of using a chemical balance or a refrigerator, a centrifuge machine or obtain absolute alcohol or acetone. The farmer has to sacrifice a few nature carps is donors of sizes compared to those of the recipient spawners, for collection of fresh glands. The pituitary gland could be macerated in a mortar wilt a pestle in little clear fresh rainwater and the extract injected immediately to the spawners.

Chapter 8

Induced Spawning of Indian Carps

Editor's note

In India, the major carps constitute the important food fishes and are cultivated in ponds. But in pond fish culture, considerable difficulties are encountered in obtaining quality fish seed for stocking purposes as these fishes do not breed in ponds. Attempts made to induce these species to breed in ponds by injection of fish pituitary hormones have met with great success. The technique developed has clearly demonstrated that it could be profitably utilized for the large scale production of fish seed. The success in inducing breeding of carps has also made it possible to try experiments for the improvement of stock by hybridization and selective breeding. A number of intergeneric and interspecific hybrids have been produced by artificial fertilization of eggs of various species of carps injected with fish pituitary hormones. The breeding as well as the hatching techniques have been described. The need for the popularization and extensive application of this technique throughout the country has been emphasized as it is hoped that the following the improved technique it would be possible to produce millions of fish seed of the important cultivated species so as to meet the great demand for fish seed in India.

Introduction

The first prerequisite of fish culture is the availability of desirable variety and required quantity of fish seed, with the extension of carp cultural practices into many States of India, the dearth of quality fish seed is felt all the more. Concerted efforts were therefore made by fisheries workers in India to study the factors responsible for the spawning of carps in natural habitats and also under semi artificial conditions in 'bundh' type of ponds of Bengal (Hora 1945).

Hamid Khan (1938) attempted for the first time to induce Indian carps to breed by hormone injections. He injected *Cirrhina mrigala* with mammalian pituitary gland extracts, but the attempt was unsuccessful. Since then, no appreciable work was done in this line till 1955. Ramaswami and Sundararaj (1956, 1957) had carried out experiments on catfishes, *Heteropneustes fossilis* and *Clarius batrachus*, and succeeded in inducing them to breed by hormone injections.

The first concerted effort to induce the Indian carps to breed by injections of pituitary gland hormones was initiated by the author in 1955 at the Central Inland Fisheries Research Substation, Cuttack. In 1957, it was possible to breed the majority of the principal varieties of Indian carps by pituitary hormone injection (Chaudhuri and Alikunhi 1957).

History of Pituitary Treatment in Pisciculture

Before going into the details of the techniques followed in bringing about the successful breeding of the Indian carps, it is desirable to make a passing reference to the history of pituitary treatment in pisciculture. It is a well established fact that the pituitary glands of higher vertebrates play an important role in the development and maturation of gonads. Studies on the effect of hormones on the reproduction of fishes were all these years confined to the use of mammalian and synthetic hormones only. These did not give satisfactory results.

In Brazillan pisciculturists were the pioneers in using fish pituitary gland for successfully inducing their indigenous fishes to breed by injecting pituitary gland hormones (Houssay 1931; Ihering 1937; Fontenele 1955). The Russians have also developed similar techniques in breeding surgeons (Gerbil' skill 1938; Kazanskii 1939). Since then a few other countries have also succeeded in making fishes breed by hormone injection (Hasler *et al.*, 1930; Fugita *et al.* 1946; Pickford and Atz 1957).

Success, however, could not be attained in inducing breeding of the important Chinese carps which are cultured in China, Japan and all over the South East Asiatic countries. The Chinese carps, like the Indian major carps, have similar breeding habits and do not normally breed in confined waters. In India, however, as already mentioned, success has been achieved in inducing the carps to breed by injecting fish pituitary extracts (Chaudhuri 1960; Alikunhi *et al.* 1960). The method followed at the Central Inland Fisheries Research Substation, Cuttack is discussed briefly in the subsequent chapers.

Fish Pituitary Gland

The pituitary gland of fish is a small body situated ventral to the brain proper in a concavity on the floor of the brain box and is connected to the brain by means of a stalk. Like higher vertebrates, fish pituitary gland also controls directly or indirectly a wide variety of physiological processes by secreting a number of hormones; the most important ones being the gonad stimulating hormones (FSH and LH) which take part in stimulating the development and maturity of the sexual organs and induce spawning in fishes.

Hormones Injection Technique

It is believed that the pituitary gland is most portent in fully ripe fish. The glands are ordinarily collected from fresh, or freshly killed, fully ripe fish for better results. It has, however, been observed that glands collected from fishes preserved in ice for 7-12 days also give successful results. Usually glands from the same species as the recipient fish are used, although glands of other fishes also gave results.

The glands after collection are immediately preserved in absolute alcohol and stored under refrigeration. Absolute alcohol dehydrates and defattens the glands. The preserved glands are taken out before injection and weighed in a chemical balance. The required quantity of gland is then macerated in a tissued homogenizer with a little distilled water or 0.3 per cent common salt solution. The gland suspension is then centrifuged and the supernatant fluid is injected intramuscularly to the carp breeders. While preparing the gland extract, usually no distinction is made with regard to glands collected from the different sexes of the donor fish. Determination of exact dose of pituitary gland to be administered depends mainly on the size and proper stage of sexual maturity of the recipient fish. Besides, the potency of glands also varies according to the time of collection, condition and stage of sexual maturity of the donor fish and also on proper preservation of the gland materials. The absence of any suitable method as the judge the exact stage of maturity of the female fish from its external appearance seems to be the main handicap in standardizing doses of injections. Successful results have, however, been obtained by administering 5-10 mg of pituitary gland (dried for 2 min) per kg body–weight of the female and 2-3 mg of gland per kg body weight of the male.

Though the Indian carps do not breed in ponds, they mature in ponds. For the induced breeding experiments, mature fishes are collected and stocked in stocking ponds and periodic examination of the condition of fish gonad is made. Artificial feeding is provided if there is scarcity of food in water or the breeders are overstocked.

With the advent of monsoon the breeders are netted out, segregated sex-wise and usually released in separate ponds. Identification of sex of the breeders in much easier at the time. A male breeder is characterized by roughness of the dorsal surface of the pectoral fin in contrast to that of the female which is smooth (Chaudhuri 1959a). Besides, ripe males ooze milt on slight pressure on the abdomen and fully gravid females have swollen vents.

Selection of good breeders is very important. Males with freely oozing milk and females with soft, bulging, rounded abdomens and swollen vents are selected for injection.

Breeding Technique

Breeders captured by drag-netting are collected in hand nets. They are next weighed and placed on a soft cushion and injected intramuscularly on the caudal peduncle through the meshes of the hand net.

After the injection, the breeders are released inside a breeding hapa. A breeding hapa is a rectangular cloth container with a cover, male of fine-meshed mosquito net cloth, and is ordinarily of 2.4 m x 1.2 m x 0.9m in size. It is fixed to bamboo poles in ponds or in the river. Usually two males are put along with one female.

Spawning usually occurs within 6-8 h after the injection. If, however, the fish does not respond, a second injection with a slightly increased dose is given about 10-12 h after the first one. Very rarely a third injection is given. Better results have been obtained in a large number of cases when a small provocative dose was given to the female and after 6 h both the sexes were injected and released together in the breeding hapa. The fishes ordinarily breed 3-6 h after the release.

Hatching Technique

After the fish has spawned, a sample of eggs is collected and kept in the laboratory for observation. As soon as the embryo displays twitching movement, the eggs are collected from the breeding hapa and transferred to the hatching hapa.

A hatching hapa consists of an outer hapa made of this meshed cloth (1.8m x 0.9m x 0.9m in size) and an inner round-meshed mosquito net hapa (1.75m x 0.75m x 0.45 m in size). The inner hapa is stretched fully inside the outer one and both are tied to bamboo poles in pond water.

Eggs after their removal from the breeding hapa are measured with a beaker of known volume and distributed uniformly in equal volumes to each of the inner hatching hapa. While measuring eggs, two samples of known volume are collected and taken to the laboratory where the total number of eggs in each sample are counted and eggs examined to find out the rate of fertilization. From this, the total quantity of eggs and the percentage of developing eggs are estimated. The incubation period of carp eggs is 14-18 h at a temperature of 24^0–31^0C. Hatchlings escape to the outer hatching hapa through the holes of the mosquito net cloth of the inner hapa. The egg-cases and bad eggs are left out in the inner hapa which is removed when all the eggs have hatched out.

On the third day after hatching when the yolk is almost absorbed, the young fry start feeding. They are measured in small perforated aluminum cups and stocked in nursery ponds.

Water and Climatic Conditions

The range of temperature in which the fishes spawn is 24^0-31^0C. Spawning occurs both during the day and the night and both in clear as well as in turbal waters. Cool rainy days have been observed to be more conducive to spawning. Hatching hapa is preferably fixed in clear water for better survival of hatchings. Thick algal bloom is harmful to eggs and hatchlings.

Results

Analysis of the results of the experiments on induced breeding of carps shows that success in 60 per cent of the cases can ordinarily be achieved. In a few cases negative results are obtained when the selected female breeders are not in the proper stage of maturity. Failure on many occasions is due to sudden rise of temperature and adverse weather conditions. In the majority of cases of successful spawning, fertilization occurs in over 70 per cent of the eggs, the maximum achieved being 99.5 per cent.

The largest fish so far bred was a Catla female weighing 14.5 kg. The breeder laid about 20 lakhs of eggs after one injection. 70 per cent of which hatched out. The maximum number of eggs laid by a Rohu (4.08 kg) was 15.50 lakhs and by a Mrigal (4.76 kg) 11.64 lakhs.

Commercial Fish Seed Production

Results of the last few years definitely shown that the method could profitably be utilized for the large scale production of fish seed. Millions of major carp spawn have been produced. During the fish breeding season, 1961 about a crore of fish seed of major carps Catla, Rohu, Mrigal and Calbasu were produced by injection of pituitary gland extracts at the Central Inland Fisheries Research Substation, Cuttack.

The cost of production of fish used is fairly low. In a fish farm where mature fishes are grown, pituitary gland materials can be obtained by sacrificing a few ripe fishes. Even otherwise, glands collected from fish markets are not costly. The fish head after collection of gland could be returned to the fish merchant for sale and only a nominal price is charged for the gland.

The fecundity of carps is considerably high. The fish seed requirements of a fish farmer could easily be met if only one pair of each species be bred satisfactorily. For instance, one Rohu female weighing 2.72 kg was bred which laid 5.84 lakhs of eggs. 92 per cent of these hatched out and 4.7 lakhs spawn were obtained for stocking. In a properly managed nursery the expected survival within three weeks of rearing should be at least 50 per cent. That means, 2.35 lakhs fingerlings are expected. The minimum selling rate of pure stock of major carp early fingerlings is Rs. 10 per 1,000. In terms of money, 2.35 lakhs of fingerlings would fetch Rs. 2,350.

Success of Induced Breeding Experiments in Various States

Importance of the induced breeding work was realized by the Central as well as by the State Governments. Immediately after the first success, a short training course was arranged in this substation by the Ministry of Fisheries and Agriculture for the training of State Officers and also for Fisheries Extension Officers. A majority of the States responded and officers were sent for the training. Subsequently, as improved methods were conducted for State Fishery Officers.

The majority of the States in India have met with considerable success in inducing major carps to breed by hormone injections and have produced spawn for stocking. In an attempt to attain self sufficiency in fish seed resources for the State of Assam, a special training course in induced fish breeding for the Fisheries officers of that State was arranged in 1958 and extensive arrangements were made in 1959 for large scale production of major carp fish seed, in collaboration with State Fisheries, Assam. Five centres were selected for induced breeding experiments and a total of approximately 40 lakhs of quality fish seed were produced. The usual practice in Assam was to collect eggs from the small rivers. The operations used to be very expensive and the percentage of major carp fish seed was very low. Further work on the production of good quality fish seed in subsequent years in Assam was very promising and it is hoped that with concerted efforts Assam will be able to become self sufficient with regard of its local demand of fish seed.

It is also expected that in the near future the majority of the States of India will be able to produce a large quantity of major carp fish seed to meet at least partially their demands of spawn.

Artificial Insemination of Eggs

Successful results, however, have been obtained in stripping carp eggs and artificially fertilizing them. Male and female breeders after injection were caught at the right moment when the eggs could be easily stripped and fertilized by the milk from the males. Preliminary experiments were very encouraging and on one occasion more than 90 per cent of the eggs hatched out and about 70 per cent of fry survival up to the fingerling stage. It is hoped that with the improvement of laboratory conditions and better facilities provided for improvement such experiments under controlled conditions, artificial raising of carp seed will be possible in the laboratory in the same manner as is being practiced in trout hatcheries. Preliminary experiments in this line have already been undertaken in the air conditioned laboratory of the Department of Fisheries, Orissa.

Selective Breeding and Improvement of Stock

Hybridization is one of the methods of combining desirable cultural qualities of selected fishes and raising hybrids of strains superior to the indigenous species.

Hybrids of better strain, more hardy and more resistant to disease, have been obtained by crossing different species or different races of fish in Russia (Gerbil skill 1952; Kazanskii 1953) and a few other Western countries (Baldwin 1956). Success in inducing breeding of Indian carps and fertilization of their eggs by stripping has made selective breeding of carps by hybridization possible. This has opened a new avenue of research with immense potentialities for the improvement of present stock.

Preliminary attempts made in the line had been highly successful (Chaudhuri 1959b). Thirteen different hybrids could be produced by crossing carps belonging to three genera and seven species. In addition another hybrid was obtained by crossing common carp male with Rohu female. (Alikunhi and Chaudhuri 1959). It is interesting to note that the majority of the hybrid species, including the intergeneric ones, have developed fully matured gonads. One of the hybrids was injected with pituitary hormone and bred successfully. Young ones of the hybrid are being grown in the ponds.

Conclusion

Success in inducing major Indian carps to breed in confined waters by injection of fish pituitary gland hormones is an important landmark in the history of fish culture in India. It has revolutionized the age old practices in vogue and has provided vast scope for the development of pond fish culture, not only in India but also in other Asiatic countries where fish culture is practiced more or less on the same lines as in India.

With further improvement of the breeding and hatching techniques and standardization of doses, it would be possible to meet to a large extent the requirements of quality fish seed in this country. The method is a simple one and can be learnt by the private fish culturists with conscientious effort. For popularization of the technique, sufficient number of demonstration centres have to be started all over the country with the help of the Fisheries extension Officers and Fisheries officers of the Community Development and National Extension Services.

The need of artificial production of fish seed will grow tremendously in the near future as there is very possibility of scarcity of fish seed in the natural habitats. With the rapid industrialization of India, large numbers of factories are being established and consequently large quantities of factory wastes are also discharged into the rivers throughout the country. These factory effluents pollute the waters and adversely affect the riverine fisheries.

Besides, construction of dams across many rivers in India have restricted to a large extent the spawning migration of fishes and brought about considerable changes in the environment and natural spawning grounds of many fishes, thus depleting the fishery in those waters.

Lastly, the frequent natural calamities brought about by flood in most of our rivers during monsoon months are affecting spawn collection from rivers considerably. Embankments constructed across many rivers to prevent flood have also destroyed many natural spawning grounds of carps.

Considering the above adverse factors, it is apprehended that in future the riverine fisheries will be improvised more and more and the chance of procuring sufficient quantity of fish seed from those areas will be remote. The need for depending on the artificial breeding of fishes will be felt all the more.

In conclusion it may be suggested that more emphasis should be given to develop the technique for induced breeding of fish and popularize it all over the country, because of the fact that the first and foremost prerequisite for successful intensive fish cultivation and development of Inland Fisheries is an assured supply of pure quality of fish seed. This method should profitably be utilized in making the important economic estuarine varieties of fishes to breed and obtain seed for brackish water fish forming.

The experiments so far conducted have been highly encouraging and have shown great promise for further research. Concerted efforts are being made to perfect the techniques, and it is believed that earnest endeavors and cooperation from all quarters would lead to the production of fish seed of desirable varieties on a commercial scale to meet the great demand for fish seed in India. Development of Fisheries in India depends to a large extent on the successful solution of its great problem of scarcity of quality fish seed for stocking all her cultivable inland waters.

References

Alikunhi, K. H., and Chaudhuri, H. (1959). Preliminary observations on hybridization of the common card (*Cyprinus carpio*) with Indian carps, Proc. 46th Indian Sci. Congr., Delhi.

Alikunhi, K. H., Vijayalakshmanan, M. A., and Ibrahim, K. H. (1960). Preliminary observations on the spawning of Indian carps, induced by injection of pituitary hormones. Indian J. Fish., 7,1-19.

Baldwin, N. S. (1956). Hybrid trout. Sylvia, Edinb., 12,1-4.

Chaudhuri, H. (1959a). Notes on the external characters distinguishing sex of breeders of the common Indian carps. Sci. and Cult., 25, 258-259.

Chaudhuri, H. (1959b). Experiments on hybridization of Indian carps. proc. 46th Indian Sci. Congr., Delhi.

Chaudhuri, H. (1960). Experiments on induced spawning of Indian carps with pituitary injections. Indian J. Fish., 7, 20-48.

Chaudhuri, H., and Alikunhi., K.H. (1957). Observations on the spawning in Indian carps by hormone injection. Curr. Sci.., 26, 381-382.

Ramaswami, L.S., and Sundararaj, B. I. (1956). Inducing spawning in the Indian catfish. Science, 123, 1080.

Ramaswami, L.S., and Sundararaj, B. I. (1957). Induced spawning in the catfish Clarius. Naturwissenschaften, 44, 384.

Chapter 9

Selective Breeding and Hybridization of Carps

Carp Breeding and Hybridization

Editor's Note

Hybridization is a method of transferring desirable characters from one species to another with a view to producing hybrids of superior qualities to those of the parent species. The usefulness of hybridization received due attention both in animal husbandry as well as in the plant breeding. Study of hybridization in fishes on the other hand is comparatively new to the fish breeders and has assumed increasing importance in recent years only. By selective breeding a new strain usually differing from their wild parents in higher productive properties is evolved by planned utilization of genetic variation inherent in a plant or an animal.

The urge of a fish culturist to have the best cultivable varieties of fish for increased production makes him select the fastest growing and healthiest fishes as breeders. In fish culture selective breeding and hybridization are mainly aimed at increasing production by evolution of a new strain or a hybrid having the following qualities: quicker growth; better food conversion; better utilization of food from the environment; better resistance to diseases and parasites; better resistance to unfavourable environmental factors caused by high and low pH, extremes of temperature, deficiency or super saturation of dissolved oxygen content of water, high salinity and last of all better taste and nutritional qualities.

Fish Hybridization in Other Countries

USSR is the pioneer country where considerable progress has been made in the field of genetic selection and hybridization of fishes. The fish culturists of Russia have employed interracial, inter specific and inter-generic crosses of sturgeons and also interracial crosses in carps. One of the most remarkable successes in recent years is the production of a hybrid between the cultured carp and the Amur wild carp. The hybrid combines the faster growth rate of the former and the cold resistance quality of the latter. This success has made carp culture possible in the cold Siberia and North-Western parts of USSR. Russians have also successfully produced a hybrid between beluga and starlet (an intra-generic cross between two economically important surgeons Huso huso and Acipensor ruthenus) which has combined the quicker growth rate of beluga and early maturity of starlet. This success led to the cultivation of sturgeons in ponds in USSR.

In USA and Canada, work on fish hybridization was mainly carried out on the various species of trout's and salmons. Splake, a progeny of the cross between speckled trout male and lake trout female proved to be a very useful hybrid. More recently, considerable hybridization work has been done on sunfishes and commercial catfishes.

In Europe, especially in Germany and Poland, a good number of strains have been evolved by continuous selective breeding of the common carp. Extensive work on the selection of common carp has also been conducted in Israel and useful strains developed.

Japan has produced several ornamental varieties of gold-fish and common carp by long time selective breeding. Hybridization work in fish started in recent years and major part of the work had been confined to the Cyprinid and Cobitid fishes.

In Malaysia and in a few of the African countries some very interesting work had been done on the hybridization of various species of Tilapia.

Fish Hybridization in India

Although fish culture is an age-old practice in India and also in China, practically no progress was made on this important aspect of fish husbandry. Unlike common carp where reproduction of the species could be controlled, the Indian and Chinese carps could not be domesticated or strains evolved by selective breeding since they failed to reproduce under captivity. This inability on the part of the Indian and Chinese carps to breed in ponds restricted any attempts on the selective breeding and hybridization of these species.

However, the development of the technique of inducing spawning of Indian carps by administration of fish pituitary hormones paved the way for making attempts on improvement of stock of those valuable cultivated species by selective breeding and hybridization. In the year 1958, the fish success in the hybridization of various Indian carps was achieved by the author at the Central Inland Fisheries Research Substation, Cuttack by artificial fertilization of the stripped eggs of injected females on one species by the sperms of injected males of another species. This achievement demonstrated the feasibility of hybridization among various species of Indian as well as Chinese carps.

Technique of Hybridization in Carps

Since the Indian and Chinese carps do not breed in confined waters the neuro-hormonal technique of breeding is employed for obtaining their sex-products. The technique followed is simple. During the carp breeding season, fully mature carps are selected for the experiments. The female fish is given a preliminary low dose of 2-3 mg of carp pituitary gland per kg body weight. After an interval of 6 hrs. a second injection of 5-11 mg/kg body-weight is given to each female. The male breeders are also given injections of a low dose of 2-3 mg/kg body-weight. Depending on the temperature of water, the female fish usually reaches the oozing condition of about 4-6 hrs. after the second injection. By occasional examination, freely oozing stage of the females could be determined, when they are taken out of water for stripping. Dry method of stripping is usually followed and this ensures higher fertilization of eggs. The water and slime is carefully wiped from the body of the female with a tissue cloth or clean towel and then eggs are stripped by pressing the belly into an enamel tray or basin. The milt from the male of a different species is next stripped similarly and spread over the eggs. The sex-products are then carefully mixed with a feather or by constantly shaking the basin or tray so as to ensure proper fertilization of the eggs. After a few minutes pond water is mixed and slowly the excess of milk is washed out. This is repeated several times after which the eggs are kept in big trays filled with water. The eggs absorb water and swell up and within 45 minutes the cleavage starts. Later on the developing eggs are transferred to hatching hapa in ponds or hatching jars in the laboratory. Further development proceeds more or less in the similar way as in Indian major carps and hatching takes place in about 15-18 hrs at a temperature ranging between 27-31^{0}C.

Since longevity of the sex-products once stripped out of the body is very short, the eggs should be stripped out in several lots and immediately fertilized with milt from males.

Hybridization in Indian Carps

The Indian carps once the sex-products are obtained, hybridization becomes very easy. Attempts made on hybridization between the various species of Indian carps met with immense success. The cultivated carps of India belong to three important genera, namely *Catla, Labeo* and *Cirrhinus*. A number of interspecific and intergeneric hybrids were produced. Morphometric characters, colour, body shape, sex-ratio, maturity and growth rate in comparison on the parental species were generally noted.

Interspecific Hybrids

Hybridization experiments conducted at the Central Inland Fisheries Research Substation produced hybrids between the four species of the genus *Labeo*, namely *Labeo rohita, Labeo calbasu, Labeo bata* and *Labeo gonius*. By crosses and reciprocal crosses the following interspecific hybrids were produced. While naming hybrids the common name of the male parent was put first followed by that of the female parent.

Male Parent	*Female Parent*	*Hybrids*
Labeo rohita	*L. calbasu*	*Rohu-calbasu*
L. calbasu	*L. rohita*	*Calbasu-rohu*
L. bata	*L. rohita*	*Bata-rohu*
L. bata	*L. calbasu*	*Bata-calbasu*
L. calbasu	*L. gonius*	*Calbasu-gonius*

Development of eggs was normal in all cases and a maximum of over 94 per cent fertilization was observed in some. The morphological characters of hybrids were intermediate between the parents although the general appearance appeared to be more like the paternal species. Of these hybrids, both the reciprocal crosses between rohu and calbasu showed great promise but the performance of others were not very satisfactory. The growth rate of both rohu-calbasu and calbasu-rohu was far superior to the slow growing parent calbasu. These hybrids attained maturity in two years. Rohu-calbasu hybrid was induced to spawn by hormone injection F_2 generation was obtained. F_2 hybrids had characters intermediate between rohu and calbasu. Rohu-calbasu hybrid when back-crossed with calbasu the progeny matured within 2 years. The sex-ratio of the progeny was 69 per cent females and 31 per cent males.

Intergeneric Hybrids

The following intergeneric hybrids were produced by crossing 7 species of fishes belonging to the three important genera of Indian carps.

Male Parent	*Female Parent*	*Hybrids*
Catla catla	*Labeo rohita*	*Catla rohu*
Labeo rohita	*Catla catla*	*Rohu catla*
Catla catla	*Labeo calbasu*	*Cala calbasu*
Labeo rohita	*Cirrhinus mrigala*	*Rohu mrigal*
Cirrhinus mrigala	*Labeo rohita*	*Mrigal rohu*
Cirrhinus mrigala	*Labeo calbasu*	*Mrigal calbasu*
Cirrhinus reba	*Labeo rohita*	*Reba rohu*
Cirrhinus reba	*Labeo calbasu*	*Reba calbasu*
Labeo fimbriatus	*Catla catla*	*Fimbriatus catla*

A few of these inter generic hybrids exhibited superior qualities than the parent species. *Catla-rohu* hybrid combined the good qualities of both the parents. The hybrid had wider body like catla and smaller head like rohu and had faster growth rate than rohu. The hybrid had the advantage over catla because of the smaller head and such had more flesh than cala of similar size. Rohu catla, rohu mrigal, mrigal-rohu and cala calbasu also showed much promise. In all these hybrids the general appearance and morphological characters were intermediate between the parental species. Majority of the hybrids had attained full maturity and were fertile. Catla rohu hybrid matured in 3 years and F_2 generation was obtained by hypophysation, F_2 hybrids grew satisfactorily and its growth was faster than rohu but slightly slower than catla.

Mrigal calbasu hybrid matured in 2-3 years. The eggs of an injected female mrigal-calbasu were stripped and fertilized with milt from catla, calbasu and mrigal males. The progeny of all the three crosses were obtained thereby producing a hybrid comprising all the three genera.

Hybridization in Chinese Carps and Common Carp

Preliminary attempts made in hybridization between the Indian carps and the three species of Chinese carps namely, *Aristichthys nobilis* (bighead), *Hypophthalmichthys molitrix* (silver carp) and *Ctenopharyngodon idella* (grass carp) met with partial success. In a few cases although eggs were fertilized the embryo either did not hatch out or the hatchlings died within a week. Hybridization between the three species of the Chinese carps on the other hand gave better result. In Taiwan, thousands of fingerlings were produced by crossing silver carp and bighead. At Cuttack, a few specimens of a hybrid between silver carp female and grass carp male were successfully obtained. The hybrid possessed rounded body like grass carp and smaller scales and mouth with protruding lower lip as in silver carp.

Hybridization between common carp and Indian carps was possible although the nature of the fertilized eggs and their incubation period are widely different. Stripped eggs of rohu could be fertilized by the milk of common carp. The hybrid had elongated dorsal fin like common carp. It grew to adult size, but no mature specimen was observed. Another hybrid between common carp male and *Labeo fimbriatus* female was also produced.

Hybridization in Carps and Viability of Hybrid Eggs

Hybridization experiments conducted with carps as described in the foregoing show that closely related fishes especially those of a subfamily can be easily hybridized than among members belonging to a different subfamily. All Indian carps fall under the subfamily

Cyprinini whereas the Chinese carps belong to the subfamily *Hypophthlmichthys*. Hybridization was easily obtained between members of the subfamily Cyprinini but when attempted between two subfamilies success was not achieved. Chromosome compatibility is a limiting factor for successful species hybridization. Incompatibility of cytoplasm of one species which the chromosome of another species may also stand on the way of success in hybridization. Detailed cyogenetic studies and study of the chromosome complexes of the parental forms are therefore essential for taking up any work on hybridization.

Hybridization in Carps and Fertility of Hybrids

The old popular belief that the hybrids as generally sterile forms in not true in case of fish. Experiments on the hybridization in Indian carps have shown that both inter specific as well as inter generic hybrids are fertile. The inter-specific hybrid rohu-calbasu and inter-heneric hybrid catla-rohu were both bred successfully by hypophysation and F_2 generation of hybrids produced. Mrigal-calbasu, another inter-heneric hybrid was hybridized by 3-way crosses producing a hybrid comprising three distinct genera. Chromosome number plays an important part in determining fertility of hybrids. The chromosome of the parental species should be able to conjugate otherwise it would result in abnormal gemetogenesis.

Hybridization and Induced Sterility

One aspect of the usefulness of selective breeding and hybridization, that is, improvement of animals through rational selection of germplasm and full development of the resulting somatoplasm has so far been dealt with. The other aspect of hybridization, namely the prevention or control of reproduction by hybridization is no less important in fish culture.

In fish culture, proper stock manipulation is essential for obtaining high production of fish. In case of a few pond breeding cultivated species, overpopulation of stocked fishes is observed due to their continuous breeding in the pond. Since Indian and Chinese carps do not breed in ponds, this problem does not arise in their culture but where Tilapia is cultivated its prolific reproduction results in poor production of fish. Mono-sex culture, however, did not meet with much success.

Induced sterility or production of mono-sex by hybridization may obviate the problem of overpopulation by reproduction. Hybridization experiments with various species of Tilapia have produced only male offspring's. This success is of considerable significance in Tilapia culture. Similarly the phenomena of andro– or gynohenesis also could suitable be applied to obtain mono-sex progeny.

In USA, a sterile hybrid has become very popular with many American bait hatcheries. These hybrids are preferred over gold fish as baits because of their sterility, thus obviating the danger of introducing unwanted fish in public waters. The sterility of certain sunfish hybrids have also made them desirable for culture in ponds with large–mouth bass since the latter is very often outstripped by the fecundity of sunfish and pond balance is lost.

Hybridization of Carps in Nature

A large number of fish hybrids have been reported from nature. Natural hybridization of the Indian carps though not very common, does occur. Carp hybrids are frequently observed in "bundh" type of ponds in West Bengal. The so called 'birbal" fish reported from the bunds of Bankura district of West Bengal is supposed to be a hybrid between catla and rohu.

The reason for hybridization of fish in nature is no definitely known. The reason attributed for occurrence of more natural hybridization among freshwater fishes than marine fishes may be because of the unstable and temporary nature of fresh-water habitat and the increased chance for the meeting of ova and spermatozoa in stream currents. In most cases hybrids in nature probably occur due to chance mixing of the sex-products. In "bundh" breeding, occurrence of natural hybrids may be due to congestion in the spawning ground. Because of limited space there is every likelihood of ova of one species getting accidentally fertilized by the sperms of another species.

The foregoing accounts give some idea about the present status of research carried out on the selective breeding and hybridization of carps in India. While some useful preliminary attempts have been made towards hybridizing various carp species and studying the performance of the F_1 and F_2 hybrids as compared to the parent species, no further work on progeny testing involving back crossings etc. could be followed up mainly because of lack of available field facilities since large-scale programme on selective breeding and hybridization requires numerous ponds. Very little work therefore, could be done towards improvement of specific traits by genetic selection. Extensive work in this important branch of genetic selection and hybridization of the cultivated carps have to be taken up in right earnest for increasing production of fish by evolving high yielding strains.

In conclusion, it is worthwhile to mention about the hazards of indiscriminate hybridization. Fish culturists should be very cautious in their selection and hybridization programme and see that useless hybrids are not propagated. Such hybrids should be destroyed and prevented from entering natural waters.

Chapter 10

Introduction of Exotics

This presentation "Introduction of exotic species of fish and their effect in the culture of indigenous species of India" though presented at a seminar long past (1964), has immense significance. Harmful effect of such introduction has been highlighted by Professor Chaudhuri in various lectures and scientific notes.

The following technical articles are dealt with in this treatise.

1. Introduction of exotic species of fish and their effect on the culture of indigenous species of India.
2. Record Fish Production with intensive culture of Indian and exotic carp.
3. Preliminary observation on the survival and growth of Chinese carp.

Editor's Note

Profitable fish culture aims at according to Professor Chaudhuri, the production of maximum quantity of fish from a given body of water. This can be achieved by culturing a number of quick growing fishes which do not compete with each other for food and are non-predatory in food habits. A judicious combination of such compatible species should raise the production of fish flesh in cultivable waters of India. The important food fishes cultivated in India are the major carps, *Catla catla*, *Labeo rohita* and *Cirrhina mrigala*. These indigenous fishes selected for culture, are quick-growing ones and non-predatory in food habits.

With a view to increasing variety with bulk of the indigenous species attempts have been made in India by planned transplantation of fish between different regions, by acclimatization of marine and brackish water food fishes and also by introduction of suitable exotic fishes for cultural purposes. The aspect of introduction of exotic fishes in India and their effect on the culture of Indian major carps have been dealt in this paper.

Exotic Fishes Introduced in India

Although introduction of foreign species was started over century ago, it is only in recent years that planned introduction of exotic fishes for fish culture has been taken up seriously. Exotic fishes have also been introduced to India as game fishes, Iarvicidal fishes and as ornamental aquarium fishes. Two species of the game fishes namely, the rainbow trout (*Salmo gairdneii*) and the brown trout (*Salmo trutta fario*) have established themselves at higher altitudes in several parts of India. Two larvicidal fishes Gambusia afinis and Lebistes reticulates have been introduced in Madras and a few other States.

The important food fishes which have been introduced to India will be discussed below in details:

Osphronemus goramy (the Gourami)

Osphronemus goramy, a native of Indonesia has been spread by the Javanese all over the Malayan archipelago and was introduced from there to Ceylon, Egypt and Sicily by persons interested in fish culture. The species could adopt itself to the local conditions in many countries and became a popular fish there.

Gourami was first introduced to India in Kolkata from Java during the early half of the last century and then into Madras from Mauritius in 1865. The fish however could establish itself only when a fresh consignment was brought to Madras in 1916. Subsequently, Gourami has been transplanted into several States in India where it is breeding normally.

Gourami is an omnivorous fish, but feeds largely on soft leaves of submerged and emergent aquatic plants, marginal grasses and algae. It also includes insects, worms, prawns etc. in its diet and can be fed on artificial food of both animal and vegetable origin. The growth rate of Gourami is rather slow, attaining 9-10 inches only at the end of the first year.

Gourami breeds almost all the year round. Parents construct nests in shallow marginal areas of ponds out of aquatic weeds and marginal grasses. About 500-3000 eggs are laid in the nest which is guarded by the parents. The incubation period is about 10 days.

Gourami is suitable for cultivation under tropical conditions. It has been reported to thrive satisfactorily up to an elevation of 2400 ft. in Mysore but could not survive the low temperature when introduced in Punjab. Natural production of Gourami in Indonesian ponds has been reported to be near about 200 kg per hectare. Gourami is an excellent food fish and is considered to be world's best freshwater table fish.

The species grows to a fairly large size, can be bred with ease in ponds, has a fine flavour and does not compete actively for food with the indigenous cultivated species of India. These qualities made Gourami an ideal fish for transplantation into all Indian waters for its intensive cultivation along with other economic food fishes.

Tinca tinca (the Tench)

The tench was first introduced from England by Maclvor in 1874 in Ooty lake. Subsequently the fish has established itself in the ponds and lakes of Nilgiri Hills. It thrives well in cold waters. Its introduction into the tropical warm waters does not appear to have met with considerable success, through the fish has been reported to have acclimatized and bred in the Sunkesula fish farm, Andhra State.

The Tench is generally regarded as a bottom fish and feeds on insects, vegetable debris, and also on plankton. Its growth is rather slow. It has grown to a maximum length of over 500mm weighing about 1.4 kg in the Ooty lake. It is a prolific breeder and breeds freely in the Nilgiris during February to April. The eggs are adhesive and are attached to leaves of aquatic plants. These hatch out in 5 to 6 days. Its cultural possibility in warm waters along with the Indian major carps has yet to be studied.

Carassius carassius (the Crucian Carp or the Golden Carp)

The crucian carp or the golden carp, a native of Central Europe was first introduced by Maclvor in 1874 in the Ooty lake along with the tench. It has established itself to the waters of the Nilgiris. Its growth is slow and attains a maximum size of about 450 mm and a weight of about 1.4 kg in the Ooty lake. It feeds mainly on copepoda and water fleas. The crucian carp breeds in ponds all the year round. The eggs are sticky and are attached to the submerged water plants. Through the species has bred at the Sunkesula Fish farm, Andhra State, attempts made to introduce the fish to the plains did not meet with much success.

Cyprinus carpio (European Carp or Common Carp)

Originally a native of China, the common carp, *Cyprinus carpio* now enjoys a wide distribution and is cultured in almost all the countries of the world. It has well established itself in Europe where the breeding of Common carp is a fairly old industry. This fish was first introduced to American waters by the early German settlers. The species has built up a huge population there and are available in all waters including the rivers, reservoirs and lakes. In Israel, Japan and South-East Asian countries, *Cyprinus carpio* comprises one of the most important economic food fishes.

Depending on the texture of the skin and size of scales, *C. carpio* has been divided into three main varieties. They are:

1. Scale carp (*C. Carpio* Var. *communis*) body completely covered with smaller scales.
2. Mirror carp (*C. carpio* Var. *specularis*)–with large size shiny scales over the body.
3. Leather carp (*C. carpio* Var. *nudus*)–body almost without any scales; has a leathery appearance.

In India, the German strain of Mirror carp was first introduced in 1939 from Ceylon. The species was stocked in the Ooty lake and since then has established itself very well in Nilgiri waters. The fish has been successfully transported to the Kumaon Hills in the Uttar Pradesh, Bombay and Bangalore and also to Madras. The mirror carp is quite hardy. It could withstand the strain of direct transport from the cold waters of Nilgiris to the warm waters of the ponds of Madras. Subsequently the fish has been distributed to the various parts of India and thriving quite satisfactorily in different Indian waters. Through the growth is much quicker and sexual maturity is attained earlier in warm waters, the fish was not observed to breed in the lower elevations. In recent years it has been possible to breed this strain in the plains at Cuttack and obtain sufficient seed for stocking and propagation.

As the German stock of the Common carp was not breeding freely in tropical waters in India, a consignment of the Chinese stock of the common carp (variety scale carp) was brought from Bangkok to Cuttack in August 1957 so as to study the suitability of the species for cultural purposes along with the indigenous species. With this view, intensive experiments have been carried out at the Central Inland Fisheries Research Sub-station, Cuttack, on the various aspects on the biology of the species. Valuable information gathered from these experiments are summarized below.

Growth of the Bangkok strain of common carp is highly satisfactory. In larger ponds, specimens have been observed to have attained a weight of about 10 kg in 28-30 months. Under identical conditions of environment and density of stocking, fingerlings and yearlings of common carp have been observed to grow faster than those of Rohu and Mrigal and almost as good as those of Catla.

The common carp is an omnivorous fish. Its feeding habits are largely similar to those of Mrigal and resemble those of Rohu to a certain extent. While the German stock breeds only once or twice in a year, the Bangkok stock breeds almost throughout the year. Intensive breeding, however, takes place from the middle of January to the end of March and also during July-August. Breeding is very limited during November-December when the water temperature is the lowest. A single female can spawn at least 4-5 times is a year.

As regards the quality of the fish flesh, it is considered to be a fairly tasty fish. The Bangkok strain of Common carp is highly domesticated and is much easier to handle than the Indian carps.

As a result of the thorough investigations made by the Central Inland Fisheries Research Sub-station, Cuttack, the Fisheries Research Committee, Government of India recommended in April 1959 that the common carp (Bangkok stock) be introduced in Indian waters because of its adaptability to diverse habitats, quick growth, omnivorous feeding habit, compatibility with the indigenous cultivated species, ability to breed all the year round in ponds and having fairly tasty flesh. The committee also desired that the various State Fisheries Departments should take up breeding and cultivation of the species. Following this recommendation there had been demands from almost all the States, and consignments of common carp fry and fingerlings have been supplied from Cuttack to those States. The species is thriving satisfactorily in many of the States by now.

Tilapia mossambica (Tilapia)

A native of South Africa, the Tilapia has been introduced all over the South-East Asian counties where the fish is extensively cultured. In India the first consignment of Tilapia was brought by the Central Marine Fisheries Research Institute, Mandapam in August 1952. In the same year a few fingerlings were brought to Madras. For making detailed investigations as to the probable effect of introduction of Tilapia on the culture of major carps, a small stock of Tilapia from Mandapam was received by the Central Inland Fisheries Research Sub-station, Cuttack in December 1953. Detailed study on the biology and culture of Tilapia alone and also in combination with indigenous species carried out at the Cuttack sub-station indicated the following:

Presence of Tilapia in carp nurseries very seriously affect the survival and growth of carp fry as it feeds extensively on carp fry and its fry directly compete with young carp fry for food. Though they are omnivorous in feeding habits, the adult Tilapia when stocked in combination with major carps has an adverse affect on the growth of carps especially that of Mrigal. Tilapia breeds throughout the year and because of the protection afforded by the mother to the young ones it quickly establishes itself in the ponds and its large population consisting of fishes of all sizes is able to exploit all the food resources available in the pond. Tilapia is a hardy fish and has a capacity to resist quite a high dose of fish poison like rotenone and as it is very difficult to capture Tilapia by drag netting complete eradication of the fish is practically impossible or it becomes very costly. Tilapia is susceptible to low temperature and does not survive at a water temperature below 10°C. As the water temperature in the plains does not fall to such low temperatures. Tilapia once introduced in any part of the country may spread all over the country. In view of the probable adverse affect of Tilapia on carp culture in India, The Fisheries Research Committee had recommended in its first meeting that Tilapia may be introduced only in areas where scope for carp culture is limited and from where this fish may not spread to the adjoining areas. The recommended places were the Western slopes of the Western Ghats and the coastal strip between Cape Comorin and South of and excluding river Tapti, and also in the District of Tinnevally, Madura and Ramnad in Madras State south of and including river Vaigai. In its third meeting the committee observed that "there appears to be no harm in extending the area of Tilapia culture northwards up to the Pennar river."

During the earlier years of introduction, highly encouraging reports regarding the suitability of Tilapia for culture were obtained from the areas where Tilapia was introduced. This was revealed by the fact that in the Madras State there was steady demand for Tilapia young ones and several thousands of acres of fallow land had been brought under fish culture and some areas showed considerable increase in the output of fish. In Kerala, Tilapia had become so popular that its culture no longer required any departmental effort and about 4000 tanks, reservoirs and lakes had been stocked with the fish. But statements made at the Fisheries Research Committee meeting at Trivandrum in April, 1959, indicated that the results of Tilapia culture in Kerala and Chennai have been disappointing and that the initial enthusiasm for its culture has appreciably gone down. The Research Committee however, after studying the results of investigations on Tilapia had resolved that the fish should not be introduced into areas which have so far been excluded.

In view of the recommendations made by the Fisheries Research Committee in its 5th meeting, the Central Inland Fisheries Research Sub-station, Cuttack, started investigations on the yield of Tilapia when cultured alone and also on the suitability or otherwise of combined culture of on the suitability or otherwise of combined culture of Tilapia and murrel. Sustained high yield of marketable size Tilapia could be obtained by keeping down the population by regular harvesting of the large size fishes at fortnightly intervals. If no manuring or feeding was done the yield showed a slight decline in the third year. With manuring and artificial feeding the production in the pond can be appreciably increased. Monosex culture of male Tilapia definitely yielded much larger specimens for marketing but he experiments invariably get spoiled due to accidental entry of female Tilapia into the ponds which start breeding. Preliminary experiments conducted on the use of Tilapia as a forage fish for the murrel *O. striatus* gave encouraging results.

Chinese Carps

Chinese carps are the important economic food fishes cultured in ponds in China, Japan and some of the South-East Asia countries. A few species of these Chinese carps are very fast growing and attain considerable large size. Like the Indian carps, these fishes also do not breed in confined waters.

With a view to finding out the relative cultural qualities of the important Chinese carps as compared to the Indian Major carps, experimental consignments of fingerlings of the Silver carp (*Hypophthalmichthys molitrix*) and the Grass carp (*Ctenopharyngodon idellus*) were brought to Cuttack during September and December 1959 from Tokyo and Hongkong respectively. A second consignment of Grass carp fingerlings was brought in December 1961. The fishes were released in the experimental fish farm at Cuttack and investigations are being made. The results of investigations made so far are mentioned below:

Hypopthalmichthys molitrix (The Silver Carp)

The growth of Silver carp fingerlings has been very satisfactory. They are primarily plankton feeders. Under comparable conditions, the silver carps were found to grow faster than Catla. The maximum weight attained by Silver carp in 31 months (33 months-old) was 8.4 kgs. Two year-old Silver carp attained full maturity in Cuttack ponds by June-July. Three year-old specimens of Silver carp have been successfully induced to breed by pituitary hormone injection at the Cuttack Sub-station. This is the first instance of pond breeding of any Chinese carp in the World though repeated attempts were made in several countries to induce these to breed did not meet with any success. The fry of the induced bred silver carp have been stocked in nursery ponds where they are growing very satisfactorily.

Ctenopharyngodon idellus (The Grass Carp)

Grass carp feeds mainly on aquatic macro-vegetation. The species was brought with the idea to ascertain whether this fish could also be utilized for the biological control of the aquatic submerged weeds, an overgrowth of which in detrimental to the growth and production of fish. Preliminary experiments at using grass carps to control the natural growth of submerged weeds like Hydrilla in nursery ponds have been very encouraging. The fish relished aquatic weeds like Hydrilla, Lemna and Azolla which were regularly supplied to the Grass carp ponds. Growth of Grass carps have been observed to be fairly satisfactory. In about two years the average size attained by grass carp was about 2 kgs. Two years-old male Grass carps were found to ooze in June-July. Three year-old Grass carps were fully mature and were successfully bred by pituitary hormone injection. Like the Silver carp, this has also been the first instance of pond breeding of a Chinese carp. The fry are being reared in the nursery ponds.

From the results obtained so far, on the study on these two Chinese carps, there is great hope that these two species would make welcome additions to pond culture in India. Both the species have a rapid growth rate and attain very big size. Grass carp will be an asset to out fish ponds for its quality to control aquatic vegetation. The main apprehension against the introduction of these species namely their inability to spawn in captivity, has been removed by the successful spawning of these carps by hormone injection by the Central Inland Fisheries Research Substation, Cuttack. With more detailed investigation and building up of sufficient stock in coming years it may be possible to get the fishes introduced throughout India.

Conclusion

It will be seen form the foregoing account that a good number of exotic fishes have either been introduced in Indian waters of are under investigations awaiting introduction. Some of these exotic fishes have been found to have high cultural qualities and are regarded as welcome additions to the cultivated indigenous species. However, it should always be borne in mind that indiscriminate introduction of exotic fishes is very dangerous and may bring disastrous effect on the indigenous cultivated species. Transplantation programme should be carried out in a planned manner and strict quarantine restrictions have to be taken. There are scores of instances when haphazard introduction of fishes from outside areas had been responsible for harmful effect on the fisheries Industry in a country. Besides, there is every chance of spreading disastrous fish diseases and parasites in the new areas. In certain cases an exotic fish may even be introduced unintentionally. As for instance, in India, while investigations are being carried out in Tilapia, the species has already made its entry into West Bengal from East Pakistan now Bangladesh where the fish was introduced a few years back.

In conclusion, it may be suggested that while judicious introduction of desirable variety of exotic fishes with cultural qualities would help in increasing production of fish, detailed investigations should be made before any transplantation so as to study the probable effect of these fishes on the indigenous cultivated species and sufficient precaution should be taken so as to check spreading of infections diseases unknown to the country.

Record Fish Production with Intensive Culture of Indian and Exotic Carps

Fish culture has a long history in India and the association of a species of Indian major carps in pond fish culture has been known to the fish farmer. The productions, however, have been low because food niches of the pond remained under-exploited, and dependence was mostly on the natural fertility of the ponds to produce fish food. With the introduction at Cuttack of exotic species of fishes, the Chinese silver carp and grass carp and the common carp of Indonesian strain by the Pond Culture Substation of the Central Inland Fisheries Research Institute, true multi-species culture of the three exotic carps and indigenous major carps were undertaken in experimental ponds. Experience gained through experimentation led to the gradual increase of fish production in still-waters and a production as high as 7.500 kg/ha/annum has been recently obtained which is reported in the present communication. The production was achieved by intensive multi-species culture of fish employing a high stocking density in desirable proportion supported by judicious application of fertilizers and supplementary feedings. The highest production (net) recorded from India earlier to this was 6,286.1 kg/ha/yr (Jhingran, unpublished. Bradach *et al.* (1972) mentioned that under favourable circumstances productions of 7,500 to 8,000 kg/ha may be attained by polyculture in the Far East. Yashouv and Halevy (1972) have reported of productions as high as 10,620 kg/ha/yr with poly-culture of tilapia, common carp and silver carp. Chaudhuri (FAO, 1971) achieved fish production of 10,390 kg/ha in year in Burma by culturing the Indian major carps together with small numbers of carps hybrids and gourami.

The experiments reported herein were undertaken in two ponds each 0.25 ha in area which resulted from renovation of a section of the moat of the Killa Experimental Fish Farm at Cuttack. This section had a poor record of fish production as the water sheet was rendered unsuitable for fish culture due to accumulation of coal-tar flowing into the water from drums dumped nearby.

The ponds were poisoned with mahua oil cake applied @ 250 ppm which was followed by application of quicklime @ 100 kg/ha. Because of entry of some murrels from nearby sources the pond 6B had to be poisoned a second time after a month of the initial poisoning. Hereafter, the fingerlings for culture were released.

The ponds were stocked in September, 1972 with 2-3 month old fingerlings raised from fry of induced bred fish. The stocking density used for both the ponds was 10,540 fingerlings per hectare of the cultivated cyprinids which included the Indian major carps catla, rohu, mrigal, the Chinese silver carp and grass carp and the common carp stocked in the ratio of 1: 3, 1: 2, 1: 2 respectively. To this was added 544/ha of miscellaneous fish in one and in the other 568/ha. The former pond has been denoted as 6A and the latter as 6B. Species comprising the miscellaneous category included *Notopterus chitala, Ompok bimaculatus, Mystus seenghala* and *Pangasius pangasius*. The purpose of introduction of these species was to further increase production by utilizing the insects, shrimps, mollusks and weed fishes (inadvertent entry) which are not made use of by the cultivated carps but compete with them for food.

Fertilization was effected by means of inorganic (a mixture of urea and triple superphosphate) and organic (cowdung) fertilizers, the former applied once a month and the latter once a quarter. Fertilization was stopped when algal blooms persisted and fish appeared to be in distress particularly during early hours of the morning. The quantum of inorganic fertilizer used was 1,530 kg/ha/yr for 6A and 1,140 kg/ha/yr in 6B. Each of the ponds received cowdung @14,400 kg/ha/yr. The stocked fish were daily fed with artificial feed, which was a mixure of groundnut oil cake and rice bran, a fortnight onwards after stocking. The quantity of feed put in pond 6A was 12,852 kg/ha/yr and 13,340 kg/ha/yr in 6B. Feeding was suspended or reduced during periods of low water level in the ponds or when the ponds were having a thick bloom of Microcystis.

Feeding was undertaken by broadcasting the feed from the pond margin during the first half of the rearing period whereafter the feed was made into adough and a number of balls of this material was placed in trays hung at different depths in the pond. Chopped bits of *Enhydra fluctuans* were mixed with the feed in the latter half of the culture period. Grass carp were given weeds particularly *Spirodela, Najas* and *Hydrilla* regularly.

Periodical removal of fish which had reached the marketable size of 1 kg was carried out in both the ponds and the remaining finally harvested after one year of culture. The earliest to be harvested were some grass carp and silver carp and a few common carps from both the ponds at the end of 5 months rearing. Some catla could be harvested after 6 months.

Table 1: Average Weights of Fishes as Computed from the Total Numbers Recorded and Rotal Weights Realized of Each Species

Species	*Pond 6A*			*Pond 6B*		
	Average Weight in gm	*Survival (%)*	*Contribution by Weight (%)*	*Average Weight in gm*	*Survival (%)*	*Contribution by Weight (%)*
Silver carp	1,152	30.85	13.07	1,032	65.90	19.04
Catla	1,197	67.92	15.00	1,179	71.69	11.94
Rohu	754	94.93	39.42	854	97.72	35.14
Grass carp 1st lot	1,106	76.60	15.63	1,219	68.68	11.84
Grass carp 2nd lot				1,542	89.00	7.30
Mrigal	766	94.33	13.39	654	100.00	9.54
Common carp 1st lot	1,516	0.41	2.30	1,772	1.17	0.86
Common carp 2nd lot				977	92.59	2.61
Ompok	79	32.00	0.07	102	60.00	0.01
Chitala	163	80.0	0.83	238	91.91	1.17
Mystus	330	36.36	0.12	398	75.00	0.02
Pangas	733	60.00	0.14	742	100.00	0.03

Second lot of 100 nos. of grass and 54 nos. of common carp were introduced in pond 6B, in which fish were found to be growing better. The grass carp were introduced with the idea of raising a second crop of the species, which was realized; whereas the common carp were put as it was observed from the periodic samplings that the earlier introduced numbers of the species had disappeared either through mortality or poaching.

The experiments were concluded after one year when the fish were finally harvested. Table 1 denotes the average weights attained by the fish in the two ponds along with their survival percentages and also the contribution of each species as percentages of the total yield. The high survival despite the high stocking density obtained with rohu puts this species in the first place as regards species wise contribution to the total gross production which in the case of pond 6A is 2,261.6 kg/ha and for 6B 2,638.6 kg/ha. The species that belied all expectations was the common carp which were almost absent from the fish harvested.

The gross and net production recorded in pond 6B was 7,500 kg/ha/yr and 7,343 kg/ha/yr respectively and that in 6A 5,734 kg/ha/yr and 5,652 kg/ha/yr respectively and is the highest recorded yield in pond culture operations in India. The former figure can be considered as outstanding for a pond not having facilities for recirculation of water. This could truly be considered as intensive fish culture since the stocking density used was more than double the rate ordinarily followed in composite fish culture in India.

References

Bardach, John E., Ryther, John H., and McLarney, William O., Aquaculture: The Farming and Husbandry of Freshwater and Marine Organisms, Wiley-Interscience, New York, 1972.

FAO, "Report to the Government of Burma on Fish Culture Development. Based on the work of H. Chaudhuri, Inland Fishery Biologist (Fish Culture)," Rep. FAO/UNDP(TA). 1971 (2954), 45p.

Jhingran, V.G., "Present status of aquaculture in India," (Mimeo), 1973.

Yashouv, A. and Halevy, A., "Experimental studies of polyculture in 1971 Bamidgeh, 1972, 24, 2.

[The work was done in association with co-scientists R. D. Chakrabarty, N.G.S. Rao, K. Janakiram, D. K. Chatterjee and S. Jena].

Preliminary Observations on the Survival and Growth of Chinese Carps in Burma

Editor's Note

Chinese Grass carp, *Ctenopheryngodon idella* (Val) and Silver carp, *Hypophthalmichthya molitrix* (C. and V.), recently brought from India to Burma were introduced at the Hlawga fish farm. The paper describes in details the observations made on the survival and growth of these exotic carps in Hlawga ponds. The survival as well as growth of both the species were very satisfactory and a high production of fish during the growth period of 6 months was obtained. The fishes, however, did not attain full maturity in the first year. It is expected that the number which survived and have grown to adult size would be adequate for subsequent propagation of these species to all over the country. The history of earlier attempts made to introduce Chinese carps in Burma is also discussed.

Chinese Carp in Burma

Introduction

The phytophagous Chinese Grass carp, *Ctenopharyngodon idella* and Silver carp, *Hypophthalmichthys molitrix* have drawn the attention of fish culturists all over the world in recent years. These fast growing fishes, especially the former species, have been introduced in many Asian, African, East European and also in North and Central American countries. (Alikunhi and Sukumaran, 1964; Avault, 1965; Hora and Pillay 1962; Vinogradov, 1968 and Woynarovich, 1968). These species have proved to be welcome additions to the culture fisheries of majority of the countries of their introduction. The Silver carp is predominantly a phyto-plankton feeder and the Grass carp feeds on higher aquatic vegetation. Because of their shorter food chain, these fishes are able to convert food into fish flesh very efficiently resulting in high yields. Besides, Grass carp could suitably be utilized as a biological control agent for many noxious weeds in fishery waters.

In Burma, the Indian major carps, namely, *Catla catla, Labeo rohita* and *Cirrhina mrigala* are the important indigenous species cultivated in ponds. However, to increase variety with the bulk and also to take advantage of the voracious feeding habit of Grass carp for the control of aquatic vegetation in lakes and other cultivable waters, attempts have been made in recent years to introduce these Chinese carps in Burma.

Information on the history of introduction of Chinese carps in Burma and details of the preliminary observations made on the survival and growth of these exotic carps in Burmese waters, are presented in this communication.

History of Introduction of Chinese Carps

Several attempts have been made since later fifties to introduce Chinese carps in Burma. Tubb (FAO/UN,1957) recommended introduction of Grass carp for the biological control of aquatic weeds in Inlè lake. Chaudhuri (FAO/UN, 1966) also stressed the utility of introduction of Grass carp, especially in Inle lake for control of the luxuriant growth of submerged aquatic vegetation. He also suggested to explore the possibility of introducing the fast growing Silver carp in Burma.

In April, 1957, a consignment of 30,000 fingerlings of Chinese carp was brought by the Government of Burma from China. The consignment consisted mainly of Grass carp and the Bighead (Aristichthys nobilis), along with a few specimens of Black carp (Mylopharyngodon piceus). Large–scale mortality occurred after release of the fingerling in Hlawga ponds. Out of those survived, 3,000 fingerlings were transported to Inle Lake in Shan State by rail and about 700 were sent by air. But heavy mortality took place during transport and although a few were released in the lake, they could not establish themselves in the new environment. A few adult Grass carp (males) and Bighead specimens are still being held at the Hlawga Fish Farm. A second attempt was made in 1959, when about 100,000 fingerlings of Chinese carps were imported from China, but the entire consignment died shortly after release in the ponds.

In 1969, with the assistance of FAO, arrangement was made to import a gift consignment of Chinese carps from India. Accordingly, a small experimental consignment of about 1000 Grass carp and 1000 Silver carp fingerlings arrived at Rangoon by air from the Central Inland Fisheries Research Substation at Cuttack, India, on October 7, 1969. The fingerlings were released in the Hlawga fish farm and are being reared there. A few hundreds of these Chinese carps, however, were subsequently transported successfully to Nyaungshwe fish farm near Inle Lake.

Observations

Observations on the survival and growth of the newly introduced Chinese carps were made at the Hlawga fish farm. The farm is situated below the dam of the Hlawga lake, a reservoir for the supply of water to the City of Rangoon. The soil is not rich. Analyses of the farm soil showed the texture of the soil as silty-clay. The soil pH is moderately acid and the soil is low in organic carbon, low in both available nitrogen and available phosphorus and medium in total nitrogen. The pH of water in treated ponds varied from 6.8 to 7.5.

Three nursery ponds, A_3 A_4 and A_5 (each 0.08 ha in area) were prepared a month prior to the introduction of Chinese carps. Liming and manuring with cowdung was done at the rate of 250 kg/ha and 2,000 kg/ha respectively. Superphosphate at the rate of 200 kg/ha was introduced a week after the introduction of the fingerlings. On October 7, 1969, 1030 Grass carp fingerlings were released in pond A_4 and 950 Silver carp fingerlings were introduced in pond A_3. After 10 days, 400 Grass carp and 100 Silver carp fingerlings were transferred to the Nyaungshwe farm. The ponds were thoroughly netted out after about a month for restocking the Chinese carps for growth study.

In the first experiment, 370 Grass carp and 570 Silver carp fingerlings were stocked in three different combinations and at different rates of stocking in the three prepared nursery ponds the rest (65 of each species) were distributed to various other ponds. The size-range average size and average weight of each species were determined at the time of stocking. The stocking rate of fingerlings varied from 3, 375 ha/ to 5,000 ha. The fishes were fed with artificial food. The feed given was a mixture of rice bran, peanut cake and chopped green vegetation (common marshy emergent plants) in the ration 1:1:2. 3.2 kg. of the mixed feed was fed daily in ponds A_3 and A_5 while 4.8 kg. of the feed was introduced in the heavily Stocked pond A_5. The quantity of feed was increased 2 times after 3 months. Pond A_3 and A_4 developed a light bloom of the blue-green algae, Microcystis which continued till the end of the experiment.

Initially the Silver carp grew much faster than the Grass carp. Growth of Grass carp was restricted since no suitable aquatic weeds were available in the fish farm or in nearby places. By the middle of

January, Spirodela was procured and 20 Kg. was introduced in each pond. Subsequently on two more occasions, once in February and the second time in April, 40 kg. of Spirodela were introduced in each of the three ponds. The Grass carp recovered quickly and the subsequent increase in growth was remarkable.

The experiment was wound up after 6 months' duration when the ponds were required for other purposes. Final survival and growth of fishes were assessed by complete dewatering of the experimental ponds and collecting all the fishes. Total gross production of fish within a growth period of 6 months was calculated to be 1,911, 2,046 and 1,396 kg/ha in ponds A_3, A_4 and A_5 respectively. The average production of Grass carp came to 809 kg/ha. The corresponding figure for silver carp was 975 kg/ha.

The survival of fingerlings, in general, was very high. The average survival in Grass carp and Silver carp was 97.03 per cent and 94.74 per cent respectively. The recovery percentage was least in Silver carp in A_3. It is quite probable that a few fish might have jumped into the adjacent A_4 pond where the percentage of survival of Silver carp was as high as 99.3.

After the closure of the first experiment, a second experiment was initiated with the Chinese carps. The rate of stocking in the second experiment was given very low so as to obtain faster growth for studying the size and age at maturity of these exotic carps in Burmese waters. At the beginning of May, 1970, 30 Grass carps and 30 Silver carps were stocked in each of the ponds A_1 and A_2. These two ponds were prepared 10 days earlier by draining out water completely and refilling with fresh water. Some water from ponds A_3 and A_4, having a bloom of Microcystis was also introduced in these ponds. Subsequently both the ponds developed a thick bloom of Microcystis. Daily 6.4 kg. of the mixed feed consisting of rich bran, peanut cake and chopped green vegetation were fed in each pond. In addition, 3.2 kg. of cut grasses were also supplied on every third day. The growth of the Chinese carps especially in pond A_2 wee observed to be highly satisfactory.

For studying the maturity of these Chinese carps, the fishes were examined regularly from the month of May. The fishes did not show any sign of maturity excepting a few males of Silver carp which developed slight roughness on the pectoral fins. On examination, gonads of both Grass carp and Silver carp were found to be immature. Testes of some of the Silver carps have only reached II–III stage of maturity.

Discussion

The observations made in the foregoing indicated that the Chinese carps could be successfully reared in Burmese waters provided proper care is taken from the very beginning. The earlier attempts in acclimatizing the Chinese carps in Burma did not meet with much success probably because of insufficient knowledge of the biology of these carps and also due to inadequate facilities for transporting the fingerlings over long distances. The Chinese carps can tolerate wide range of temperature as is evidenced by their successful acclimatization in various tropical (Alikunhi and Sukumaran, 1964, Hora and Pillay, 1962) as well as in various temperate regions (Vinogradov, 1968; Woynorovich, 1968).

Tubb (1968) included both the Grass carp and silver carp among the cultivated species is Burma, but information available to the present authors show that Silver carps were not previously introduced to Burmese waters although the attempts were made earlier to introduce Grass carp.

Both the Chinese carp species are thriving very well after their introduction into the Hlawga fish farm. The survival rate after the initial stocking would have been higher but for the accidental entry of a number of juveniles of the murrel, *Ophicephalus striates*, which wee responsible for predation of small fingerlings of Chinese carps. However, when the fingerlings grew larger, the survival rate was very high as observed in Experiment 1.

The growth rate of Chinese carps in Burmese water is quite satisfactory. Grass carps would have grown much quicker if aquatic weeds, such as, *Wolffia, Lemna, Spirodela, Hydrilla* and *Najas* were readily available for feeding in large quantities. Alikunhi and Sukumaran (1964) have given an account of the growth rate of Grass carp and Silver carp in India. Compared to this, the growth rate of Chinese carps in temperate water is much slower. Woynarovich (1968) has compared the growth of Grass carp in different regions such as, China, USSR, Rumania, and Hungary. The weights achieved in first 2 years very from 200-1,200 g only. According to him, the increase in weight per year comes to 1 kg only after the second Summer. The silver carp in the same regions grow to 120–2,030 g in 2 years.

Sukumaran *et al.* (1970) have studied the comparative growth of Silver carp and Catla and have observed that, in general, the former recorded better growth than Catla. Although no experiment was conducted by the present authors to study the comparative growth of Chinese carp and the indigenous carps of Burma, the growth rate of Chinese carps as observed in the two experiments seem to be much faster than the observed growth rate of Catla, Rohu and Mrigal in Burma. Besides, observations made in Indian (Alikunhi and Sukumaran, 1964) show that the Chinese carps grow much faster than Rohu and Mrigal and the Silver carp grows even better than the fastest growing of Indian major carps namely, Catla. Both the Chinese carps are quite compatible with the Indian carps although some competition exists between the predomination phytoplankton feeder silver carp and the predominantly zooplankton feeder Catla.

Sukumaran *et al.* (1970) carried out a series of experiments on combined stocking of Silver carp and Catla and also in combination with Grass carp. They obtained a total gross production varying from 510.7 to 650.75 kg/ha/6 months. These production figures seem to have been obtained by manuring alone, since there was no mention of any feeding being done. At the Hlawga fish farm in Burma, by proper manuring and feeding the net production of fish obtained, varied from 1283.8 to 1877.5 kg/ha/6 months.

Regarding the study on the maturity age of Chinese carps, there are host of references suggesting wide variations in the maturity age of these fishes in different regions and under varying climatic conditions. The Chinese carps attain sexual maturity in their natural habitats in China at the age of 4 years (Chen and Lin, 1935). According to Lin (1965), the Silver carp becomes mature in 2-3 years and Grass carp in 4-5 years, under the climatic conditions in Taiwan. As a rule the males mature one year earlier than the female. Ling (FAO/UN, 1965) mentions that the Silver carp females need three-plus years and Grass carp four-plus years to attain full maturity. Alikunhi and

Sukumaran (1964) observed Silver carps to attain full maturity in 2 years at Cuttack, while in Grass carp, the males only attained full maturity during the same period. Alikunhi *et al.* (1963) succeeded in breeding 3 years old Grass carp and Silver carp. The induced-bred Indian carps as well as Chinese carps have been observed to attain maturity earlier than to be obtained from natural habitat. It was expected that the Chinese carps introduced and reared in Burmese waters would also mature in the first year since these are the progeny of few generations of induced bred fishes. However, excepting the appearance of slight roughness in males of Silver carp, there was no sign of maturity in others. Even the males of Silver carps have not attained full maturity although several males of both the species have exceeded the minimum size of maturity attained by Chinese carps at Cuttack (Alikunkhi *et al.*, 1963). This difference in the maturity age between the Chinese carps at Cuttack and the Chinese carps of the same stock in Burma may probably be due to the effect of seasonal change of temperature which is believed to influence maturation of gonads in fish. Lower Burma does not have extremes of temperatures and as such the seasonal change of temperature is not very much marked. This may be the probable reason why the indigenous Indian carps take much longer time to attain full maturity in Burma than their counterparts at Cuttack in India.

Conclusion

The observations made is the foregoing have demonstrated that the Chinese carps, introduced recently in Burma have survived the critical period and a considerable number have already attained the adult size. These fishes are expected to attain full maturity by the next fish breeding season. However, it is still necessary to take proper care to grow them to suitable spawners and induce them to breed so as to obtain sufficient stock for propagation to all over the country. The Chinese carps would definitely be able to increase the production of fish in Burma, if judiciously cultured along with other compatible species. The Grass carp would have additional demand for its ability to control aquatic vegetation. There would be the necessity for initiating a programme for systematic stocking and culture of Grass carp in Inle Lake. Once the Grass carp establishes itself in the lake, it would enrich the water by its faecal matter containing semi-digested plants and thus pave the way for introduction of other economic cultivated species, resulting in high yields of fish from the lake.

References

Alikunhi, K.H. and K.K. Sukumaran. 1964. Preliminary observations on Chinese carps in Indian. Proc. Indian Acad. Sci. (B), 60 (3): 171–89.

Alikunhi, K.H., K.K. Sukumaran and S. Parameswaran. 1963.Induced Spawning of Chinese carp, *Ctenopharyngodon idellus* (C. &V) and the Silver carp, *Hypophthalmichthys molitrix* (C &V) in ponds at Cuttack, India. IPFC Procs., 10 (2): 181-204.

Alikunhi, K.H., K.K. Sukumaran and S. Parameswaran. 1965. Observations on growth, maturity and breeding of include-bred, pond-bred silver carp, *Hypophthalmichthys molitrix* and grass carp, *Ctenopharyngodon idellus*, in India during July 1962 to August, 1963. Bull. Cent. Inst. Fish Educ. (2): 19 pp.

Avault, J. W. Jr., 1965. Preliminary studies with grass carp for aquatic weed control. Progr. Fish-cult., 27 (4): 207-9.

Chaudhuri, H. 1968. Breeding and selection of cultivated warm water fishes in Asia and the Far Mast–a review. FAO Fish Rep. (44) Vol. 4: 30-66.

Chaudhuri, H., S.B. Singh and K. K. Sukumaran, 1966. Experiments on large scale production of the Chinese Grass carp, *Ctenopharyngodon idellus* (C and V) and the Silver carp, *Hypophthalmichthys molitrix* (C and V) by induced breeding in pond in India. Proc. Indian Acad. Sci. 113 (2) Sec. B: 80-95.

Chen, S. B. and S. Y. Lin, 1933. The fish fry industry of China. Chekiang Prov. Fish Exp. Stn., 4 (4).

FAO/UN 1957. Report to the Government of Burma on inland fisheries development. Based on the work of J. Alan Tubb. FAO Expert in Inland Fisheries. Rep. FAO/EPTA, (776): ii + 64 p.

FAO/UN 1965. Report to the Government of the Republic of China on the production of fry and fingerlings of Chinese major carps by induced spawning. Based on the work of S. W. Ling, Rep. FAO/EPTA, (2044): I + 12 P.

FAO/UN 1966. Report to the Government of Burma on the development of fish culture. Based on the work of Chaudhuri, FAO/UNDP/TA Inland Fishery Biologist. Rep. FAO/UNDP (TA), (TA 2298): (i) + 14 p.

Hora, S. L. and T. V. R. Pillay 1962. Handbook of Fish culture in the Indo-pacific Fisheries Region. FAO Fish. Tech. Pap., 14: 204 pp.

Lin, S. Y. 1965. Induced spawning of Chinese carps by pituitary injection in Taiwan. Chinese-American Joint Commission on Rural Reconstruction. Fisheries Series No. 5:31 p.

Sukumaran, K. K., S. B. Singh, D. S. Murty and P. C. Chakrabarti. 1970. Studies on compatibility and competition between silver crap, *Hypophthalmichthys molitrix* (Val) and Catla catla (Ham). IPFC. Procs. 13 (2): 185

Tubb, J.A., 1968. Status of fish culture in Asia and the Far East FAO-Fish. Rep. (44) Vol. 2: 45–53.

Vinogradov, V. K., 1968. Techniques of rearing phytophagons fishes, FAO Fish. Rep. (44) Vol. 5: 227-32.

Chapter 11

Aquaculture and Global Fish Seed Production

Editor's note

The subjects discussed are as follows:

This vital issue covers broad subjects for analysis and discussions. Of many technical write ups only six (6) could be retrived for presentation in this treatise. The following technical papers have been considered for edition:

1. Large scale production of fish seed of the Chinese carp (*Cteopharyngodon idellus*) by induced breeding.
2. Recent advances in Global Aquaculture.
3. Seed production–its problems.
4. Record seed production with intensive culture of Indian and exotic carp.
5. Fish seed production–a prequisite to intensive freshwater fish culture.
6. Problems of warm water fish seed production.

Experiments on Large-scale Production of Fish Seed of the Chinese Grass Carp, *Ctenopharyngodon idellus* (C. and V.) and the Silver carp, *Hypophthalmichthys molitrix* (C. and V.) by Induced Breeding in Ponds in India

Introduction

With a view to study the cultural possibilities of these carps in Indian waters and their compatibility with the cultivated species in India, namely, the Indian major carps, experimental consignments of fingerlings of the Chinese Grass carp, *Ctenopharyngodon idellus* (C. and V.) and the Silver carp, *Hypophthalmichthys molitrix* (C. and V.) were brought in 1959 (Alikunhi and Sukumaran, 1962) and released in the Experimental Fish Farm at Killa in Cuttack, India. These herbivorous, quick-growing species are likely to be welcome additions to the culture fisheries in India. The silver carp is predominantly a phytoplankton feeder and as such efficiently converts the food into fish flesh resulting in high yields. The Grass carp, on the other hand, feeds on various aquatic macrovegetation which otherwise remain mostly unutilized and create 'weed problem' in cultivable waters. The voracious feeding habit of Grass carp on submerged weeds and on small floating weeds is a point of strong recommendation for its culture and propagation as a biological control agent for the noxious weeds.

In view of the aforesaid qualities of the above species, there has been a heavy demand for their seed. Considerable difficulty is experienced in obtaining the desired quantity and quality of seed of these species for stocking as these fishes do not breed naturally in ponds. The need to evolve a dependable technique to procure adequate quantity of their seed has been strongly felt by a large number of research workers. This led to extensive efforts in the South-East Asian countries and the USSR towards induced spawning of these species in confined waters. The success of the technique of induced breeding of Indian major carps of similar breeding habit, by administration of fish pituitary hormone has been well established (Chaudhuri and Alikunhi, 1957; Chaudhuri, 1960; Alikunhi *et al.*, 1960) and the technique has been further developed for attaining commercial production of fish seed of these economic species (Chaudhuri *et al.*, 1962; Vijayalakshmanan *et al.*, 1962; Chaudhuri,

1963). In the case of Chinese carps, successful spawning of the Grass carp and Silver carp by hormone injection has been reported by Aliev (1961), Alikunhi *et al.* (1962) and Tang *et al.* (1963). However, only limited success has been obtained by the above workers. It has been expressed that further standardization of the technique of hormone treatment is necessary to achieve success on a large scale. Attempts towards this have been continued at the Central Inland Fisheries Research Substation, Cuttack, and have met with success. During June to August 1964, intensive efforts were made to obtain sufficient quantity of fish seed by inducing spawning in Grass carp and Silver carp by administration of fish pituitary hormone. The details of the experiments conducted and the success obtained in breeding these carps on a productive scale are presented in this communication.

Experiments and Results

The induced breeding experiments conducted with Grass carp, *Ctenopharyngodon idellus* and Silver carp, *Hypophyhalmichthys molitrix* are described below separately. Since a large number of experiments were carried out, it may be difficult to accommodate the details of all of them in the text. Therefore, only the successful experiments have been described in the text.

Grass Carp

Altogether twenty-eight sets of Grass carp were injected with heteroplastic fish pituitary hormone, and the donors of glands were *Cirrhina mrigala* (Mrigal), *Labeo rohita* (Rohu) and *Hypophyhalmichthys molitrix* (Silver carp). Out of these, four sets were injected with Mrigal glands, seventeen with Rohu glands and seven with Silver carp glands. Successful results were obtained in eight sets (2 sets receiving Mrigal and 6 sets Rohu glands) of which seven sets yielded healthy fry.

The female spawners of Grass carp in sets 1 and 2 received the hormone in two doses at an interval of seven hours and were stripped six to seven hours after the final injection. In set 1, the female receiving lower dose displayed no sex play, but yielded better results as compared to set 2 which received a slightly higher dose and exhibited distinct sex play. The eggs hatched out after 19 hours, and the percentage of hatching was found to be quite good.

Set 3 which received more or less the same dose as sets 1 and 2 did not give satisfactory results. The percentage of hatching was very low obviously due to hot weather and high water temperature.

The females in sets 4, 5 and 6 were administered the hormone in three gradually increasing doses at three-hour intervals and in all cases natural fertilization of eggs was obtained. In set 4 delay of three to four minutes occurred in obtaining milt for the stripped eggs, but no adverse effect was noticed.

The female in sets 7 and 8 received only two doses of hormones at an interval of four hours, and high percentage of fertilization of eggs was observed.

The total dose of hormone administered in the above instances varied between 10 and 12 mg./kg. body-weight of the recipient fish.

Silver Carp

Altogether forty-five sets were injected with homoplastic as well as heteroplastic glands from Mrigal and Rohu. Out of these in nineteen sets (13 sets receiving homoplastic and 6 sets Rohu glands) there was successful spawning and twelve sets yielded healthy fry.

Three female breeders of Silver carp in sets 2 to 4 received only one dose of pituitary hormone at the rate of 3–4 mg./kg. body-weight and released eggs freely in about eight hours which were fertilized by milt from uninjected males. The weather was ideal throughout.

The females in sets 1, 5 to 12 received the hormone in two doses after six to eight hours, and in set 16 after four hours, the total dose administered varying from 8 to 11 mg./kg. body-eight. High water temperature and breaking up of eggs membrane caused heavy loss of fertilized eggs. On several occasions release of insufficient milt by the males appeared to be responsible for the low percentage of fertilization.

To fix female spawners in sets 13–15 and 17–19 the hormone was injected in three gradually increasing doses at three-hour intervals. In the first three some sex play was noticed. On one occasion (in set 15) natural spawning inside the breeding hapa was noticed from which 200 fry were obtained, but the same was rendered slightly doubtful owing to the splashing of some stripped eggs by the male breeder.

Physico-chemical Condition of Water

The general physico-chemical condition of water in Padmapukuri and Farm Pond II were studied for some of the days when the experiments were carried out.

In the case of both Grass carp and Silver carp the negative results were mostly due to slight loss of condition of the female breeders which on examination were found with resorbing eggs. Other adverse factors were high water temperature, unsuitable water conditions possibly resulting in breaking up of egg membrane, and lack of sufficient milt in the male breeders.

Discussion

With the success in inducing spawning of the Indian carps by injection of fish pituitary hormones (Chaudhuri and Alikunhi, 1957), it was thought that probably the same technique would hold good in breeding the Chinese carps also which have similar breeding habits. Attempts have been made in several countries including USSR, China, Japan and some of the South-East Asian countries to induce these carps to breed by hormone injections. As reported earlier, success has been reported by Aliev (1961) in USSR, Alikunhi *et al.* (1962) in India and Tang *et al.* (1963) in Taiwan. Hickling (personal communication, 1964) observed successful ovulation in Grass carp, but the eggs could not be fertilized. The experiments conducted by Aliev (1961) were not strictly in confined waters since the spawners after injection were released in sections of a canal maintaining flow of water where fishes naturally spawned. However, artificial fecundation of eggs by stripping was not very successful and only stray hatchlings were obtained by him. Lin (1949) reported that spawning carps from the river could be caught and stripped successfully. This method is regularly practiced in Japan (Kuronuma, 1958) where the Chinese Grass and Silver carps have established in rivers (Inaba *et al.*, 1957). Alikunhi *et al.* (1962) have demonstrated successful spawning of Grass carp and Silver carp in ponds at Cuttack by hormone treatment. They obtained success with three-year-old pond-grown fishes which were observed to have attained sexual maturity in the Cuttack ponds in two years. In 1963, Chaudhuri successfully bred two-year-old pond-grown Grass carp by hormone injection at Kalyani Research Station, West Bengal, and obtained healthy fry; and Alikunhi *et al.* in 1963 induced to breed one-year-old Silver carp in ponds at Cuttack (Annual Report of the Central Inland Fisheries Research Institute, 1963–64). Tang (1960) recorded natural spawning of Grass carp and Silver carp in the Ah Kung Tien reservoir. Tang *et al.* (1963) have reported spawning of the Chinese carps–Big head, Grass carp and Silver carp by pituitary injections. While Alikunhi *et al.* (1962) achieved success with pond-grown breeders. Tang *et al.* (1963) collected fully ripe breeders for their experiments from the Ah Kung Tien reservoir where fishes have been naturally spawning. In all the instances mentioned above positive response was obtained but the success was limited and the production of fish seed was meager.

In the present experiments the authors have attempted to evolve a technique so as to produce fish seed of the Chinese Grass and Silver carps on a productive scale to meet the heavy demand, especially that of the Grass crap, for stocking in cultivable waters throughout India. The technique of breeding adopted by Alikunhi *et al.* (1962) as stated, was essentially the same as those followed in the case of Indian carps. However, it was observed by the present authors that though there was great similarity in the response of Chinese and Indian carps to similar hormone treatment, the results obtained were not usually strictly identical. In the case of Indian carps, the usual practice is to give a preliminary low dose or the hormone to the female breeder followed by another higher dose to the female and only one dose to the males after six hours when breeders of both the sexes are introduced together in breeding hapas. Spawning is usually obtained within three to six hours after the second injection (Chaudhuri *et al.*, 1962). The same procedure when adopted in the Grass and Silver carps, successful natural spawning was not obtained and in several instances eggs were released inside the hapa but were not fertilized. In a few other cases, the breeders were not disturbed and natural spawning was awaited up to eight to nine hours after the last injection but without any success. Such conditions, therefore, necessitated stripping and artificial fecundation when the females were ready to ovulate. This, however, has certain disadvantages, because a little delay or early stripping may culminate in very poor results and furthermore the whole process is very cumbersome. To obviate these drawbacks, attempts were made to synchronise proper stimulation in the male and female breeders to obtain natural spawning by regulating doses, duration and frequency of hormone administration. In order to gradually bring up the condition of the females, the hormone was injected in the course of three slightly increasing doses at closer intervals of three hours. Also, in the case of males of Silver carp which were not giving very satisfactory release of milt, similar treatment was given to them to obtain better results. With this slight modification in the technique, encouraging response was obtained. On three occasions, Grass carps were observed to have naturally spawned inside the hapa. The eggs obtained were naturally fertilized and produced healthy fry, though the percentage of hatching was not very satisfactory. In the Silver carp, however, only on one occasion, a small percentage of fertilized eggs was observed from naturally released eggs, but the findings was rendered doubtful owing to accidental splashing of some stripped eggs into the breeding hapa by the male fish when it was being stripped.

To our knowledge, the instances mentioned above, especially of Grass carp, producing viable eggs and healthy fry after injection, seem to be the first record of a natural spawning in a pond. Aliev (1961) obtained natural spawning after releasing the injected breeders in a canal maintaining flow of water.

The total dose of hormone varying from 9 to 12 mg./kg. body-weight of the recipient female, and 2–4 mg./kg. body-weight of the recipient male, administered in two to three courses within six to eight hours, was found to be effective ordinarily in the case of both Grass and Silver carp. The dose of hormone administered by Aliev (1961) and Tang *et al.* (1963) are difficult to compare with the above, because they expressed the dose in terms of number of glands and did not take into account the weight of individual glands. The time interval given by Aliev (1961) between two injections was twenty-four hours, but in our Indian climatic conditions the complete course of two or three injections were given within six to eight hours, and the entire operation of injecting and stripping the spawners was usually over within eight to fourteen hours. An interesting observation was made on July 9–10, 1964, when three Silver carp spawners released eggs freely within eight hours after the administration of a first low dose of hormone at the rate of 3–4 mg./kg. body-weight. The weather on the particular day was very favourable with cool breeze and continuous drizzling. The eggs could be successfully fertilized by milt from uninjected males and healthy fry were obtained.

During the experiments the sexual behaviour of males of Silver carp on a number of occasions was found to be poor which resulted in yield of little or no milt at the time of stripping. Detailed examination of the gonads revealed that out of as many as 107 males, 60 males showed deformed testes which were either beaded or lobed and in some cases very thin. The release of scanty milt by the males of Silver carp could also be responsible for somewhat poor results in several cases. Release of insufficient milt at the time of stripping was also recorded by Aliev (1961).

It is a well-known fact that the success of the induced breeding experiments primarily depends upon the selection of spawners in their prime stage of maturity. In the present experiments considerable difficulty was experienced in the selection of suitable spawners of the Chinese carps because most of them attained full maturity as early as in April-May, and so during the monsoon season the properly ripe females got mixed up with those where resorption had set in. It is apprehended that this could be the main reason for the unsuccessful results in many instances. Added to this, delayed monsoon during the year was another adverse factor. The prevailing high water temperature during the summer months prevented the taking up of any induced breeding experiment before the monsoon season. It has been noticed in the present as well as earlier induced breeding experiments that water temperature is a very vital factor, and rains are very conducive but not essential. During the present experiments the water temperature ranged between 27°C and 35°C., but above 30°C, the results were usually poor. On the 15th July 1964, for instance, one set of Silver carp when bred showed a high percentage of fertilization of eggs, but subsequently during the day heavy mortality of development eggs occurred due to high water temperature and only stray hatchlings could be obtained. Aliev (1961) reports loss of Silver carp spawn in Russia when the water temperature was 26.2° to 28.5°C., which is however an optimum range in tropical climates.

During the experiments the phenomenon of breaking up of egg membrane and dropping of embryos was noticed which affected very adversely the yield of healthy fry on several occasions. The possible factors responsible for it could not be determined, and investigations in this line are being pursued at the Research Station. Tang *et al.* (1963) suggests that possible reasons for mortality of eggs during hatching could be overdose of hormone which results in the release of large number of immature eggs, and also sudden fall in water temperature at times.

From the experiments described in the foregoing a total of 2.4 lakhs (24 million) of fry of Grass carp and 3.57 lakhs (357 million) of fry of Silver carp were obtained. The advanced fry and fingerlings of Grass carp and Silver carp produced has already been distributed all over India on a limited scale and a gift consignment has also been sent to Philippines. The results of this year indicate that in the near future fry of these Chinese carps could be produced on a commercial scale to meet the increasing demand of their seed.

1. The experiments to induce spawning in the Chinese Grass carp, *Ctenopharyngodon idellus* and the Silver carp, *Hypophthalmichthys molitrix*, grown in ponds, were successfully conducted during June-August 1964. The breeders consisted of induced-bred as well as of the original stock. Large-scale production of fish seed was achieved yielding 2.4 lakhs of fry of Grass carp and 3.57 lakhs of Silver carp. The results obtained suggest the possibility of commercial production of seed of these Chinese carps.
2. Induced fish breeding techniques as developed for the Indian major carps were adopted in general; however, the response was not found to be quite satisfactory, because stripping and artificial insemination of eggs had to be resorted to in most of the cases. The above technique was, therefore, modified and in place of two injections at an interval of six hours, the female breeders received the hormone in three gradually increasing doses at a closer interval of three hours. With this slight modification in the technique, encouraging response was obtained. On three occasions, Grass carp were observed to have naturally spawned inside the hapa and healthy fry were produced.
3. The total dose of hormone administered varied from 9 to 12 mg./kg. body-weight of the recipient female and 2–4 mg./kg. body-weight of the recipient male, and was ordinarily found effective in the case of both Grass and Silver carp. On one occasion when ideal weather condition was prevailing, a single low dose of hormone 3–4 mg./kg. body-weight of female spawners induced successful

spawning in three sets of Silver carp. The donors of glands were the Indian major carps (Rohu and Mrigal) and Silver carp.

4. The entire operation of injecting and stripping the spawners usually took eight to fourteen hours during the present series of experiments.
5. The negative and poor results mostly were due to such female spawners where resorption of eggs had set in and also in several instances due to adverse physico-chemical conditions of water resulting in heavy mortality of developing eggs.

References

Aliev, D.C. "Experiments of breeding from Far eastern vegetarian fish in the conditions of Turkmenia (in Russian)," Abstract from Voprosi lehtiologii, Vol I, 1961, Edition 4(21).

Alikunhi, K.G., Vijaya-Lakshmanan, M.A. and Ibrahim, K.H. "Preliminary observations on the spawning of Indian carps induced by injection of pituitary hormones," Indian J. Fish., 1960, 7(1), 1–19.

—— and Sumumaran, K.K. "Preliminary observation on Chinese carps in India," Proc. Indian Sci. Congr., Cuttack, 1962: Proc. Indian Acad. Sci., 1964, 60 B(3), 171–188.

—,— and Parameswaran, S. .."Induced spawning of the Chinese Grass carp, *Ctenopharyngodon idellus* (C. and V.) and the Silver carp, *Hypophthalmichthys molitrix* (C. and V.) in ponds at Cuttack, India," IPFC Proc. 10th Session, Sec. II, 1962, 181–204.

Chaudhuri, H. "Experiments on induced spawning of Indian carps with pituitary injections," Indian Jour. Fish., 1960, 7(1), 20–48.

Chaudhuri, H. "Induced spawning of Indian carps," Proc. Nat. Inst. Sci. Ind., 1963, 29B(4), 469–87.

Chaudhuri, H. and Alikunhi, K.H. "Observations on the spawning in Indian carps by hormone injection" Curr. Sci., 1957, 26(12), 381–82.

Chaudhuri, H. Bhowmick, R.M., Pal, R.N. and Munshi, A.B. "Observations on induced breeding of major Indian carps during 1960 at Cuttack," Proc. Indian Sci. Congr., Cuttack, 1962.

Inaba, D., Nomura, M. and Nakamura, M. "Preliminary report on the spawning of Grass carp and Silver carp in the Tone River, Japan, and the development of their eggs," Jour. Tokyo Univ. Fish., 1957, 43(1), 81–96.

Kuronuma, K. and Lin, S. Y. "Spawn taking of Chinese carp in the Tone River, Japan (Record of 1957)," IPFC Curr. Affairs Bull., 1958, No. 22 "Life-history of Woon Ue, *Ctenopharyngodon idellus* (C. and V.)" Lingnan Sci. Jour., 1935, 14(1) and 14(2).

Tang, Y.A. "Rreproduction of the Chinese carps, *Ctenopharyngodon idellus* and *Hypophthalmichthys molitrix* in a Reservoir in Taiwan," Jap. Jour. Ichthyology, 1960, 13(1/2), 1–2.

Tang, Y.A., Hwang, Y.W. and Liu, C.K. "Preliminary report on injection of pituitary hormone to induce spawning of Chinese carps," IPFC occasional paper, 1963, 63/14, 1–7.

Vijayalakshmanan, M.A., K.V. "Experiments on inducing spawning of Indian carps by Pal, R.N., Sukumaran, injection of fish pituitary hormones during 1959 at K.K. and Ramakrishna, Cuttack," Proc. Indian Sci. Congr., Cuttack, 1962.

[The work was done in association with Co-Scientists S.B. Singh and K.K. Sumumaran. Published in the Journal "The Proceedings of the Indian Academy of Sciences" Vol. LXIII, No. 2, Sec. B, 1966.]

Recent Advances in Global Aquaculture Fish Seed Production

Aquaculture has been recognized in recent years, as one of the most important strategies to accelerate global fish production. Fish, as a group is the final result of a complex biological cycle in an aquatic ecosystem. Fish culture had been an age-old practice in Asia, especially, in China and India, but the average production had been generally low because of the traditional type of culture. After the second world war, the aquaculture technology was gradually improved and more and more scientific methods were introduced. As a result, the fish production was increased from a meager 500–600 kg/ha/yr to 2–4 mt/ha/yr. However, during the last few decades, all round development in the aquaculture industry had taken place. Consequently, the yield of fish has gone up considerably due to intensification and expansion of modern aquaculture throughout the world. Such remarkable achievements could be possible mainly because of the application of the latest technologies in the two very important aspects of aquaculture, namely (A) Fish Seed Production and (B) Food-fish Production.

Fish Seed Production

The availability of adequate quantity of quality seed of the cultivated fish species, is an essential pre-requisite for obtaining large-scale success in the culture of any fish. Aquaculture industry in the world suffered tremendously due to the extreme dearth of seed for culture, since the majority of the major cultivable fishes either fail to reproduce in captivity or do not attain full maturity when reared in ponds or other enclosed waters. They generally breed in their natural habitats under optimum environmental conditions. The Asiatic carps (Indian and Chinese carps), cultured for centuries in ponds, belong to this category, as also the important euryhaline species such as, mullets, milkfish and sea-bass, the catadromous freshwater eels and majority of the commercially important marine fishes. For decades in the past, concerted efforts were made by fishery scientists to breed these fishes under artificial conditions. The success was finally achieved in late mid-1950's when members of Asiatic carps were induced to breed by injection of fish pituitary hormones. A dependable technique was subsequently developed to produce millions of seeds of these economic varieties of Indian and Chinese carps and induced fish breeding become a routine procedure in the various countries of Asia and Europe to produce quality seed to meet the demands of seed of these food-fishes for culture. Following this success, in subsequent decades, it was possible to induce breed the popular grey mullets, milkfish, sea-bass, several catfishes and a few other species. More recently, a number of marine commercial species have been induced to spawn by administration of fish pituitary and other hormones. Hormones of mammalian origin, both natural and synthetic as well as a few other hormones are progressively being used in recent years for inducing spawning in fishes. A few important hormones used in regular seed production programme, are mentioned below:

1. HCG (Human Chorionic Gonadotropin obtained from human pregnancy urine) alone or in combination with FPE (Fish Pituitary extract) gave varying success in breeding fishes and obtaining their seed.

2. Luteinizing Hormone-Releasing Hormone (LH-RH) and its analogues. Here attempts had been made to intervene at the hypothalamic-hypophysis interface in the endocrinal chain instead of injecting exogenous gonadotropins to stimulate hypophysis directly.
3. The mammalian (mGnRH) and/or piscine (sGnRH) gonadotropin-releasing hormone analogue combined with dopamine antagonists, is a further improvement over the previous one.
4. Salmon gonadotropin-releasing hormone analogue (sGnRH-A) in combination with domperidone, known as 'Linpe Method' is largely used in China in the large-scale seed production of Chinese carps. 'Ovaprim' its trade name, prepared by Syndel Lab., Canada, is used for induced breeding in several countries.

However, the poor fish-farmers of majority of the developing countries of the world still prefer to use the whole pituitary gland extract mainly because of the lower cost, simpler technique and ready availability of the hormones.

The induced breeding technique has also opened a new line of research on the genetic selection and hybridization in these fishes for producing superior strains of fishes.

Food-Fish Production

With the expansion of cultural areas and intensification of culture, the overall production of fish through aquaculture had increased considerably. Besides, by the introduction of scientific cultural technology including improved management practices, such as, stock manipulation, pond fertilization, supplementary feeding, disease control etc., the fish production could be increased from a meager 500–600 kg/ha/yr to about 2 to 4 mt/ha/yr.

Further, with the introduction of exotic species of superior qualities and faster growing Chinese bighead, silver carp, grass carp etc., an elaborate system of culture is being practiced, known as poly-culture or mixed culture of a number of fast growing compatible species of fishes having different feeding habits and different feeding zones. By polyculture of a large number of species, it is possible that the different varieties of food that are produced and are available in the pond environment, are fully utilized and converted to fish flesh. By mixed fish farming the production could be increased from 8–10 tons to as high as 20 mt/ha/yr.

Also, introduction of a faster growing species of tilapia (*Oreochromis niloticus*) in many countries of Asia and the Pacific, the fish production had been increased tremendously. Further by genetical research and intensive farming technology, a red hybrid strain has been produced which could give a production as high as from 600–800 mt/ha/yr.

The fish production could further be enhanced considerably by following Non-conventional Methods of Culture, such as, (*a*) Cage and Pen Culture (*b*) Integrated fish-livestock-crop farming systems and Recycling of Animal Wastes, and (*c*) Running water fish Culture.

Cage and Pen Culture

Cage and pen culture, although an age-old practice in a few countries of South-east Asia, the modernization of the cultural techniques have in recent years given very high yields in fresh, brackishwater as well as through mariculture. In Japan, millions of metric tons of yellow-ail and sunbreams are produced every year in marine cages.

Integrated Fish-livestock-crop Farming Systems

This is very high yielding aquaculture system where aquaculture is combined with livestock raising and vegetable/crop farming. In this system there is mutual benefit among the different commodities and the wastes and bye-products of one is utilized by another and practically nothing is wasted. Fishes get the maximum benefits since they utilize the wastes and bye-products of livestocks and crops thereby saving expenditures on feeds and fertilizers. Recycling of animal wastes is very very important since it converts the animal wastes to fish flesh by fertilizing the pond waters which otherwise would have been a nuisance and a health hazard. By simple recycling of wastes 5–10 mt/ha/yr fish can be produced without using inorganic fertilizers or supplementary feeding.

Running Water Fish Culture

The principle behind water fish culture is that the flowing water constantly removes the nitrogenous and other metabolites from the culture waters, thereby eliminating these toxic products which when accumulated inhabit fish growth. Running water culture can give unbelievable production, such as, over 4000 mt/ha/yr (estimated) of common carps in Japan and over 2000 mt/ha/yr of rainbow trouts in Idaho, USA. Such high production is possible because of intensive stocking, elimination of metabolites and feeding ad libitum in oxygen-rich flowing water.

To sum up, it may be confidently said that recent advances in aquaculture have shown that the industry has reached such a stage that it could provide a substantial portion of fish protein to the diet of protein-deficient poorer masses of the third world countries.

Seed Production: Its Problems

Introduction

Production of quality fish seed in adequate quantities is the first and most important phase of pond fish culture. India has vast freshwater resources which could profitably be utilized for fish culture. Of these, about 6 lakh ha water areas are at present utilized for fish culture, 4 lakh ha are readily available for culture and nearly 6 lakh ha water areas need reclamation. If all these areas are brought under cultivation the demand for seed would be tremendous. According to the report of the Fish Seed Committee (Govt. of India 1966), the requirement for optimum stocking of the readily cultivable waters only would be an additional quantity of 58 millions of spawn, calculated on the basis of the seed production in the country during 1964–65. This shows that India is hopelessly deficit in its seed production. With the development of the technique of fresh water aquaculture in recent years, the Central Inland Fisheries Research Sub-station, Cuttack has obtained production as high as an average of 7500 kg/ha/yr with the maximum of 8631 kg/ha/9½ months as against an average production of 600 kg/ha/yr in India. Such high production was obtained by the culture of Indian and exotic carps, providing adequate supplementary feeding and fertilization of ponds. This breakthrough in fish production would raise aquaculture to the status of an industry and considering its economics, aquaculture would definitely attract more and more people to take it as a profession. Consequent to this, the demand for seed of cultivated species both indigenous and exotic would be increased tremendously. To cope up with the rising demand, an all out efforts have to be made to tap all available sources for enhancing seed production in the country.

Methods of Seed Production and their Problems

Considering the great importance of production of quality fish seed for enhancing fish production, the Fist Seed Committee made several recommendations. This was followed by holding of a seminar on "Production of quality fish seed for fish culture" during November 1 and 2, 1968 at the Central Inland Fisheries Research Institute, Barrackpore. The seminar was participated by renowned fishery scientists and workers in the line. Recommendations of the seminar include making an all out drive for enhancing seed production in the country by (*i*) collection of seed from rivers, (*ii*) production of seed from bundhs and (*iii*) by induced breeding.

Seed Production by Collection from Rivers

Production of seed from riverine sources was till recently the only source of carp seed in the country and even now is considered to be its main source. According to the report of the Fish Seed Committee, contribution by riverine spawn collection to total spawn production in the year 1964–65 was as high as 91.67 per cent. The age-old practice of collection of spawn from rivers was mainly restricted to the river Ganges and its tributaries in Bihar and pre-partitioned Bengal. In recent past, a few more State Fisheries Department had also initiated collection of riverine spawn. The Seed Committee as well as the Seminar on "Production of quality fish seed" laid great emphasis on the necessity for intensive spawn prospecting in rivers all over India for location of new centers which could be profitably exploited. Work on spawn prospecting started by Central Inland Fisheries Research Institute in 1964 and continued till 1970 in various parts of the country could locate several new spawn collection centers. Considering the importance of the problem of All India Co-ordinated Research Project on "Investigations on riverine spawn prospecting and collection techniques" was initiated in 1971 having headquarter at Allahabad and three centers in Assam, Bengal and Bihar.

Although rivers are the main source of spawn production the quality of spawn in majority of the rivers excepting the Gangetic river system, seems to be very poor, comprising 70 to 80 per cent or even more of uneconomic species. Even in the Ganges proper the percentage of spawn of cultivated varieties are going down every year. In Mahanadi, quite a substantial quantity of spawn is collected by the Orissa State Fisheries Department every year, but the quality of spawn is very unsatisfactory consisting mainly of the minor carp species bata and reba. Percentage of catla is often very negligible. The poor quality of spawn appears to be mainly due to indiscriminate destruction of large fishes from the riverine pools where fishes take shelter during summer months when the river bed practically dry up. Fishing from riverine pools and large scale destruction of brood fish in the shallow breeding grounds where they migrate for breeding, is largely responsible for poor quality of fish seed in rivers. Besides, construction of dams and embankments in rivers restricting migration of brood fish upstream and to the shallow breeding grounds in adjacent areas are also responsible for rapid fall of spawn production in some rivers. Industrial pollution and washings of highly toxic insecticides brought from the catchment area are depleting the stock of fish and fish seed in rivers.

Heavy mortality occurs during transport of spawn, fry and fingerlings from remote and out of the way place riverine centers to the seed markets. Seed production in rivers depends largely on the timely rainfall and optimum weather conditions. Very often production of seed in rivers is poor due to late rain and prolonged drought condition. Floods also restrict collection of fish seed from rivers. The problems, as mentioned above, are largely responsible for the fall in quality seed production from rivers. It is apprehended that more and more industrialization of the country and large scale use of insecticides and pesticides in Agricultural field would be a real danger to our riverine fish fauna and especially to the fish seed population of our rivers.

Seed Production from Breeding Bunds

Breeding bunds is yet another source of fish seed which according to the Fish Seed Committee report contributed 5.38 per cent of spawn production in the year 1964–65. Bunds are special type of ponds where natural condition is simulated by the sudden rush of fresh rainwater from the catchment area when carps get stimulated and spawn. Bundh breeding is an age-old practice in the Midnapur and Bankura districts of West Bengal. More recently, successful bundh breeding for raising quality spawn were followed in Madhya Pradesh by the State Fisheries department. Stray bundh breeding is also reported from a few other states. Concerted efforts have to be made to popularize bundh breeding and extend its scope to all other states where topography suitable for construction of dry bundh is available as in West Bengal and Madhya Pradesh. It is hoped that the State Fisheries Department would initiate bundh breeding for production of quality fish seed in their respective states.

This source of fish seed has also its limitations and problems since it is dependable to a great extent on the climatic and other environmental conditions. Inadequate rainfall is largely responsible for failure of spawning in bundhs. Colossal waste of fertilized eggs is a common occurrence in wet bundhs where spawning sometimes is very heavy. Facilities for collection and hatching of large quantities of eggs are usually inadequate which results in heavy mortality of hatchlings. Since major carp breeders are not selectively stocked in wet bundhs, miscellaneous species of fishes are also observed to breed along with the cultivated ones resulting in poor quality of spawn. In wet bundh, often very small size and unhealthy breeders are observed to spawn thereby producing weaker and slow-growing progenies. Such problems in bundh breeding can be rectified by selective breeding of healthy breeders of cultivated species alone in dry bundhs. Lastly, there are limitations since 'bundhs' can only be constructed on a topography of undulating terrain. The construction and maintenance of bundhs is also very expensive and beyond the means of an average fish farmer.

Seed Production by Hypophysation

Success in inducing spawning of Indian major carps by injection of fish pituitary hormones, one and a half decade age, has opened a new source of quality fish seed. The technique has been standardized and is being used for the large scale production of seed of the cultivated species. Since the first success in 1957, Central Inland Fisheries Research Sub-station, Cuttack conducted several special training courses on induced breeding for the benefit of State Fisheries Officers and Extension workers, besides giving regular training to the trainees of Central Institute for Fisheries Education, Bombay and Barrackpore every year. As a result, at present almost all the State Fishery departments in India and other agencies are producing quality fish seed by this method. While some of the States have shown excellent progress in utilizing the technique for commercial production of quality spawns, in a few others, due to several factors, the progress made was not upto the mark. According to Fish Seed Committee's report the production of seed by hypophysation technique during 1964–65 was 1.5 per cent of the total production as compared to 91.7 per cent contributed by riverine sources. Since 1964–65, the progress had been very rapid and the production of seed by this technique has been manifold. A fresh statistics on the seed production would have given a figure at least 10–15 times of the previous estimated production. This technique appears to be the most dependable one although some difficulties are encountered when experimented under varying agro-climate conditions in different parts of the country.

During 22–27 February 1971, an workshop on 'Induced breeding (Hypophysation) of carps' was held at the Central Institute of Fisheries Education, Bombay with a view to study the present status of induced fish breeding programme in the various states and to elucidate the various problems confronting the field workers and suggesting remedial measures so as to help enhancing production of quality fish seed in the country. The main problems encountered by various workers can be categorized as the following:

1. Non-availability of adequate numbers of right type of breeders.
2. Unfavourable climatic and physio-chemical conditions of the environments.
3. Poor hatching of eggs and mortality of hatchlings.

Non-availability of Adequate Numbers of Suitable Breeders

Failure in induced breeding experiments is largely due to the poor condition of breeders selected for breeding. In general, very little attention is paid towards proper rearing and maintenance of a stock of brood fish. Breeding experiments are conducted by collecting some adult fish from various source just prior to the experiments. Such breeders are bound to give very little response and consequently seed production suffers. Very often fishes collected from weed infested water bodies are found to possess ill-developed gonads. This may probably be due to insufficient sunlight and heat that is required for the proper development of gonads. Accumulation of sufficient quantity of fat in the abdomen is sometimes responsible for ill-formed gonads in fishes. Adiposity in May carp breeders especially in catla in certain bodies of water is a problem which needs thorough study.

However, when such fishes were transferred to shallow ponds in the month of March and April and kept in crowded conditions, these have been observed to attain maturity. In some waters males and females breeders have been observed to mature at different times. All these difficulties can be obviated if adequate number of breeders are raised and maintained in ponds long before the onset of the breeding season. This would ensure proper development, growth and maturity in breeders. A large stock of brood fish would facilitate the fish culturist to pick and choose the right type of breeders while making a selection for breeding purposes. Selection of good breeders in prime condition is very important for attaining success in induced fish breeding. The standardized doses, when injected, should give good results under favourable weather conditions if ripe breeders are selected.

Dearth of Pituitary Glands

Shortage of pituitary glands has been observed to be one of the important problems limiting large-scale seed production by induced breeding. At present, Kolkata fish market is one of the main sources of pituitary gland collection. Many states get their requirements from the said market although effectiveness of the glands collected from the markets are also doubted by some workers. The need for a 'pituitary bank' is therefore greatly felt. As a first step towards establishment of a 'pituitary bank', Central Inland Fisheries Research Sub-station, Cuttack produced large number of ampoules containing pituitary extract preserved in glycerine and supplied to fish breeding centers in various States, Agricultural Universities and also to private fish farmers who were in genuine need of pituitary glands. The supply was continued for several years and the potency of the hormones in ampoules were tested under different climatic conditions. However, it is advisable that each fish breeding unit or center should be self-sufficient in its requirements of pituitary glands. To solve this problem it is suggested that the required number of donor fish may be raised in the fish farm itself. The common carp is an ideal donor fish from which pituitary gland can be collected almost all the year round since common carp is found mature almost throughout the year. This procedure should be the best for ensuring a ready supply of most potent and dependable stock of pituitary glands.

Unfavourable Climatic and Physico-chemical Conditions of the Environment

Unfavourable weather condition is another important limiting factor in large-scale seed production. It is well known that favourable weather conditions, rain and optimum water temperature (26°–29°C) are conducive to successful spawning. Prolonged drought and failure of rain fall to bring the gonads to a prime condition. Spawning may sometime occur at water temperatures above the optimum range but the percentage of fertilization and hatching is normally poor. In some regions favourable weather conditions are available for a very limited period. The short period should therefore be fully taken advantage of for producing maximum quantity of carp spawn. Otherwise it may by necessary to have an air-conditioned laboratory with arrangements for lowering down water temperature and facilities for circulation of water for breeding and hatching purposes.

Poor Hatching of Eggs and Mortality of Hatchlings

Problems relating to hatching of eggs and survival of hatchlings are very many. Poor hatching may be due to the texture of hapa cloth being very thick, deposition of silt on the eggs and hatchlings in case the hatching hapa is fixed in flowing waters, sudden rise and fluctuations in water temperature, bad fixing of hatching hapa, presence of algal bloom when the developing eggs are beyond up by the oxygen bubble produced by photosynthesis and get spoiled. Destruction of eggs by Tilapia and common carp, and hatchlings by copepods are also common. Loss of eggs due to holes made by crabs in hapa are also reported. Precaution at every step has to be taken for ensuring high survival of developing eggs and hatchlings. Better hatching and higher survival of hatchlings are obtained if running water is provided. In recent years, a glass jar hatchery complex has been set up at the Central Inland Fisheries Research Sub-station, Cuttack with facilities for circulation of water which is definitely an improvement over the hatching hapa fixed in stagnant ponds. An indoor hatchery dispenses with the hazards of nature and also many of the hatching problems mentioned earlier and ensures better hatching and higher survival of hatchlings.

Establishment of Composite Fish Seed Farm

Various problems confronting large-scale fish seed production in the country by hypophysation have been discussed above. It is high time that every State should start at least one composite fish seed farm for production of quality fish seed. The fish farm should preferably be located at a place where maximum rainfall in the state is recorded. This would ensure better rainfall in the state is recorded. This would ensure better success in induced breeding experiments and better hatching success. A composite fish seed farm should be self-sufficient in its requirement of breeders as well as the donor fishes for pituitary gland collection. It should preferably have a source of water which could be tapped. The farm should be provided with adequate number of nursery and rearing ponds for raising fry and fingerlings and stocking ponds for rearing breeders and donor fishes. The fish seed farm should also be provided with a glass jar hatchery complex with facilities for circulating water for hatching large quantities of eggs at a time and if possible under controlled temperature. Large-scale breeding of the Chinese Silver carp, grass carp and common carp should also be taken up at the seed farm. Such a farm would be the nucleus from which quality fish seed in the form of spawn, fry and fingerlings would be distributed all over the state and would go a long way to meet the requirement of quality fish seed in the state. But above all, with a view to take advantage of a fish seed farm and utilize it properly, a batch of suitably trained, devoted scientific workers are necessary.

In the foregoing, an account of the various means of seed production has been provided and the problems discussed. Considering the fact that the country's present seed production is far below its requirement, there is the urgent need to augment seed production. So all the three important sources of seed have to be developed simultaneously and tapped properly so as to make up the huge deficit. Riverine seed collection though contributes to the bulk of the seed produced in the country, if the overall percentage of quality fish seed is taken into account the actual percentage will go down considerably. As discussed earlier, the future of riverine fisheries does not seem to be very bright and consequently the fate of seed also cannot be better. It is therefore, necessary that greater attention should be given to quality seed production by hypophysation, the most dependable technique we have, so far. An all out drive should be made by the Central as well as State Governments to further its scope to all over the country and open a number of breeding centers in each state in addition to the establishment of composite fish seed farms in each state. Unless the seed production of the mix cultivated species of Indian and exotic carps are increased manifold, the breakthrough recently achieved towards obtaining high production of marketable fish by composite culture of these species, cannot be taken into the field all over the country by fish farmers and thereby fish production in the country would have a serious set back.

Fish Seed Production: A Pre-requisite to Intensive Freshwater Fish Culture

Introduction

One of the greatest problems that faces the fish culture industry of our country today is the non-availability of quality fish seed in adequate quantities. The cultivable catla, rohu and mrigal, known collectively as the Indian major carps, are naturally distributed in the Indo-Gangetic river system. As such, the seed of these species is available mainly in the rivers of the northern plains where they breed during the monsoon, when the rivers are in floods. Good-quality seed is available in the rivers in West Bengal and Bihar and also in Uttar Pradesh. However, the seed collected from the river systems, which by itself is a major industry in at least Bengal and Bihar, is not adequate to meet the demands of the whole country in terms of spawn, fry and fingerlings, which run into 5,510 million, 1,650 million and 830 million respectively. These estimates were made about 10 years ago. The requirements have grown considerably during this interval. It is imperative that the riverine collections are made extensively and other modes of speed production are intensified.

The eggs and young ones of fish are collected from the rivers with the help of a funnel-shaped net which is fixed against the current in flooded rivers. The net has a tail piece attached to its cod-end, where the drifting young ones of fish, known as spawn, get collected. In this tailpiece certain other organisms, including insects and miscellaneous fishes and their young ones, also get collected. The river collections are thus a mixture of the seed of several economic and uneconomic varieties of fish. The CIFRI has surveyed several riverine stretches, and new centers for collection of the seed of major carps have been located during the past ten years. Several new nets have been designed to improve the efficiency of the catching gear.

Seed Production

The other methods of seed production of cultivable varieties include their production in bundhs. There are two types of bundhs: (*i*) wet bundhs, which retain water throughout the year; and (*ii*) dry bundhs which, as the name suggests, dry out during the summer months. In wet bundhs, during the course of years, the fishes get acclimatised to the pond conditions and breed during the monsoon months when riverine conditions are simulated therein. One would find several wet bundhs where the fish breed. There may be others where they do not breed at all though they have been in existence for several years. The location of breeding grounds in large-scale wet bundhs is a big problem but once the breeding ground is located it is possible to collect the eggs where the cultivable species have bred. The collection is usually made with the help of a net cloth which is dragged after disturbing the water so that the eggs which settle down at the bottom are made afloat. The eggs, thus collected, are hatched in pits with circulation of water or in double-walled hapas (cloth enclosures fixed with the help of bamboo poles in ponds for hatching the eggs). The outer hapa is made of muslin cloth while the inner one is that of mosquito netting to enable the young ones to wriggle out into the outer hapa leaving the dead shells behind in the inner hapa. In case of fishes breeding in wet bundhs, where the breeding grounds have not been located, the eggs are either fed upon by predatory and weed fishes or else covered by mud or destroyed in many other ways resulting in their poor hatching and survival. It is, therefore, necessary to locate the breeding grounds in large-size wet bundhs and collect the eggs and hatch them out artificially to increase the seed production.

In dry bundhs, the brood fishes are stocked during the monsoon months soon after fresh rain water is collected there. Such bundhs are very small in area, ranging from 0.25 acre to 2.5 acres, and are mostly situated in the laterite soils of West Bengal and Madhya Pradesh. The fishes breed in these bundhs within a few hours of stocking. If they do not breed within two days of stocking they are removed and a fresh stock of breeders is put in these bundhs. In West Bengal, in recent years the fish farmers have started injecting the brood fish with pituitary glands to induce the entire stock to breed. While the seed collections from the river and wet bundhs include several other varieties of fishes, the seed from dry bundhs is almost cent per cent pure. It is possible to breed a single species in a dry bundh and, if desired, to have seed of a few varieties in a known combination.The eggs are collected by de-watering the dry bundhs through an outlet which is provided on the lower-most level in the bundhs. The method of collection is the same as that for wet bundhs. The eggs are hatched in earthen pits or double-walled hapas and the spawn made available for sale. The dry bundh has an edge over the wet bundh in that it can be used to breed the fish 3 to 5 times in a season whereas in a wet bundh breeding generally occurs only once when almost all the ripe carp breeders participate in spawning.

Injection Method

The most recent method of producing the seed of cultivable varieties is through injections of pituitary gland extract. This is the most dependable method of production of pure seed on a commercial scale. The glands is removed from ripe fishes, macerated and dissolved in a small quantity of distilled water which is injected into the body of a ripe fish generally in the tail region through a hypodermic syringe. The fish chosen for injection is selected with great care. A fully ripe male oozes milt when a slight pressure is applied on its abdomen while a full ripe female has a swollen abdomen and protruding reddish vent. It is also possible to select a male and female by feeling the inner side of their pectoral fins. The pectoral fin in case of a male is rough while that of a female smooth. The injection is generally given when the weather is cool which is especially so towards the evening and at night. It is generally seen that the male-female ratio is 1: 1 by weight but 2: 1 by numbers. In most cases a preliminary dose is given to a female and the male is injected only once along with the female which is given the second injection after a lapse of 6 hours. The injected fishes are put in breeding hapas where they usually breed within 4 to 6 hours of the second injection. The eggs are collected after the embryo starts twitching movements inside the egg shell. The eggs are laid for hatching in double-walled cloth hapas in ponds or in glass jars in the hatchery.

The exotic grass carp (*Ctenopharyngodon idella*) and silver carp (*Hypophthalmichthys molitrix*), which have more or less similar breeding habits as the Indian major carps, are also bred through administration of pituitary gland extract. While the Indian major carps release the eggs and milt easily after injections, the two exotic carps have usually to be stripped. The eggs are first stripped from an oozing female, about 4 to 6 hours after the second injection, and immediately fertilized by spraying the milt from a ripe male. The entire operation is a matter of a few minutes and no water is allowed with the eggs or milt. However, the eggs are washed several times with water after fertilization and then allowed to swell by keeping in an outer hapa fixed in the pond. The hatching is done in double-walled hapas or glass jars in a hatchery.

The third exotic species, the common carp, *Cyprinus caprio* of which there are three varieties in India, is a pond-breeder. It breeds without any difficulty, upsetting the population density. The survival is also very poor as the eggs and the young ones are destroyed by predatory insects and fishes. A simple method to breed this fish has been developed and is now in use throughout the country. Ripe males and females are selected and kept in breeding hapas along with a submerged aquatic weed, Hydrilla or Najas. The eggs released by the female are sticky and get attached to the leaves of the weed. The breeders are removed after spawning and the weed is distributed to a few outer hatching hapas for hatching. The eggs hatch out in 48 to 72 hours depending upon the temperature.

Though no reliable figures are available, it is perhaps true that even the present fish seed production in the country is not commensurate with the estimated demands for 1964–65. If we consider the trade turnover which the fish seed industry would have (based on 1964–65 requirements), one would simply wonder at the potentialities of this industry. Taking the cost of quality spawn to be Rs. 350 per hundred thousand, the cost of 5,510 million of spawn would be about Rs. 19.3 million. This does not include the costs towards packing and transportation which gives rise to ancillary industries. If the fry requirements of 1,650 million are sought to be met at Rs. 12 per 1,000 fry, a turnover of Rs. 19.8 million is involved which again does not include the packing and transportation costs. 830 million of fingerlings have a ready market at Rs. 150 per 1,000 and would cost Rs. 124.5 million. To meet the fish seed demand of the country all the available sources should be fully and judiciously exploited. The riverine collections which contribute largely to the total spawn production in the country deserve to be judiciously exploited, being a natural resource.

Prospecting surveys in new stretches of different rivers need to be conducted as a regular programme to locate new centers of spawn availability.

The cost of spawn production in a dry bundh differs from place to place depending upon the cost of construction of the bundh, cost of breeders and their transportation and labour charges. The cost of spawn production has been calculated to be Rs. 112 per hundred thousand during 1973–74 from the dry bundhs in West Bengal. The sale price of quality spawn being Rs. 350 per hundred thousand, a profit of Rs. 238 per hundred thousand is possible by developing the dry bundh breeding industry. The cost of production of one hundred thousand spawn is Rs. 100 to Rs. 150 in case of induced breeding depending upon the experience and skill of the fish farmer. This method of seed production is the most reliable with a hundred per cent guarantee regarding the quality of the seed. It is also the only means which could be followed almost everywhere and could be depended upon.

Rearing the spawn to fry and the fry to fingerling stage in nursery and rearing ponds, respectively, are yet other aspects of fish seed production. These, however, involve considerable expenditure on land, pond construction, nets, spawn, fertilizer, feed, labour, etc. The cost of fry production in West Bengal has been estimated to be Rs. 800 per hundred thousand whereas its selling price is Rs. 1,200 per hundred thousand. However, the cost of production of 1,000 fingerlings in sewage treated ponds is Rs. 30 to Rs. 40 whereas its selling rate is Rs. 150 per 1,000 fingerlings. It is from sewage-irrigated pond that one could earn a net profit of Rs. 7,000 to Rs. 10,000 per hectare in 3–4 months' time by way of fingerling production alone. The new technology developed at the Central Inland Fisheries Research Substation, Cuttack, for spawn production through induced breeding and the management techniques evolved for achieving a higher survival rate in nursery and rearing ponds would result in far greater profits.

One of the ways of meeting the increasing demands on animal protein is through intensive fish culture. The industry thus has bright prospects. The fullest utilization of cultivable water resources would call for a major fish seed industry in the private sector. With further development and popularization of intensive freshwater fish culture in India the requirements of quality fish seed would increase as also with the addition of new water resources in the form of ponds and tanks, and irrigation or multipurpose reservoirs. Fish seed production is an evergrowing industry–where else would one find such a lucrative vocation?

[H. Chaudhuri; Published in 'Indian Farming' 1975]

Problems of Warm Water Fish Seed Production

Introduction

Many of the important cultivable species of fishes spawn only in their natural habitats making it necessary for a fish farmer to collect the stocking material from nature. While the breeding requirements of some of these species have been partially understood, much remains to be known before they could be bred in confined environments. Progress has been made in the recent past in inducing some of the Indian and Chinese major carps (Chaudhuri and Alikunhi, 1957; Soin, 1959), Aliev, 1966, grey mullets (Tang, 1964; Yashouv, 1966, and other cultivated species, Clemens and Sneed, 1962; Tongsanga *et al.* 1963; Shelbourne, 1964) of the world to breed in captivity to obtain the necessary stock for culture, yet it is difficult to discard the traditional methods as production of tablesized fish must go on, and especially so, when aquaculture is fast becoming an important industry in several countries.

A number of hitherto uncultivated species are now in great demand and the problems connected with their seed production and rearing are being investigated. The present review, however, deals with the problems of seed production and distribution of only warm water species of fish, which are already being cultivated extensively.

Collection and Production of the Seed of Carps

Collection of Seed from Natural Waters

Spawn

Both in India and China, the fish fry trade is a well-organized industry (Lin, 1949; Hora and Pillay, 1962; India, Ministry of Food and Agriculture, 1966) providing employment to hundreds of people. Considerable expertise has developed over the years in the collection and handling of spawn and fry but the lucrative nature of the industry has drawn several unskilled personnel in recent years and this has affected the catches and their survival.

Spawn (post-larvae) of Indian carps, *Catla catla* (Catla), *Labeo rohita* (rohu) and *Cirrhinus mrigala* (mrigal) and of Chinese carps *Hypophthalmichthys molitrix* (silver carp), *Ctenopharyngodon idella* (grass carp), *Aristochthys nobilis* (bighead), *Cirrhinus molitorella* (mud carp) and *Mylopharyngodon piceus* (black carp) are collected during monsoon months from flooded rivers by means of large conical begnets. While the general technique of collection is the same, the material, design and dimensions of the gear vary in different places both in India and China. In Indonesia, *Puntius gonionotus* (tawes) breeds in rivers during the monsoon months and spawn and fry are collected with scoop nets made of fine cloth.

In India, several studies have been carried out with a view to improving spawn collection methods. Alikunhi, Chaudhuri and Ramachandran (1952) identified the causes of heavy mortality of spawn either during collection or in nursery ponds and suggested suitable improvements. Proper fixation of the net which avoids much resistance to fast flowing waters, a change in the constructional design of the cod end of the net, periodic removal of spawn to avoid overcrowding in the receptacle, proper sieving of the collections to filter off predatory and competitor species, avoidance of overcrowded conditions during temporary storage, were some of the measures suggested to better the yield from natural sources.

In order to meet the increased requirements of carp seed several riverine stretches in India have been prospected from 1949 onward and a large number of new spawn collection centres located (Jhingran, 1972). With a view to assessing the spawn-yielding potentiality of various centres on different rivers, new measures of spawn abundance and indices of spawn quality and quantity have been developed (Barrackpore, 1965). A standard spawn-collection net has been devised and proved to be 200 per cent more efficient than the gear in use in different parts of the country (Shetty, Malhotra and Ghosh, 1972). Catching efficiency has been further increased by developing a new gear similar to 4-seam trawl instead of the commonly employed 2-wing nets.

Chinese traders make a preliminary estimate of the catch composition of spawn and fry of Chinese carps on the basis of certain characters in the field itself. The fry are sorted according to size groups and species using graded sieves and their oxygen requirements (Hora and Pillay, 1962). The elaborate process of sorting, though of great ingenuity, is however costly and time-consuming, involving high labour and transport costs and heavy losses during handling and distribution (Wu and Chung, 1964). Alikunhi *et al.* (1951) also tried to segregate the spawn of Indian major carps from a mixed collection of riverine spawn on the basis of their developed field method for distinguishing the Indian major carp spawn from spawn of minor carps based on size and the shape and size of the yolk-sac. While this method helps in judging the quality of spawn as a whole, it does not help in segregating the desired spawn. Because of this, valuable nursery space is wasted by stocking mixed collections, often comprising a high percentage of undesirable species. The riverine collections are also very undependable as monsoon vagaries and consequent heavy flood or drought often result in poor collections. These problems emphasize the necessity for locating new centres for the collection of quality spawn on the one hand and the need for a more dependable method of production of quality fish seed on the other.

Fry and Fingerlings

Mass collection of fry and fingerlings of cultivable species of carps from natural sources is not ordinarily practiced excepting in India. Several fry and fingerling collection centres which are exploited regularly are located on the Gangetic and Indus river systems in the northern and Godavari, Krishna and Cauvery river systems in peninsular India. Castnets, dragnets and basket traps are commonly used as the collection gear.

The segregation of cultivable species from uneconomic varieties is fairly easy. The catch per unit of effort at several fry and fingerling centres is very high and the operations are economically viable in many areas.

Breeding in Confined Waters Through Controlled Natural or Seminatural Conditions

Breeding of Common Carp

Breeding of common carp in confined waters is very common. They are bred in special ponds (Dubison type) with grassy raised bed in the center as in Europe or ponds with hard or cemented bottom where "egg collectors" like "Kakabans" in Indonesia are provided from outside. In China, Eiohhornia, Ceratophyllum and Myriophyllum are used. In Japan, egg collectors made of polythene are being increasingly used. In India, however, breeding is usually undertaken in cloth "hapas", fixed in ponds preferably without any fish population, and Hydrilla provided as the substratum for egg deposition. This is an improvement over the "kakabans" method as selected breeders alone are used and a high percentage of fertilization is ensured. In Hungary, hypophysation is employed for mass production of fingerlings. Fish are stripped after injection and fertilization brought about artificially. The stickiness of the eggs is eliminated by use of urea and tannin and the eggs are incubated in jars. Hypophysation facilities complete and advance spawning and a high rate of fertilization and hatching irrespective of meteorological conditions (Woynarovich, 1969).

Breeding Indian and Chinese Major Carps

The traditional practice of breeding Indian major carps in special types of ponds, known locally as dry "bundhs" has undergone several improvements in recent years (Dubey and Arjun Singh and Pande, 1972). As the factors responsible for the spawning of carps are not yet fully elucidated the practice of "bundh" breeding is still not widely practiced. Though the soil type is considered to be of prime importance, breeding has been observed in "bundhs" constructed on clayey, silty and stony soils (Dubey and Tuli, 1961. Kulkarni (1972) calls it eco-stimulus breeding as the required stimulus for breeding is provided by changes in the environmental conditions such as provision of more diluted and oxygenated water, change in p^H, salts, etc. However it is one of the most important means of obtaining pure seed of cultivated carps.

The hatching techniques adopted are generally poor and result in great loss of numbers. Use of cement nurseries and glass hatching jars would be economic and enable handling a very large number of eggs at a time. The cloth "hapas" which are not only expensive but also difficult to operate in the limited space in the "bundh" have to be avoided to get better results.

Breeding the Tawes *Puntius gonionotus*

P. gonionotus spawns in ponds with hard bottom in Indonesia, Thailand and Malaysia. Spawning ponds are specially prepared by leaving them dry for several days till the bottom soil cracks. A part of the pond bottom is sometimes covered with sand and gravel. The pond is filled with well oxygenated water and the spawners introduced; reproduction follows immediately. The eggs are demersal and hatch out in two or three days, the fry are collected when they are three weeks old (Hofstede, Ardiwinata and Botke, 1953).

Another method of breeding *P. gonionotus* resembles the "bundh"-breeding techniques of India. The construction of the spawning pond (0.1–0.2 ha) is undertaken in such an area that sufficient rain water from the catchment gets easily collected in it. The pond is also provided with a small pit (4 × 4 m) at the bottom, where the breeders are released. The fish are induced to spawn with the entry of fresh rain water from the catchment after a heavy shower (Schuster, 1952).

Artificial Propagation

While artificial propagation of salmonids has been known for over a century, artificial reproduction of cyprinids is recent. Intraperitoneal or intramuscular injections of the pituitary gland extract are necessary for this purpose. Pituitary glands are removed from mature or maturing carp and preserved in acetone or alcohol. Since the Indian and Chinese major carps mature only once during a year, it is during this short period that the glands have to be collected. Inadequate knowledge of the maturity stage, especially of a maturing fish, from its morphology makes the choice between mature and immature fish difficult and glands on a commercial scale are collected from immature fish also on several occasions. Common carp, which are comparatively large glands and remains mature for a greater part of the year, should be considered a good donor. In India, the problem of preparing and preserving the extract has been solved by developing techniques for ampouling and making it readily available to fish farmers (Ibrahim and Chaudhuri, 1966).

Administration of a single large dose of pituitary hormone was a common practice among Chinese fish breeders (Soin, 1959) which, though resulted in ovulation, affected the quality of eggs. Fractional injections of hormone approaching the threshold dose, which give good results, are, however, widely used throughout (Chaudhuri, 1960; Wu and Chung, 1964) and effect saving of pituitary material. While natural spawning and fertilization is a common feature in injected Indian major carps, it is rarely so with Chinese major carps (Wu and Chhung, 1964; Chaudhuri *et al.*, 1967). Artificial fertilization by the "dry" method is adopted (Chaudhuri, Singh and Sukumaran, 1966; Vinogradov, 1968).

The fractionation and purification of carp (Singa, 1974) and salman pituitary (Donaldson *et al.*, 1972) have been attempted successfully in recent years. However, since the effective doses of lyophilized second fraction of carp pituitary and that of SG-G 100 to precipitate spawning in Indian major carps and the catfish (*Heteropneustes fossilis*) are the same as that of the crude pituitary extract, purified preparations would not find much use in commercial breeding unless the inhibitory factor, if any, is eliminated (Chaudhuri, 1975).

An essential requirement for success in induced breeding is the selection of good and healthy breeders in prime condition. The stage of maturity of female breeders can be determined by collecting ovarian eggs by means of a catheter and examining them under a microscope. Visceral fat, common in grass carp and catla from weedy ponds, affects the development of gonads, especially the ovaries, and such breeders do not respond to pituitary injections. Catla has been found to shed its adiposity and develop proper gonadial condition if transferred to shallow ponds by March-April and reared in crowded conditions. Three methods have been developed in Malaysia (Malacca, 1967) to bring about full maturity in grass carp. Malformation of the male gonads is a very common feature in silver carp and unless a large number of breeders are available for selection, fertilization of the ova becomes a major problem. It is generally advisable to rear an adequate number of breeders and maintain them in ponds much before the onset of the spawning season.

Hypophysation is still rain-dependent as the changes brought about in the physico-chemical and climatic conditions of the environment are largely responsible for the large-scale success or failure of induced breeding. Failure of rains and prolonged drought result in poor development of the gonads. Rains being erratic, especially in a tropical country like India, prevailing air and water temperatures range too high and affect spawning. Spawning may sometimes occur at water temperatures above the optimum level but the percentage of fertilization and hatching is greatly affected and results in poor yields. The use of short periods of favourable weather conditions for successful breeding suggests that air-conditioned breeding and hatching centres may be a solution to this problem (Alikunhi *et al.*, 1964).

One of the problems encountered in seed production in India is the high mortality during incubation and hatching, due to sudden fluctuations in water temperature, presence of algal blooms producing oxygen bubbles during photosynthesis resulting in buoying up of the eggs to the water surface, depositing of silt and poor oxygenation inside the "hapa" direct destruction of the eggs by miscellaneous fishes in the pond. Hatching in weiss apparatus, Zoug jars (Woynarovich, 1964) or the glass jar hatchery complex (Bhowmick, 1975) ensures quicker hatching and higher survival. The incubation net frame developed at the Saitama Fisheries Experiment Station for hatching Chinese carp eggs also has a very high efficiency (Kuronuma, 1968).

The same individuals of Indian major carps have been bred twice instead of only once as is normal, within an interval of two months, during a single monsoon season, producing almost an equal number of eggs at each spawning (Barrackpore, 1972). This indicates the possibility of doubling the production of spawn from an individual fish. The use of artificial light for simulating longer photoperiods, higher temperatures and provision of nutritive food may help in bringing about early maturity and result in increasing the spawn production.

A major problem faced by aquaculturists in carrying out large-scale induced breeding operations is the lack of a suitable alternative to the fish pituitary. There is an urgent need to develop a natural or synthetic steroid or chemical hormone to meet the increasing demand for commercial production of fish seed all over the world.

Production of the Seed of Catfish and Air-breathing Fishes Seed Production of Danubian wells (*Silurus glanis*)

Danubian wells (*S. glanis*) are bred in ponds in Hungary, Yugoslavia and Czechoslovakia. The breed easily in storage basins (0.03–0.08 ha in area and 1.0–1.5 m in depth) if proper nests are provided. Spawning takes place when the water temperature rises to 20°C, hence filling of the ponds, placement of "tent-nests" and stocking of selected breeders should be done the same day. Hatching of the eggs could be done in hatching crates of Zoug jars. The young ones are average to strong light and this must be avoided (Huet, 1970).

Seed Production of Channel Catfish (*Ictalurus punctatus*)

Channel catfish breed when the temperature is around 21–24°C. Commercial culture of the species was limited due to the non-availability of the young. Production of channel catfish in hatcheries, though not difficult, was erratic and necessitated the development and refinement of hatchery techniques. It could be spawned in ponds, pens or aquaria with or without the use of hormones. Wire-meshed spawning enclosures (3 × 1.5 m) are fixed in spawning ponds and wooden legs (1 × 0.25 m sides) or 50–1 milk cans submerged in 60 cm of water to be used as spawning containers (Crawfood, 1958). An equal number of males and females are placed in the pond at the rate of two pairs for each container. In the pen method, a male and a smaller female are placed in each spawning pen provided with a single spawning container. As the female might devour the eggs it is necessary to remove it from the enclosures after the eggs are laid. Males can be used two or three times for spawning.

The pen method has an edge over the pond method as it protects the spawning pair from intruding fish and provides an opportunity for selection of each pair besides delaying or hastening spawning, handling of the fish needing hormone injections and immediate removal of spawners after egg-laying to enable stocking the pen with another pair (Clemens, 1968). While the pond method could be adopted by an unskilled farmer, the pen method requires skilled personnel capable of proper sexing and pairing the fish. The latter method also requires better quality brood stock.

Sneed and Clemens (1960a) have described the techniques of hypophysation in channel catfish. A single pair is kept in a suitable tank with running water and the female injected intra-peritoneally with 4–5 mg of acetone-dried whole ground pituitary per kg of body weight of female. Selection of gravid spawners is essential.

The eggs are removed to a mechanical hatching through provided with running water and a system of moving paddles for agitation (Nelson, 1960).

Production of the Seed of *Clarias* spp., Other Catfishes and Air-breathing Fishes

Clarias are indigenous to Pakistan, India, Bangladesh, Burma, Malaysia, Vietnam, Thailand and the Philippines. *C. batrachus* in India and both *C. batrachus* and *C. macrocephalus* are popular in Thailand.

In Thailand, the fry (25 mm) are collected from natural sources such as canals and ditches where they breed in comparatively calm waters of the marginal shallows about 50 cm below the water surface. The fry are collected by means of small fine-meshed hand nets from spawning holes and transferred into a fry container. Generally, 2,000–15,000 fry are found in each hold. This method of collection yields approximately 50 million fry annually (Sidthimunka, Sanglert and Pawaspootanon, 1968).

Pond breeding methods were developed by Wanichakorn (1957) to meet the increasing demand for *Clarias* fry by making holes in pond banks and provision of some weeds therein for attachment of eggs. Tongsanga, Sidthimunka and Menasaveta (1963) and Sidthimunka, Sanglert and Pawaspootanon (1968) induced breed *C. macrocephalus* in small aquaria for commercial production of seed.

While *Clarias* was breed in India as early as 1956 by using homoplastic pituitary glands (Ramaswami and Sunderaraj, 1957a), mass production was possible only through induced spawning by a massive dose of carp pituitary gland extract (Khan and Mukhopadhyay, 1972).

Several other species of catfish and air-breathing fishes (*Heteropneustes fossilis, Pangasius pangasius, P. sutchi, Channa* spp., *Anabas testudineus* and *Osphoronemus goramy*) are cultured in the Indo-Pacific region. Their seed are generally collected from nature though they are all amenable to induced spawning. The fry of *Pangasius sutchi, P. larnaudi* and *P. mioronemus* are caught in Cambodia in the Mekong river during the late monsoon months. They live on the bottom but regularly come to the surface to take in air. They are caught in slow-moving water shaded by bushes and trees. Advanced fingerlings and juveniles are caught with dipnets (Hora and Pillay, 1962).

Earlier spawning of *P. sutchi* is reported to have been accomplished by hormone injection. Recent attempts in India to breed the indigenous *P. pangasius* by pituitary treatment have met with some success.

A low water column is the most essential requirement of the breeding environment of air-breathing fishes to enable easy aerial respiration. In *Anabas, Heteropneustes, Claris* and *Channa* spp. Aerial respiration starts five to eight days after hatching (Banerji, 1974; Thakur, Pal and Khan, 1974). Rising to the surface for aerial respiration involves considerable loss of energy and in deeper waters may lead to exhaustion or mortality due to "drowning". Pandian found the energy loss due to vertical trips of young *Channa punctatus* reared in 40 cm deep water to be 161 cal/day as compared to 92 cal/day for those which were reared in 2.5 cm deep water.

In India, fingerlings (30–100mm) of *Clarias magur, H. fossilis, A. testudineus, Channa marulius, Channa striatus* and *Channa punctatus* are collected from flooded fields, swamps and paddy fields by scoopnets, dragnets and traps and marketed alive for culture.

Osphronemus goramy breeds naturally both in streams and ponds. The species is not so common in India as in the Philippines and Indonesia. In India and the Philippines, no special preparation is made for spawning the fish except that aquatic weeds are planted in the ponds for construction of nests and egg deposition. In Indonesia, however, the fish is bred both in "growing" and spawning ponds. The growing ponds are planted with *Hydrilla* for binding the bottom soil and providing shelter to the fishes and branching stems of common bamboo (*Bambusa vulgaris*) for nesting. The depth of water is maintained at 70 cm. Four to eight year old brood fish, maintained separately in special ponds about 50 cm deep and fed on *Carica* and *Colocasia* leaves and rice bran, are introduced in the growing pond at 1 female per 100 m^2 in the proportion of 3 females to 2 males. As soon as spawning begins, a fishy smell emanates from the nest and an oily substance comes to the surface The eggs hatch out approximately 10 days after the breeders are introduced.

Spawning ponds for goramy measure 100 m^2 and are 60 cm deep at the inlet and 1 m at the outlet with muddy bed and steep banks. Branching stems of bamboo and fibres of *Arenga saccharifera* are provided in the ponds for nest building. Ten female and five male breeders are stocked in such a pond and fed on *Colocasia* and *Carica* leaves at 5 kg per day and about 21 of rice bran twice a week. The females build nests in about 10 days and spawn in the nest 3 days. Hatching takes another 3 days. The hatchlings are fed white ants at one tea cup per nest. When about 1 cm in size, the fry is transferred to the rearing pond. If it is desired to reuse the pond, the nests could be taken out and the eggs hatched separately in earthenware jars. It is necessary to change the water regularly.

Production of the Seed of Cichlids (*Tilapia* spp.)

The tilapias, though a native of Africa, are now widely cultivated the world over. They can be divided into two groups: one herbivorous and macrophytophagous comprising *T. melanopleura* and *T. zillii,* and the other, microphagous and omnivorous group comprising *T. macrochir, T. mossambica, T. nilotica* and *T. nigra.* They have lower fecundity but this is compensated for by buccal incubation of eggs. Of these *T. mossambica* and *T. nilotica* can withstand brackish water also.

Seed production is not a problem in tilapias but it is their prolific breeding which requires control. Two methods, *viz.,* association of tilapias with a predator and monosex culture, were tried to control overpopulation with varying degrees of success.

Since monosex culture of tilapias gave good results, the problem of obtaining all male or all female fry was sought to be solved by hybridization. Hickling (1960) obtained all male fry by crossing the males of an African strain (*T. hornorum zanzibarica*) with Malayan strain to *T. mossambica.* Fast growth of the males is an obvious advantages as there is no fear of overstocking through wild breeding but they cause great damage to the dikes by building nests even in the absence of females. All female hybrids would therefore be better despite their poor growth and lesser production. One "Malayan" male (*T. mossambica*) when crossed with two "Malayan" females (*T. mossambica*) gave 97 and 88 per cent females (Malacca, 1963). Crossing male *T. hornorum* zanzibarica with females *of T. nilotica* in Uganda also resulted in the production of 100 per cent male fry (Pruginin and Kanyike, 1965). All male hybrids were also obtained, though irregularly, by crossing male *T. macrochir* with female *T. nilotica* in Ivory Coast by Lessent (1968).

All hybrids produced so far, however, belong to the microphagous group. Huet (1970) observes that production of herbivorous hybrids would be of great interest considering the ease with which they could be fed artificial food. It is also suggested that fry raising must be separated from production of fish as these are two distinct operations.

Production of Elevers (*Anguilla* spp.)

The catadromous eels pass from one to three years of their life in the sea depending upon the area they come from. The European eel (*A. vulgaris*), the American eel (*A. rostrata*) and the Japanese eel (*A. japonica*) are economically very important.

Eel culture is practiced intensively in Japan and a very high rate of production is obtained. Elvers (60–70 mm/0.16–0.2 g) are captured in winter at the months of water courses with different types of nets and stocked in ponds for culture until autumn whence they attain a size of 150–200 mm and a weight of 5–8 g. Cannibalism is a highly characteristic feature and young eels need to be graded 5 to 6 times before they reach a size of 100–150 mm. These are stocked again but in different ponds according to their size. These are known as seed fish and used for stocking large ponds which produce eels for the table (Huet, 1970).

The Indian eels, *A. bicolor* and *A. bengalensis* have also been reared recently and found to have both high growth and production potential. It has been found that the elvers attain marketable size (350 mm in length and a weight of 100 g) in running waters in one year (Dorairaj, 1976). However, an elver resourse survey is urgently needed to determine the areas and period of their abundance to help locate the centres for their collection on a large scale.

Production of Seed of Brackish Water and Marine FIshes

Production of Mullet Seed

The mullets are raised in fresh and brackish water ponds in China, Hawaii, Hongkong, India and Israel. In India, China and Hongkong, they are cultured along with major carps; in Hawaii with milkfish and in Israel with carp and tilapias. The most important species of mullets are *Mugil cephalus, M. capito, M. tade* and *Liza macrolepis.*

The mullets migrate to the sea for spawning though they do attain maturity in the brackish water lakes and estuaries. Hendrickson (1969) reports natural spawning in captivity when pond-grown mature mullets were transferred to tanks where pure sea water was being circulated. It is believed that mullets need waters of high salinity for proper fertilization of their eggs.

In China, dipnets are used for catching the mullet fry during low tides. Fixed dipnets, which are operated from the shore, are also used. In India, pieces of coarse cloth dragged by two persons constitute the gear in West Bengal; white fine-meshed dragnets and cashnets are used in Tamil Nadu. Chinese dipnets are quite common in Kerala too.

Mullet seed is quite tolerant of wide fluctuations in salinity and enters the estuaries and brackishwater lakes as fry and fingerlings where they are taken in impoundments along with tidal water and reared in association with other species, including predators and competitors. However, it is possible to identify the different species of mullets and segregate them for selective stocking (Sarojini, 1958). The seasons and areas of availability of mullets in India have been surveyed and several centres for seed collection located.

Production of mullet seed through hypophysation has been tried in Taiwan, Province of China (Tang, 1964; Liao *et al.*, 1972), India (Chaudhuri, 1966; Chaudhuri *et al.*, in press), Israel (Yashouv, 1966 and 1969); Abraham *et al.*, 1967), and Hawaii (Hendrickson, 1969); Shehadeh and Ellis. 1970). While ovulation, fertilization and hatching was carried out successfully, further rearing seemed very difficult. In Taiwan, Province of China, the problem of selecting and securing mature, uninjured mullets from the sea was solved by developing a special raft fitted with an outboard motor for quick transport. Though the effective dosages were found to be 3.5 mullet pituitary glands combined with 30 R.U. of Synahorin, it was felt that more pituitary glands and less Synahorin in the first injection and less glands and more Synahorin in the second injection would be better. The time between the two injections was cut to 12 h but it has not been possible so far to shorten the ovulating time by less than 24 h. A high percentage of hatching (over 70) is possible when a high dissolved oxygen content is maintained and minimum variations are allowed in water temperature and salinity (Liao *et al.*, 1972).

Larval rearing was found to have, at least, two critical periods, the third to fourth day and the 11th to 13th day. The survival rate was less than 2 per cent during the second critical period and the larvae continued to die till the 18th day, by which time the survival rate was less than 1 per cent. This problem could be partly solved by using freshened sea water from the sixth or seventh day onward but the timing and freshening rate require further studies.

Both natural and artificial feed for mullet larvae are still a problem are require further studies. The newly-hatched larvae cannot tolerate light of high intensity, at least, up to the 13th day, hence studies are required to be conducted on the optimum light intensity tolerance to reduce unnecessary mortality.

In Hawaii, Kuo, Nash and Shehadeh (1974) developed standard practices to breed the grey mullet under controlled conditions. The dosages of hormone to be administered have been established in relation to the stage of maturity of the female, which could be determined by the size of the intraovarian oocytes. Gravid females, having a mean egg diameter larger than 650 μ but not less than 600 μ, require a total dose of 12 μg of partially purified salmon gonadotropin per g body weight. Females with small and less developed eggs receive a higher total dose (21 μg per g body weight). The injection is always fractional: one-third of the total dose is followed 48 h later by the remaining two-thirds.

Out-of-season vitellogenesis of intraovarian oocytes of captive grey mullet has also been induced by the manipulation of photoperiod and temperature regimes (Kuo, Nash and Shehadeh, 1974). Spawning is possible to be induced in such females where the oocytes have progressed to the tertiary yolk globule stage but the egg size has remained below the effective range for injection (600–650 μ) by daily injections of pituitary gonadotropins increasing the doses from an initial 0.12 to 3.4 μg per g body weight. The manipulation of photoperiod and temperature, alone or in conjunction with the use of exogenous hormones, helps in achieving two primary objectives of aquaculture, *viz.*,

1. In increasing the individual spawning frequency, and
2. In prolonging the spawning season throughout the year.

Operational procedures for rearing the grey mullet larvae from incubation of eggs to the end of 42-day larval period have been developed by Nash, Kuo and McConnel (1974). The eggs are incubated up to the gastrula stage at high density for 12 h at 22°C. They are washed under stirred irradiated sea water (32°/oo salinity), then give a one-minute bath in sea water containing the antibiotics potassium penicillin (80 IU/ml) and streptomycin sulphate (0.05 mg/ml) and distributed into the rearing kreisels (circular rearing tanks) at 250 eggs/1. Incubation is then continued for the remaining 24 h. Continuous circulating motion in the kreisel keeps all the eggs in suspension. Both penicillin (10 IU/ml) and streptomycin (0.01 mg/ml) are added daily to control the marine bacteria. The larvae begin to fccd on the third day and are provided with cultured plankton and rotifers supplemented with available natural plankton. *Artemia nauplii* are provided on the seventh day.

Survival has been increased in the early larval stages by increasing the size of the rearing kreisel and increasing aeration and circulation. Nash, Kuo and McConnel (1974) believe that the survival could possibly be increased further by using even larger units. Whilc light control is desired throughout the entire larval rearing period, exposure of the larvae to direct sunlight during the first days is lethal and artificial lights are therefore preferred.

The problems resulting from ingestion of the empty cysts or shells of *Artemia* cause heavy mortality of the larvae up to four weeks of age. The larvae are also susceptible to the attacks from ectoparasites and predators introduced with the natural plankton and amphipods have been observed attacking even six-week old larvae causing their death.

While considerable progress has been achieved in mullet breeding, techniques of rearing the breeders to full maturity in ponds, proper dosages of pituitary extract and methods of hatching and rearing, especially the right kind of food for the hatching and young fry, need to be studied further before the hypophysation technique could be successfully applied for mass production of the mullet seed throughout the world.

Production of Milkfish (*Chanos chanos*) Seed

Milkfish, though widely distributed in the offshore waters of the Indian and Pacific Oceans, is most extensively cultured only in the brackishwater ponds of the Indo-Pacific region. Like the mullets, it is also suitable for culture in both fresh and brackishwater ponds. However, unlike the mullets it does not attain maturity in fresh or brackish water ponds. It is believed that it spawns in the coastal waters. Young ones are collected in large quantities from foreshore areas for stocking the ponds. The sites selected for collection are sandy and flat with clear water. Locations near the mouths of small rivers flowing into the sea are the best.

Different methods are adopted in different countries to collect the young larvae and all operations from collection, transportation to its marketing are undertaken by expert fishermen and traders. *Chanos* fry industry has perhaps the same status in Indonesia and the Philippines as the carp fry industry in India and China.

In Indonesia, milkfish fry (15–25 mm) is usually collected during high tides from March to May and September to December, when the sea is not very rough. Fry (10–13 mm) of milkfish is available on the Philippines coast from March to August and collected by dipnets in narrow tidal streams, where they are highly concentrated at the peak of the season.

Though milkfish culture industry in Taiwan, Province of China, is of recent origin, it is much more productive than that of Indonesia and the Philippines. The fry are collected during spring tides on full and new moons in April and May by using triangular dipnets. Fry collectors either wade or, in deep waters, float on an inner-tube. A bag seine of varying length, with an opening at the top of the bag for periodical removal of the fry, is also used effectively in deep waters where it is pulled by two men or anchored, where the current is strong, with its mouth facing the current (Lin, 1968).

Availability of the seed is the most important problem of milkfish culture industry in the entire region, the supply from natural sources being unpredictable and marked with wide fluctuations. The depletion of in fish seed in open waters is aggravated by water pollution resulting from the discharge of industrial wastes (Blanco, 1972). Considering the difficulty in collecting mature breeders from the sea and keeping them alive for induced breeding, it appears necessary to bring about maturity conditions in pond-reared fish. Attempts to breed milkfish artificially are underway on a large scale in all the three major producing countries. It is reported that ovulation of pond-grown milkfish has recently been achieved in Hawaii.

Production of the Seed of Yellow-tails

The yellow tail (*Seriola quinqueradiata*) cultured in Japan spawns off southern Kyushu in late April or early May. The larvae (about 15 mm) are brought ashore by the Kuroshio current where they are caught in floating seaweeds by encircling with fine-meshed nets.

The fry are graded immediately into small, medium and large categories as serious mortality results from cannibalism if this is not done (Furukawa, 1972). The graded fry are stocked in nylon net cages for further rearing. In natural conditions, they begin to feed when they are four days old; 20–30 mm fry feeds on fish eggs and arthropod larvae found amongst floating seaweeds and 40–50 mm fry feeds on the larvae of other fish (Tanaka, 1970). In net cages, however, the fry is fed on minced fish and/or shrimp. Feeding on anchovies resulted in 100 per cent mortality as the fry could not digest the unsaturated fatty acids of these oily fish (Bardach, Ryther and McLarney, 1972).

Yellow-tail eggs have been raised experimentally but most of the fry used for aquaculture are caught from nature (Tanaka, 1970).

Conclusion

In spite of the many advances made in the methods of fish seed production and distribution, this remains a major constraint on the expansion of fish culture. It has not yet been possible to arrange the supply of adequate numbers of pituitary glands for breeding. It would, therefore, be better to find a chemical substitute with a standard potency. Much remains to be known about the maturity condition of the fishes from external morphology or the size of the egg to determine the dosages to be injected. Methods of inducing "natural" spawning through hypophysation in Chinese carps have yet to be devised. Methods to bring about proper gonadal development and early maturity or alternatively to reduce the fat content should be studied and techniques of raising fish separately for breeding and for conservation, as in poultry production, need urgent development.

Production of milkfish and prawn seed on a commercial scale is likely to remain a problem until these species can be grown to sexual maturity under controlled conditions. As it is, collection of milkfish spawners and keeping them alive for induced breeding itself forms a major problem. Non-availability of efficient and cheap feeds for the mullet larvae are other important problems hampering large-scale production.

Since fish culture also aims at improving the product, selective breeding and hybridization for obtaining all male or all female broods, and varieties of seed resistant to fluctuations in the environment need to be developed. Continued research is also required on the diseases of the egg, larval and fry stages of almost all species.

References

Alikunhi, K.H., 1960. Practical hints on the breeding of common carp. Barrackpore, Cenrtral Inland Fisheries Research Institure, 11 p. (mimeo).

Alikunhi, K.H., H. Chaudhuri and V. Ramachandran, 1951. A possible method for segregation of species from a mixed collection of carp fry. Sci. Cult., 17(5): 220.

Alikunhi, K.H., H. Chaudhuri and V. Ramachandran, 1952. Response to transplantation of fishes in India, with special reference to conditions of existence of carp fry. J. Asiat. Soc. (Sci.), 18(1): 35–54.

Alikunhi, K.H. *et al.*, 1964. Observations on the breeding of carps in bundhs, near Nowgong, Madhya Pradesh, during July-August 1964. Bull. Cent. Inst. Fish. Educ., Bombay,(1): 22 p.

Alikunhi, K.H. *et al.*, 1964a. Preliminary observations on commercial breeding of Indian carps under controlled temperature in the laboratory. Bull. Cent. Inland Fish. Res. Inst., Barrackpore, (3): 19 p.

Alikunhi, K.H. *et al.*, 1971. Induced spawning of the grey mullet (*Mugil macrolepis*) and observations on rearing hatchlings to fingerling stage. Workshop on induced breeding (hypophysation) of carps. 22nd to 27th February 1971, Working Paper No. 33: 2 p.

Banerjee, M.K., 1970. A note on the mortality of *Cyprinus carpio* (Linn.) eggs by the gastropod *Lymnaea luteola* (Ham.). J. Inland fish. Soc. India, 2: 161–3.

Banerjee, S.R. 1974. Hypophysation and life history of *Channa punctatus* (Bloch) J. Inland Fish. Soc. India, 6: 62–73.

Barrackpore, Central Inland Fisheries Research Institute, 1965. Report on fish spawn prospecting investigations, 1964. 1. Uttar Pradesh and Gujarat. Bull. Cent. Inland Fish. Res. Inst., Barrackpore, (4): 191 p. (mimeo).

Barrackpore, Central Inland Fisheries Research Institute, 1972. Annual report. Annu. Rep. Cent. Inland Fish. Res.Inst., Barrackpore, (1972): 114p.

Bhowmick, R.M., 1975. Carp hatchery, its commercial utility and operation. Central Inland Fisheries Research Institute, Summer Institute, 1975. Lecture, 5 p.

Chaudhuri, H., 1960. Experiments on induced spawning of Indian carps with pituitary injections. Indian J. Fish., 7(1): 20–48.

Chaudhuri, H., 1975. The use of hormones in induced spawning of carps. P Pacif. Sci. Congr., 13(1): 14 (abstract).

Chaudhuri, H. and K.H. Alikunhi, 1957. Observations on the spawning of Indian carps by hormone injection. Curr. Sci., 26(12): 381–2.

Chaudhuri, H., S.B. Singh and K.K. Sukumaran, 1966. Experiments on large-scale production of fish seed of the Chinese grass carp *Ctenopharyngodon idellus* (C. and V.) and the silver carp, *Hypophthalmichthys molitrix* (C. and V.) by induced breeding in ponds in India. Proc. Indian Acad. Sci.(B)., 63(2): 80–95.

Chaudhuri, H. *et al.*, 1967. Note on natural spawning of grass carp and silver carp in induced breeding experiments. Sci. Cult., 33(11): 493–4.

Chaudhuri, H. *et al.*, 1967. Experiments on artificial propagation of *Mugil cephalus* Linnaeus in India. J. Fish. Biol. (in press).

Jhingran, V.G., 1972. Riverine fish seed resourses and their exploitation in India. In Seminar on production of quality fish seed for fish culture, 1–2 November, 1968. Central Inland Fisheries Research Institute, Barrackpore. Pp. 56–80.

Kulkarni, C.V., 1972. Production of high quality fish seed in Maharashtra. In Seminar on production of quality fish seed for fish culture, 1–2 November 1968. Barrackpore, Central Inland Fisheries Research Institute, pp. 38–52.

Tripathi, S.D., 1974. Production of carp seed from wet and dry bundhs. In Summer Institute 27 June–31 July 1974. Barrackpore, Central Inland Fisheries Research Institute, 13 p.

Tripathi, S.D., Growth, maturity and spawning in grass carp reared on a diet of terrestrial vegetation. JNKVV (Jawaharlal Nehru Krishi Vishwa Vidyalaya) Res. J. (in press).

Tripathi, S.D. and R.K. Sharaf, 1974. Predatory role of weed fishes vis-à-vis aquatic insects, tadpoles and froglets. JNKVV (Jawaharlal Nehru Krishi Vishwa Vidyalaya) Res. J., 8(2): 123–7.

H. Chaudhuri

(FAO/UNDP–1976)

Record Fish Production through Intensive Fish Culture in a Farmer's Ponds

Editor's note

Trials to augment fish production by application of composite fish culture technology in ponds with facilities for water replenishment were undertaken in two ponds owned by a fish farmer. With a high stocking density of 13 320 fingerlings/ha, the ponds recorded average gross and net productions of 7633.5 and 7444.8 kg/ha/year which was about seven times the normal production of around 1000 kg/ha/year reported by the fish farmer.

Introduction

It is by now widely recognized that fish production from still-water ponds can be increased many times through composite culture of carps (Lakshmanan *et al.*, 1971; Singh *et al.*, 1972; Sinha *et al.*, 1973). In recent years, intensive composite fish culture carried out in experimental ponds at Cuttack yielded average productions of over 8000 kg/ha/year (Chaudhuri *et al.*, 1975).

Fish culture practices at the fish farmer's level are empirical and limited to stocking of ponds with assorted varieties of fish seed collected from the river or rarely with progenies produced through hypophysation. Except for casual feeding with kitchen wastes, hardly any meaningful culture practices are followed in fish farming resulting in low and uncertain productions. The urgency of extending the composite fish culture technology to fish farmer's ponds was therefore keenly felt in order to demonstrate high production rates of fishes. This communication presents the results of one such demonstration conducted in a fish farmer's ponds where the farmer provided the necessary inputs and watch and ward, the stocking material and technical expertise were offered by the Fish Culture Research Station, Cuttack.

Observations

Physico-chemical Conditions of Soil and Water

Soil of pond No. 1 was silty loam (sand 32.68 per cent, silt 55.60 per cent and clay 11.72 per cent) and that of pond No. 2 silty-clay loam (sand 21.56 per cent, silt 57.72 per cent and clay 20.72 per cent). The pond soils were more or less similar in chemical reaction and nutrient status except for available nitrogen and organic carbon.

Following fertilization of ponds (Table 1), the chemical conditions of soil and water fluctuated considerably in both the ponds. Turbidity, however, persisted throughout the period. The initial and final conditions of soil and water are presented in Table 1.

Plankton

Plankton production was generally subnormal (below 1.5 ml in 45 litres of pond water). Zooplankton comprised mostly copepods and cladocerans in both the ponds. Their abundance was relatively more in Pond No. 1 than in Pond No. 2. Occasionally, blooms of *Euglena* sp. and *Microcystis* sp. were noticed during summer months.

Survival

Table 2 shows the percentage survival of the different species. The overall survival was more (78.4 per cent) in Pond No.1 than in Pond No.2 (69.7 per cent). The survival rate of common carp was relatively low in the ponds.

Growth

On the basis of absolute growth (Table 2), grass carp ranked first followed by silver carp and mrigal in both the ponds. While common carp registered the poorest growth in Pond No. 1, it performed better than catla and rohu in Pond No. 2. Rohu obviously was the slowest-growing species due to its high density in the total population and its equally high survival rate in both the ponds. The growth of catla was comparatively unsatisfactory. The overall average weight of fishes in both the ponds was 800 g.

Table 1: Water and Soil Conditions

Parameters	*Pond No. 1*			*Pond No. 2*		
	Initial Values	*Values After Fertilization*		*Initial Values Values*	*Values After Fertilization*	
		Range	*Average*		*Range*	*Average*
Water phase						
Temp °C	29.70	22.40–34.00	28.10	30.30	22.20–34.20	28.20
pH	7.50	7.10–7.90	7.40	7.30	7.10–7.90	7.50
Total Alkalinity (ppm)	95.00	88.00–116.00	101.82	95.00	88.00–128.00	105.55
Free CO_2 (ppm)	–	2.64–15.30	9.04	–	4.40–14.30	10.44
P_2O_5 (ppm)	0.03	0.03–0.18	0.08	0.04	0.01–0.18	0.09
$NH_{3-}N$ (ppm)	0.02	0.06–1.30	0.39	0.03	0.02–0.60	0.27
$NO_{3-}N$ (ppm)	0.03	0.02–0.88	0.24	0.05	0.01–0.63	0.15
Chloride (ppm)	–	129.60–227.90	184.65	–	112.00–202.00	156.05
Calcium (ppm)	–	22.40–29.48	26.28	–	22.00–28.00	25.56
Magnesium (ppm)	–	15.53–31.42	23.88	–	18,48–25.88	22.52
D.O. (ppm)	10.18	2.00–6.40	4.47	5.09	2.30–11.20	6.03
Electrical conductivity/ mho/cm	–	559–1186	834	–	620–885	769
Soil phase						
pH	6.40	5.80–7.00	6.50	6.40	6.00–6.90	6.50
Available Nitrogen (N) (mg/100g soil)	29.37	31.00–47.80	39.28	31.11	31.70–47.38	41.71
Available phosphorus (P_2O_5) (mg/100g soil)	6.00	2.00–6.00	3.83	6.00	2.00–12.00	5.56
Organic carbon (%)	0.84	0.87–1.62	0.95	1.22	0.82–1.35	0.93

Table 2: Absolute and Quarterly Relative Growth, Survival and Gish Production

Species	Initial Average Wt. (g)	Final Average Wt. (g)	Survival %	Relative Growth in %				Production (kg)		Contribution by Wt. (%)
				1st Quarter	2nd Quarter	3rd Quarter	4th Quarter	For the Pond	Per Hactare	
				Pond No. 1						
Catla	6.1	576.0	48.5	97.8	1.3	0.2	0.7	56.23	374.492	4.97
Rohu	24.0	455.0	98.5	49.0	35.3	11.0	4.7	268.85	1790.541	23.76
Mrigal	38.0	793.0	82.5	43.0	39.5	7.5	9.9	131.45	875.457	11.60
Silver carp	2.5	1020.0	97.7	95.1	3.9	0.5	0.6	398.20	2652.012	35.19
Grass carp	5.5	1121.0	85.0	62.2	31.0	4.4	2.3	189.93	1264.934	16.78
Common carp	7.5	315.0	58.0	*	*	*	*	73.75	491.175	6.52
Chital								13.20	88.839	1.17
								1131.61	7547.839	

Contd...

Table 2–Contd...

Species	Initial Average Wt. (g)	Final Average Wt. (g)	Survival %	Relative Growth in %				Production (kg)		Contribution by Wt. (%)
				1st Quarter	2nd Quarter	3rd Quarter	4th Quarter	For the Pond	Per Hactare	
Pond No. 2										
Catla	6.0	673.0	78.5	92.0	6.7	1.0	0.9	109.40	728.604	9.45
Rohu	24.0	551.0	92.6	49.8	34.6	10.6	5.2	308.70	2055.942	26.67
Mrigal	38.0	934.0	98.0	61.2	21.1	7.5	10.2	183.15	1219.779	15.83
Silver carp	2.5	1108.0	73.7	92.8	6.0	0.5	0.8	328.20	2185.812	28.36
Grass carp	5.5	1282.0	46.3	59.9	35.8	3.0	1.3	103.68	690.509	8.96
Common carp	7.5	789.0	29.0	81.0	9.6	3.1	6.1	106.97	712.420	9.24
Miscellaneous#								17.20	114.552	1.49
								1157.30	7719.191	

*: Data not available; # Comprised *Puntius* spp., *Rasbora* sp., *Amblypharyngodon* sp., *Wallago attu*, etc.

A gradual fall in the growth rate was observed with increasing age of fish. The quarterly relative mean growth for all the six species put together were as follows:

Pond No.	*Quarter I*	*Quarter II*	*Quarter III*	*Quarter IV*
1.	69.4	22.3	4.7	3.6
2.	73.1	18.5	4.3	4.1

Harvesting

Initiating the harvesting programme from eighth month of rearing, majority of silver carp and grass carp which had grown to about 1.0 kg were removed from both the ponds. Thereafter, monthly

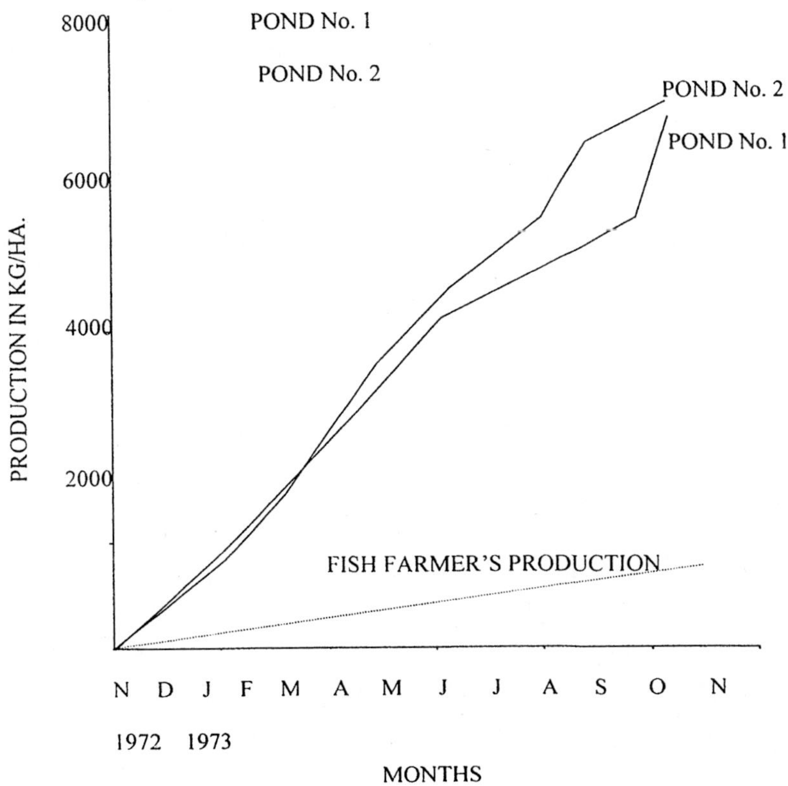

Figure 1: Month-to-month Computed Fish Production at 80 per cent Survival in Both the Ponds Against the Fish Farmer's Production

harvesting was carried out. After removing about 75 per cent of the fish population, final harvesting at the end of one year was completed by reducing the water level to the minimum followed by application of mahua oil cake.

Fish Production

The month-to-month estimated fish production assessed at 80 per cent survival level was observed true particularly in Pond No. 2 (Figure 1). Pond No. 1 recorded gross/net productions of 1131.61/1103.31 kg/year (*i.e.* 547.84/7359.08 kg/ha/year) and Pond No. 2, 1157.30/1129.00 kg/year (*i.e.* 7719.19/7530.43 kg/ha/year) with an initial stocking weight of 28.3 kg/pond (*i.e.* 188.76 kg/ha). The average gross production worked out to 7633.5 kg/ha.

Economics of Fish Production

Pond-wise economics of fish production is as follows:

	Pond No. 1 (0.15 ha) Rs.	Pond No. 2 (0.15 ha) Rs.
A. Operating cost:		
1. Rental value of land	25.00	25.00
2. Preparation of the pond cost of the Piscicide *i.e.* 150 ml of Tafdrin-20	2.50	2.50
3. Cost of stocking material at Rs. 35/- per 1000 fish	70.00	70.00
Packing charges	30.00	30.00
Transport charges	20.00	20.00
4. Cost of fertilizers		
I. Organic (Cowdung)	200.00	20.00
II. Inorganic		
Urea	–	105.00
Triple superphosphate	84.50	73.12
Ammonium sulphate	62.25	–
III. Lime	6.00	6.00

5. Cost of feed		
(*i*) Mustard oil cake	1307.00	1307.00
(*ii*) Rice bran	298.00	278.00
6. Labour charges (including watch and ward, pond preparation, daily routine work like feeding etc. 1 man @ Rs.3/- per day)	540.00	540.00
7. Harvesting charges for 8 men on 6 netting days	72.00	72.00
8. Other expenditure including marketing, dewatering the pond, etc.	100.00	100.00
Total	2627.25	2648.62
B. Income:		
1. Fish production	1131.61 kg	1157.30 kg
2. Revenue realised by sale of fish @ Rs. 5.00/kg.	Rs. 5 658.05	Rs. 5 786.50
C. Profit:		
1. Ratio to operating cost	115.3%	118.4%
2. Ratio to gross income	53.6%	54.2%
D. Cost of fish production	Rs. 2.32/ kg	Rs. 2.28/ kg
E. Average cost of fish production	Rs. 2.30/kg	

Discussion

The highlight of the experiment was the realisation of average fish yield of 7633.5 kg/ha in a year in a rural fish farmer's pond, demonstrating the technical soundness and economic viability of composite fish culture technique. The total yield thus recorded was more than seven times the normal production of about 1000 kg/ha/year reported by the fish farmer.

A special feature of the present experiment was the replenishment of water which helped in maintaining a higher level of dissolved oxygen besides flushing out growth-inhabiting metabolites (Brookway, 1950). However, it had the disadvantage of washing the rich nutrient water. Kowamoto (1957) showed manifold increase in yield of carps from running waters over those from still-water ponds. Rabanal (1967) suggested heavy stocking with intensive feeding as one of the biological means for increasing fish production in ponds with running water facility. In the present work, however, continuous flow of water was not resorted to considering some important factors: (*i*) That in a pond with steady flow of water, the resources of the pond in the form of natural fertility and natural food cannot be advantageously utilized; (*ii*) that the fish will have to rely almost entirely on costly artificial food for its growth; (*iii*) that the pond stands exposed to the risk of receiving highly polluted (with agricultural insecticides and pesticides) water during continuous irrigation; and, (*iv*) that the chances of entry of unwanted fishes from outside and escape of cultivated fishes from within the pond greatly increase due to continuous inflow of water.

However, without essentially altering the cultural technique, high density stocking was done to take advantage of water replenishing facility.

The relatively low survival rate of common carp was probably due to initial mortality, and later predation by *Wallago attu* which had inadvertently entered Pond No. 2. However, it is interesting to observe that the surface feeders grouped together yielded almost identical production (454.4 and 437.6 kg) though the survival and growth rates of both catla and silver carp differed considerably in the two ponds.

Taking an overall account, it is seen that the lower growth rate (713.3 g) in Pond No. 1 compensated by higher survival rate (78.4 per cent) as against higher growth rate (889.5 g) and lower survival rate 69.7 per cent in Pond No. 2 thereby indicating identical production potential of the two ponds.

Harvesting of marketable fish periodically reduced the pressure of density on the pond and thus facilitated growth of other fishes. Such a phased removal by the farmer would help in easy marketing of the produce in local markets.

As the feed requirements increase with the increase in size of the fish (Gaudet, 1967; Swingle, 1967), adequate provision was made to provide the feed in progressively increasing quantities as can be made out by the relation of daily ration to the initial weight of fishes. It has, however, to be admitted that the level of feeding should have some relationship to the financial abilities and resources of the individual fish farmer, as supplementary feed is the single input item costing the maximum to the farmer. In this case, the poor quality of rice polish was compensated by supplying a large quantity in view of it being cheap and easily available to the farmer.

Turbidity in the ponds was mainly due to abundance of silt and suspended particulate organic matter. The phenomenon of gaseous release on account of decomposition of organic matter at the bottom of the ponds probably caused the persistence of turbidity throughout the period of study. The role of such persistent turbidity on the growth of natural food item and of the fish population is not clear at present. However, the large number of fishes did not indicate any sign of distress, obviously due to higher level of oxygen being maintained by the replenished water.

H. Chaudhuri**, N.G.S. Rao, G.N. Saha,
M. Rout and D.R. Kanaujia

Freshwater Fish Culture Division of Central Inland Fisheries Research Institute, Cuttack (Orissa)– in Journal of the Inland Fisheries Society of India, Vol. 10, December, 1978.

References

Anon., 1955. Standard method for the examination of water, sewage and industrial wastes. American Public Health Association, New York: 522 p.

Brockway, D. R., 1950. Metabolic products and their effects. Prog. Fish Cult., 12 (3): 123–29.

Chaudhuri, H., R. D. Chakrabarty, P. R. Sen, N. G. S. Rao and S. Jana, 1975. A new high in fish production in India with record yields from freshwater ponds. Aquaculture, 6 (4): 343–56.

Gaudet, J. L., 1967. The status of warm-water pond fish culture in Europe. FAO Fish. Rep. (44) Vol. 2: 70–87.

Kowamoto, N.Y., 1957. Production during intensive carp culture in Japan. Prog. Fish Cult., 19 (1): 26–31.

Lakshmanan, M.A. V., K. K. Sukumaran, D. S. Murty, D. P. Chakraborty and M. T. Philipose. 1971. Preliminary observations on intensive fish farming in freshwater ponds by the composite culture of Indian and exotic species. J. Inland Fish. Soc. India, 3: 1–21.

Piper, C. S., 1947. Soil and plant analysis. The University of Adelaide, Adelaide: 268 p.

Rabanal, H. R., 1967. Stock manipulation and other biological methods of increasing production of fish through pond fish culture in Asia and the Far East. FAO Fish. Rep. (44) Vol. 4: 274–88.

Singh, S. B., K. K. Sukumaran, P. C. Chakraborty and M. M. Bagchi, 1972. Observation on composite culture of exotic carps. J. Inland Fish. Soc. India, 4: 38–50.

Sinha, V. R. P., M. V. Gupta, M. K. Banerjee and D. Kumar, 1973. Composite fish culture in Kalyani. Ibid., 5: 201–7.

Swingle, H. S., 1967. Estimation of standing crops and rates of feeding fish in ponds. FAO Fish. Rep. (44) Vol. 3: 416–23.

Chapter 12

Some Outstanding Successes on Short Duration Consultancies

Editor's note

Professor Chaudhuri's few, out of many, consultancy reports have been presented in a nut shell. The reports had been very productive, realistic and successful in the field and the concerning countries got the benefit of his advice which brought them immense sustained protein resources. The most successful, however, was his Burma assignment as the country's fishery resources got a boost on his successes.

The Project documents available for edition below are presented mostly in professor's language:

A. IDRC, Aquaculture Sudan Project
B. IDRC, Carps (Malaysia) Project.
C. IDRC, Carps (Malaysia) 2nd Project
D. Grass Carp Project–Fiji
E. Fish Culture Development in Burma (Myanmar) Project.

A. IDRC Aquaculture (Sudan) Project

Editor's Note

The duration of the assignment was only 16 days. Yet his observations and findings enabled Sudan Government to very successfully implement the directives given by Prof. Chaudhuri. Within a short time the poor people of the country could get protein food cheaply.

Prof. H. Chaudhuri was engaged as consultant in April 1978. Some of his observations are presented in his own words with some modifications.

Aquaculture (Sudan) Project

Objectives

The general objective of the project is to increase the production of fish in rural communities through aquaculture in freshwater ponds and specifically:

1. To adapt and develop improved systems of polyculture (*i.e.* the rearing simultaneously in a single pond several species that are not mutually competitive);
2. To develop suitable techniques for spawning Nile fishes (*e.g. Labeo niloticus* and *Barbus bynni*) that are locally well accepted;
3. To develop improved nursery rearing and management techniques; and,
4. To test and apply the above experimental results in government ponds at Essilaat.

Experimental Fish Farm at El Shagarra

The El Shagarra fish farm is located at a distance of 8 km south of Khartoum. Although the farm was reported to have been established in 1953, the major construction work was completed during November, 1956 to 1960. The fish farm has a total water area of approximately 12 ha. with the total land area as 60 ha.

Originally 8 large-size growing ponds, each 1.5 ha. in area were constructed. Subsequently, one of these 1.5 ha. pond was converted to eight 0.15 ha. experimental rearing ponds. Since the total number of ponds are not adequate for experimental purposes, the Research Station is contemplating to convert some of the large ponds to smaller ponds by partitioning. Some work has already been inititated to that effect.

Out of the present seven (7) big ponds, one has been allotted to the Extension section, two ponds will remain as such, one big pond will be divided into four (4) ponds, one into eight (8) ponds and each of the remaining two ponds will be divided into two ponds. The construction work when completed will provide sufficient number of ponds for carrying out experiments on brood-stock development, polyculture and also for rearing fingerlings of different species.

Besides, the above mentioned ponds, there are six (6) smaller rearing ponds of three hundred (300) sq.m. each. There are also three (3) more earthern ponds (one 200 sq.m. and the other two 120 sq.m. each). All these ponds could be used as nurseries or nusery-cum-rearing ponds. In addition, there are twelve (12) cement cisterns having an area of forty (40) sq.m. each. These cisterns can be used for short-term rearing of larvae before these are transferred to nurseries. These can also be used for live-feed culture.

The space for the construction of hatchery appeared to be too narrow and congested. The hatchery should have been made bigger with sufficient floor space for accommodating spawneries and spawning tanks under the roof.

During the consultant's visit, netting was arranged in all the ponds which were under various experimentation or stocked with fish. The three hundred (300) sq.m. small rearing ponds (5 numbers) were stocked separately with twenty (20) would-be spawners of grass carp and common carp in each, segregated sex-wise. While a few of the grass carp males (2–2.5 kg) were oozing, the females (2–3 kg) were not fully ripe. Many mirror carps were observed to be in fully ripe conditions. Besides, there were stray specimens of *Tilapia nilotica* and a few *Labeo niloticus*. Four other smaller ponds were dry. Eight of the cisterns were under experimentation on plankton production with different doses of poultry manures.

Large size growing ponds were having stocks of *Tilapia nilotica, Labeo niloticus, Lates niloticus, Clarias lazera* and scale and mirror carps. These, however, were not under any experimentation. In one of the eight 1.5 ha. larger rearing ponds (8/8) an experiment was being conducted with polyculture of grass carp, common carp, tilapia and nile carp. Another 0.15 ha. pond (8/4) was stocked in 1976 with four thousand (4000) fingerlings of *Labeo niloticus*. The growth of the fish was very slow. In two years the fingerlings have grown hardly to two hundred (200) to two hundred twenty five (225) mm. The stocking rate with a single species without any proper management technique, was too high. The fingerlings from the same stock stocked at a very low rate of stocking in a 1.5 ha. pond, have grown to 1.5 to two (2) kg. A few other 0.15 ha. ponds were having stocks of common carp. Some of the ponds have developed very thick bloom of blue-green algae and one of such ponds had a fish kill when about thirty (30) big size common carps died. The consultant examined the bloom and found to be mainly consisting of *Anabaena* and *Anabaenopsis*, with stray specimens of *Microcystis*. He suggested drawing down of water and then diluting it by replenishment of freshwater whenever a bloom of blue-green is observed to appear in a pond.

Programme of Research

As regards the programme of research for the project and schedule of activities, the consultant had detailed discussion with the Project Leader who requested him to prepare a detailed methodology on the various research programmes to be carried out. Accordingly, the consultant prepared a programme of research based on the objectives of the project with details of the methodology to be followed and handed over to him.

Brood-stock Development

For developing a suitable technique for spawning nile fishes such as *Labeo niloticus* and *Barbus bynni*, initially an adequate stock of spawners have to be developed. This is possible either by collecting juveniles and growing them in the ponds or collecting adults from the natural habitat and rearing those to spawners.

The El Shagarra fish farm has small stock of *Labeo niloticus* adults and a good number of juveniles but practically no stock of *Barbus bynni*. The farm has adequate number of common carp adults and also about one hundred or so grass carp adults. These fishes need proper care for developing into good spawners. Initially, it would be required to judiciously stock the adults of these species in prepared ponds and provided them with supplementary feeding for building up of a good stock of brood fishes.

Induced Spawning

Grass Carp

The present stock of grass carp adults at the El Shagarra fish farm and in the ponds of Wad Medani may not exceed 150–200 specimens. So far developing an improvement system of polyculture with grass carp as a component, it would be necessary to induce breed the present stock of grass carp for obtaining fingerlings for polyculture experiment. Some oozing male specimens of grass carp were examined and it is expected that during the spawning season fully ripe female grass carp would be available. Proper facilities for spawning grass carp by hypophysation during the ensuing breeding season may be developed before hand. A stock of pituitary glands for the breeding experiments may be collected by sacrificing some common carps from the fish farm. If the hatchery is not ready, the injected fishes may be released in breeding hapa fixed in the adjoining river, White Nile and eggs hatched in double hatching hapa. The specifications and instructions for making a few breeding and hatching hapa have been given to the researchers of the station.

*Labeo niloticus**/*Barbus bynni*

Labeo niloticus adults at present available in the growing pond no. 5 were examined. They have grown to 1.5–2 kg in size. Fifteen to twenty fishes were caught in one haul and were examined. No oozing or gravid fish was present. One fish (1.7 kg) was killed and gonads examined. It was found to be an immature female (stage II). So, it is not definite that fully ripe spawners would be available from this stock during the next spawning season. It would be advisable to collect ripe specimens from the river/reservoir and induce those to spawn. The breeding hapa may be fixed at the spot of collection. The fishes may be bred simply by change of old water with sufficient rain-water/freshwater from the river together with maintaining a continuous flow or recirculation of the same water by use of a pump. In case of failure, the technique of hypophysation may be tried. The same method may also be tried with *Barbus bynni*.

*[N.B.: *There had been a report of Labeo niloticus spawning in El Shagarra pond but no details of the breeding including information on the nature of eggs, incubation period, etc. were provides.]*

Sexual Dimorphism

The specimens of *Labeo niloticus* examined at the Jebel Aulia reservoir/river although were not oozing or in fully ripe condition, were maturing specimens and gave some indications of sex differentiation. The pectoral fins of males have just indicated little roughness. Besides, in males, the dorsal fin tip is slightly longer than that of the females and the pelvic and anal fins in the former were comparatively longer and had concave outer edge (tips of rays). The males were slightly brighter in coloration than the females. The findings are very tentative and should be confirmed by examination of large number of fully ripe specimens of both sexes.

Common Carp

The mirror carp (line carp?) strain of *Cyprinus* carpio may be breed by transferring ripe healthy specimens of male and female in the ratio 3: 1 to freshly introduced water in a cement tank provided with suitable materials for attachment of eggs.

Tilapia nilotica

The spawning of *T. nilotica* is very easy. On the contrary, the problem with the species is to control its reproduction so as to check overcrowding of the population for obtaining optimum growth. Monosex culture of *Tilapia* has been recognized as the best solution to this problem of overpopulation and since males grow faster than the females, the present trend in *Tilapia* culture is to produce all-male progeny by selective hybridization. All-male hybrid broods have been produced in several interspecific crosses. Of these, *Tilapia nilotica* females have produced 100 per cent males when crossed with the males of *T. aurea* (indigenous to Israel), *T. macrochis, T. variabilis* and *T. hornorum*. The Fisheries and Hydrobiological Research Section in Sudan is contemplating to import males of *T. hornorum* for the production of all-male offsprings for inclusion as a component of the species selected for polyculture research.

The main limitation to the commercial use of the all-male hybrid *Tilapia* is the difficulty of maintaining pure parental stocks that produce 100 per cent male offsprings. Contamination of the breeding stocks has to be avoided. In recent studies on hybridization, this is being done by labeling with genetic markers.

However, in the absence of all-male hybrids, *T. nilotica* culture can be tried by stocking it in combination with a predatory species such as Nile perch and by stock manipulation.

Lates niloticus

Attempts on inducing spawning of the Nile perch either grown in ponds or collected from the river may be made. The ripe spawners may be introduced in cement tanks provided with continuous flow of water or recycling the same water with splashing by a small pump. Hormone injections may be tried if such treatments fail to induce spawning.

Male *Lates niloticus* about two (2) kg in weight and grown in the fish farm was observed to be fully mature and was oozing milt. One of the large size Nile perch (3–5 kg) specimens appeared to be a mature female as the vent was swollen and reddish in colour. The consultant tried to demonstrate the technique of induced spawning in this pair by injection of HCG/pituitary gland, but facilities for keeping the injected breeders either in breeding hapa fixed in the river or in cement cisterns with continuous recirculation of water by the use of a pump was not available. Since the water temperature was very high continuous splashing of water was necessary.

Sexual Dimorphism

In male Nile perch a small papilla was visible in the coastal region which was absent in the female. Besides, the pelvic fin in the male was comparatively longer than the larger size female. This finding is very tentative and needs confirmation.

Nursery and Rearing Pond Management

Proper preparation of nurseries and rearing ponds and their management is very important for obtaining higher survival and better growth of the fry and fingerlings. The technique is same for majority of the species selected for polyculture such as grass carp, Nile carps and common carp.

Weed control in small nurseries is not a problem. In rearing ponds, however, most of the noxious submerged weeds, filamentous algae and floating weeds (excepting water hyacinth and *Pistia*) may be biologically controlled by grass carp.

Air breathing aquatic insects (Notonectids and others) can be killed by the application of soap-oil emulsion. Other larger predatory insects and their larvae may be eradicated by repeated nettings.

Nursery and rearing ponds should be fertilized with available organic manures (Poultry droppings/cowdung) and may be supplemented with low dose of triple superphosphate. Suitable artificial feed may be provided especially when the rate of stocking is high and the production of zooplankton is either low or depleted.

Harvesting of fry in proper time is essential for obtaining high rate or survival. The fry ordinarily is expected to grow to 25–30 mm size in about 2–3 weeks after which the number should be thinned out and stocked in rearing ponds.

Rearing pond management is similar to that of nursery ponds. The fingerlings are grown to 100–150 mm size in about 3 to 6 months and then used for polyculture experiments.

Seed Collection from Nature

In the first year of the project, it may be necessary to collect fry and fingerlings from the river/reservoir for obtaining materials for polyculture experiments as well as for studies on the rate of growth and food habits of different species. Eventually, the juveniles could be reared to build up a stock of spawners for induced breeding experiments.

Transport of Fry, Fingerlings and Spawners

Standardized technique may be developed for the transport of fry, fingerlings and adult fishes. Transport of fry and fingerlings in plastic bags under oxygen packing is a regular practice in many countries. For the local species it may be necessary to determine the optimum number to be transported depending on the size of the fish and duration of the transport. Larger fishes may have to be transported by using some suitable anesthetics into the water.

Polyculture Experiments

For selecting suitable species as components in polyculture research for higher yields, it is essential that the species should satisfy certain basic conditions. The species should be quick growing and compatible to each other; should have different feeding habits and utilize the different ecological niches in the pond. With a view to get sufficient information on the growth rates, food habits and feeding zones of the indigenous species selected for the polyculture experiments, the consultant referred to available literature but there was not much authentic information available on these parameters. He then discussed with the M. Sc. students of the University of Khartoum who were having their thesis work on the biology of Nile fishes and on other related topics. The following informations were gathered on the indigenous species.

Labeo niloticus

L. niloticus which grows to seventy two (72)cm in length and 4.5kg in weight is common in sheltered waters. It usually feeds on plant debris, epilithic algae and mud. It is good eating but bony when less than 25 cm (Amrithalingam and Khalifa). According to George, *L. niloticus* is bottom feeder subsisting largely on molluscan shells. But Miss Asma Abdel Rhaman Ahmed, M. Sc. student having the thesis work entitled "On the biology of Cyprinid fish in the Northern part of the White Nile", is of the opinion that the species is a column feeder, feeding on decaying vegetation, detritus and sometimes molluscs. The breeding season is June-August. Pectoral fin in males is harder than that of females. Fecundity is two (2) million (4.5 kg fish) and the ovarian egg size is 1–1.5 mm.

Barbus bynni

Barbus bynni grows up to eighty two (82) cm and 5.5 kg. It feeds on mollucs (snails), insects and aquatic plants. According to some, it is a bottom feeder and feeds on detritus and molluscs (more bivalves than gastropods). It's fecundity is about 1 million (4 kg) and the ripe ovarian eggs measure 2–2.5 mm.

For polyculture experiments these two indigenous species are combined with two exotic species along with *Tilapia nilotica*. The species provisionally selected for experimentation in the project are:

Labeo niloticus–a column feeder/bottom feeder (?)

Barbus bynni–a bottom feeder

Tilapia nilotica–a plankton feeder

Grass carp–macro-vegetation feeder

Common carp–omnivorous bottom feeder

Lates niloticus–a predator (as a police fish)

These fishes do not form an ideal combination for polyculture research. There is no efficient phytoplankton feeder (surface feeder) like silver carp in the combination.

In absence of all-male hybrid Tilapia, the *Nile perch,* a predator is introduced in order to keep the population of Tilapia under control. Preliminary trials may be given with 25 per cent of plankton feeder, 20 per cent of column feeder, 40 per cent of bottom feeders, 10 per cent of grass carp and 5 per cent of predator and at two rates of stocking. Based on the performance of the different species, the number may be increased or decreased on the basis of the results obtained in the first year on the overall production, the experiment may be modified in the second and the third year and some more efficient species may be included in the group.

Pituitary Bank

For conducting experiments on induced spawning of fish by pituitary hormone injections, the gland materials may be collected from common carp raised in the farm. Common carp is an ideal donor and mature specimens are available all the year round. The glands may be preserved in absolute alcohol or as acetone-dried or extracts made in glycerine and water in the ratio 1: 2 and ampouled.

During the visit, the consultant demonstrated collection of pituitary gland from common carp and its preservation in absolute alcohol. About 25–30 glands of common carp and one from *L. niloticus* were then collected by the research staff.

Concluding Remarks

The agreement on the initiation of the Aquaculture (Sudan) project was signed in July, 1977, but up till now the project has made very little progress. Several factors might be attributed to the reasons for the delay in implementation of the research project.

At first, the Project Leader who is also the Administrative Head of the Fisheries and Hydrobiological Section and in-charge of El Shagarra Station, is loaded with both administrative and technical responsibilities which necessitate his frequent visits on missions outside the country as well as to in-country travels. In the absence of a second man with adequate experience in aquaculture, the progress of work suffers when the Project Leader is away from the Headquarters.

Secondly, the present research staff of El Shagarrs station, although some are well-trained in Limnological and hydrobiological work, do not have sufficient practical experience to carry out research on the induced breeding of fish, hatchery and nursery management, and on the techniques of research in polyculture.

Thirdly, there is insufficient informations available on the biology of the indigenous species selected for induced breeding and polyculture work.

Lastly, the research station is at present ill-equipped and lacks essential field and laboratory equipments and appliances for successfully carrying out the research programmes as envisaged in the objectives of the project.

Regarding recruitment of the project staff, it is understood that both the Fisheries Research officers selected for the project have been sent to Auburn for their M. S. course studies. Of the two post of Fisheries officers, only one post has been filled up and the other post is still lying vacant. Three technicians' posts are also not yet filled up. So till the officers sent to Auburn return, and start actively participating in the project work, the progress of the research project is expected to be slow. Besides, these officers may not have sufficient practical experience in the induced breeding and rearing of carps.

It is therefore, suggested that one or two enthusiastic researchers connected with the project may be sent immediately for short term practical training in induced fish breeding, hatchery and nursery management, and on polyculture techniques, either to Central Inland Fisheries Research Station at Cuttack, India or to any other similar institute. This is absolutely necessary for the successful implementation of the project.

As stated earlier, the station is badly in need of essential equipments for fish breeding or hatchery work. It is requested that IDRC may expedite procurement of the equipments allotted to the project.

The consultant felt that the time for his visit would have been more appropriate if it had coincided with the breeding season of the cultivated species. By the time, the essential equipments also would have arrived and he could have given some practical demonstrations of the induced breeding and hatchery as well as nursery management techniques.

It is hoped that the project would be able to make appreciable progress when all the staff are recruited and some of the staff start working on the project after receiving training in the techniques of induced breeding of fishes and management of nurseries and rearing ponds, and also in polyculture of carps.

B. IDRC Carps (Malaysia) Project

Editor's Note

The consultancy was for a short duration executed in 1975. For detail of work done the contents may be referred to. It was a complicated objective to be executed successfully. The success of the project indicated super-class efficiency of the consultant.

The consultant demonstrated the technique of pituitary injection and gave instructions for proper maintenance of the brood stock and on all other aspects of induced breeding including hatchery techniques. He suggested training of staff in induced breeding and nursery management.

The content reads as follows:

1. Introduction
    - 1.1 Background Information
    - 1.2 Objectives
2. Breeding of Chinese Carps at Malacca: Past Studies and Present Status
3. Meeting on Project Projection
4. Observations
    - 4.1 Present stock of spawners and their condition
    - 4.2 Collection, handling and transport of spawners
    - 4.3 Selection and injection of spawners
    - 4.4 Breeding and hatching techniques
    - 4.5 Procedures for preparation of hormone extracts
    - 4.6 Pond-rearing of post-larvae
    - 4.7 Feeding fish in the pond
    - 4.8 Instructions through lecture and slide projection
5. Recommendations
6. Conclusion

Introduction

Background Information

The Chinese carps, especially the grass carp (*Ctenopharyngodon idella*) and the bighead (*Aristichthys nobilis*) are the important food fishes cultivated in Malaysia. Though these species are being cultured for several decades in Malaysia, no effective technique for obtaining the seed of these difficult-to-spawn fishes has been evolved. Consequently, the seed have to be imported from outside every year. It is reported that Malaysia imported forty two (42) million Chinese carp fry in 1970 and about twenty five (25) million in 1971. The foreign exchange expenditure on such imports every year is quite considerable. It is, therefore, imperative that the country should try to meet the increasing demand of its requirement of fry by evolving a dependable technique of inducing spawning in these carps.

With a view to helping to develop a standardized technique of obtaining the seed of Chinese carps by neuro-hormonal treatment, IDRC, Canada has recently sponsored a research project in collaboration with the Government of Malaysia primarily at the Freshwater Fisheries Research Station of MARDI at Batu Berendam, Malacca. Since induced breeding of the Chinese carps needs expertise in the proper selection, handling and standardizing breeding and hatching techniques, IDRC thought it might be useful to get the services of a research worker from a neighbouring country where such results have already been achieved, for assisting the project in Malaysia. With this objective in view and also for ensuring regional collaboration and mutual exchange of results, the services of Dr. H. Chaudhuri, Head of Fish Culture Division and Chief, Central Inland Fisheries Research Station, Cuttack, India was sought and made available for about a month as a consultant at the Freshwater Fisheries Research Station, Malacca.

Objectives

The main objective of the collaboration programme as laid down by IDRC is to assist in standardizing the procedures for induced breeding and hatchery rearing of carps in Malacca using salmon gonadotropin SG-G100, for spawning, while established an effective liaison between the two IDRC projects in Malaysia and India, with the aim of comparing results in similar fishes at different latitudes.

Specific research activities as indicated in the terms of reference of the consultant were to compare practical methods in use, advise on the collection, selection and handling of spawners, compare criteria for determining maturing or ripe adults; to provide advice on hatchery systems, pond rearing procedures for post-larvae, feeding methods, etc. and also to compare the procedures for ampoule extracts of pituitary developed at Cuttack with the methods being developed for the standardized dosage of salmon gonadotropin SG-G100, with a view to developing the most simplified standard and effective spawning procedure.

Breeding of Chinese Carps at Malacca: Past Studies and Present Status

Studies on the induced spawning of Chinese carps especially of grass carp started as early as in 1959 at the Malacca Research Station. Experiments carried out by Slack, Prowse and Hickling did not give successful results. Hickling obtained ovulation in 19 out of 48 grass carps injected with *Puntius gonionotus* pituitary glands but the eggs could not be fertilized. According to him, lack of the stimulus of climatic changes in the tropical environment of Malacca might inhabit full maturation of grass carp. Kurth obtained partial success but Chen and his associates were the first to obtain successful spawning of all the three major Chinese carp species in 1968–69. According to them, proper selection of both spawners and donors was important and that "food is the over-riding factor and not the local tropical climate" by which the spawning cycles in these carps could be controlled and staggered.

With the initiation of the IDRC project, some preliminary breeding experiments were carried out by the spawning group on the experimental species (*P. gonionotus*), index species (common carp) and on bighead and silver carp by using salmon gonadotropin SG-G100. Viable eggs were obtained from an injected ripe common carp and from three (3) out of five (5) injected ripe *P. gonionotus*. No controls were kept in these experiments. Negative results were however obtained from SG-G100 injected bighead and silver carp. In one experiment maturing females of *P. gonionotus* with slightly distended belly were given two injections of SG-G100 at the rate of five (5) mg/kg at an interval of twenty four (24) hours. No swelling of the genital opening or any marked change was observed in the treated fishes as compared to the controls.

Meetings on Project Projection

Upon his arrival the consultant was requested to participate in the meeting being held under the chairmanship of Mr. Lee for formulation of the programme of work for the remaining period of the project and give his advice whenever necessary. The meetings were initially held separately with each of the spawning group (Mr. Tajuddin and Mr. Jim Funk) and the nutrition group (Mr. Pathmasoothy and Dr. Bayley) and finally jointly with both the groups. Dr. Davy was also present in the meetings. While discussions were held as to whether breeding work should straightway be started with the target species grass carp or not, the consultant suggested that the number of grass carp adults being limited, it is advisable to get some experience initially with *P. gonionotus* which is abundant, smaller in size and easier to handle and breed. This work could be completed within a few days. He also suggested that since silver carp is available in good numbers, the species could be profitably utilized as an index species for preliminary trials in standardization of doses as spawning behaviour of all the major Chinese carps is more or less identical.

Regarding the programme of the nutrition group, the consultant was of the opinion that the nutritionist should concentrate more on studies on the nutritional requirements of the target species for formulating an ideal feed for optimum development of gonads. However, regular feeding with supplementary feeds to both grass carp and bighead has to be initiated immediately since majority of the fishes are in unsatisfactory conditions and a few are even imaciated and diseased.

Observations

Present Stock of Spawners and their Condition

The present stock of grass carp in the farm was ascertained as little over 300 or so. Since the number of mature fish in the above stock was very limited, very few specimens of grass carp could be sacrificed for studying the gonadal conditions. Examination of external appearance revealed that the condition of the majority of the fishes were poor. A few female grass carp died, which were cut open and examined. The ovaries were found to be immature in some, while in a few others resorption of eggs had progressed considerably. It was understood that no special care was taken for raising the spawners and no regular feeding was done. Adult grass carp and bighead carps were scattered in various ponds. These were being collected by seining and transferred to a set of 18 one-acre ponds (B-series). These fishes were examined. Ripeness of the eggs in females was studied by collecting a few eggs from the ovaries by means of a catheter. A good number of mature grass carp and bighead females were observed to be in advanced stage of regressing gonad and were unsuitable for experimentation. Excepting a few stray specimens, the males were not oozing milt even on pressure on the abdomen. As reported, Malacca appears to have two peak spawning seasons, one during June-July and the other in Nov-Dec. Considering this fact, the consultant's visit was rather little late for the spawning work. Even *P. gonionotus* females were not in fully ripe condition. While the males were freely oozing milt, the females were either having resorbing or maturing ovaries. A few grass carp purchased recently from a fish farmer were well-fed and were having better gonadal condition.

Collection, Handling and Transport of Spawners

The methods of collection, handling and transport of spawners as in practice, needs improvement at every stage for minimising stress and injuries to the delicate spawners. As the water level in ponds is drawn down considerably during seining, the captured fish are kept in very shallow muddy water from where they are collected with *tango* and then transferred to the stretcher and held with hand for examination. These repeated handlings provide too much stress and also injury to the fish. For transporting fish from one pond to the other, the fishes are released into a metal tank where they are liable to receive injuries by dashing against the sides of the tank. The fishes are netted again from the tank for releasing in the pond. Such repeated handlings could be minimized to a great extent.

Necessary instructions were given as to the methods to be followed for hastening the whole operation with least stress to the fishes. A few net hapa and scoop nets of assorted size might be necessary which would be very helpful for the purpose.

Selection and Injection of Spawners

While grass carp and bighead carps were being collected and transferred to the B-ponds (18 one acre ponds) almost every fish was examined and a few selected females were cathetered for examination of eggs. The consultant brought with him samples of catheters and got a few made in the chemical laboratory of the research station. The use of catheter for drawing a few eggs from the ovary for examination of the stage of ripeness of the eggs were taught to the spawning group. Cathetering, however, should be done for final selection of spawners for injection after making a preliminary selection by external examination. Instructions for determination of ripeness of females from the external criteria as well as by the examination of gonads were given. Identification of males even during off-season can be made by the stouter and longer pectoral fin. Sexes of *P. gonionotus* was distinguished by this method. A rough idea of determing the seven different stages of ovaries and testes was also given to the spawning group. Weighing of spawners was usually done at the station on a dial balance which was quite heavy for taking to the field every time. It is more convenient to weigh the selected spawners held in a scoop net by means of a good spring balance and finally noting the correct weight by subtracting the weight of the scoop net. Methods of administering intra-muscular injections in a more suitable site in the body (caudal peduncle) were demonstrated to the spawning group (including the laboratory attendants) and trained them thoroughly in the correct way of drawing hormone extract in the syringe without drawing any air bubble and giving quick injection to the fish through the mesh of the *tango* or scoop net. A dunlopillo was used for placing the breeders during injection to minimize stress. It requires only a little practice for a worker to get properly trained so as to complete the whole process of selecting, weighing, injection and releasing the breeders in the hapa within a very short time. This obviates the trouble of anesthetizing the fishes every time.

Breeding and Hatching Techniques

Demonstration of induced breeding of grass carp or bighead could not be taken up due to the unsatisfactory gonadal condition of the breeders. In the majority of ripe spawners examined, gonadal regression had already commenced. Such fishes do not give successful results. Besides, since the number of spawners were limited it was not thought proper to sacrifice the valuable would-be-spawners for demonstration purpose. As mentioned earlier, a few grass carp procured some time back from a fish farmer, were found to be in better condition although majority of them were either badly injured or diseased. One pair was selected for demonstration of the breeding technique. The female appeared ripe. Since the consultant did not have any previous experince with SG-G100, the dose was determined on the basis of information available as to the successful doses in mullets. He tried 15 mg/kg of SG-G100 in two successive doses to the female. Identical doses with crude carp pituitaries usually give successful results. The female however, could not be stripped after two injections. A third dose was given but the female died probably because of previous injuries. The environmental conditions were ideal and it was expected that the fish would breed. Although the fish did not respond, practical training in various operations of induced breeding were given. Fertilisation of eggs by stripping was, however, demonstrated in a common carp and some fertilized eggs were hatched out in the hatching trays in the laboratory.

Instructions on the hatching technique were given both for hatching eggs in ponds with outer and inner hatching hapa, as well as in the laboratory glass jar hatchery. Detailed information on the working principles and operational details of glass jar hatchery were given and all particulars of the hatchery complex supplied.

Procedures for Preparation of Hormone Extracts

The consultant gave thorough training to the spawning group in the method of selection of donors, collection of pituitary glands, their preservation, weighing, preparation of extracts for immediate use, as well as preparation of glycerol extract for ampouling. Ampoules brought from Cuttack were shown and the method of ampouling was explained.

Pond-rearing of Post-larvae

Detailed instructions on the various phases of pond-rearing of post-larvae and fry were given. A demonstration of control of aquatic insects (Notonectids) by an emulsion of soap and vegetable oil was given in a glass jar in the laboratory and in a small pond in the farm. The highly predatory backswimmers were killed within 15–20 minutes without affecting zooplankters. Fish fry can be released within 18–10 hours after the application.

Feeding Fish in the Pond

Feeding of grass carp and bighead for raising good spawners were taken up by the nutrition group. It was decided to feed grass carp with Napier grass and also with soyabean seedlings/peanut oil cake in addition to Napier grass. The consultant advised feeding other grass carps (not included in the nutrition experiment) with aquatic weeds like *Hydrilla, Najas, Lemna* or *Spirodela* in addition to supplementary feed such as, rice bran and peanut oilcake. The problem of mode of feeding was considered. The consultant suggested the use of trays for feeding with artificial food. This would provide information whether the fishes are consuming the food or not and also about the quantity consumed. It is also essential that for regular supply of above-mentioned weeds, these should be collected and planted in a few ponds, especially in those which are at present lying fallow. A few trips were made to collect aquatic weeds. Identifying characters of a few weeds such as, *Najas, Utricularia, Lemna* and *Spirodela* were pointed out. One common unidentified weed growing on the shallow areas of ponds having yellow flowers, was tentatively identified as *Limnocheris* sp. belonging to the family Butomaceae.

Instructions through Lecture and Slide Projection

A lecture accompanied with projection of coloured slides was delivered to the research staff of the Fisheries Research Station, Malacca on "Induced fish breeding" and "Enhancing fish production by polyculture". Sixty-one slides depicting the details of various operations involved in induced breeding of Indian and Chinese carps for large-scale production of their seed were shown and explained. The slides (50) on polyculture depicted the various phases of fish farm management including nursery and rearing ponds and raising of large number of marketable fish by polyculture of 6 to 10 species with fertilization and supplementary feeding thereby enhancing fish production in ponds considerably. During and after the lecture a number of questions were asked and important points discussed.

Recommendations

1. Raising of adequate number of good spawners is essential for successful induced breeding programme. It is, therefore, recommended that at least 300–400 good spawners of each species may be raised for each spawning season by collecting them well in advance of the season and feeding them with adequate food. For raising spawners for 1976

and 1977 seasons, yearlings and fingerlings may have to be procured immediately.

2. Utmost care should be taken in handling spawners during collection, selection, transport, injections and also during inspections for stripping. While seining spawners, pond water need not be drawn down to low level; batches of spawners may be examined from small sections of the net and quickly transferred to one or two net hapa. Handnets or scoop nets of assorted sizes with handles made of cane or any light metal (aluminium) and suitable netting (smooth) material, may be made for collection, transport, examination, weighing and injection of breeders. If fishes are released in the tank during transport, it is advisable to provide a padding on the sides so that fishes may not get injured by dashing against the sides. While transporting fish, potassium permanganate or preferably acriflavine should be put in water in the transport tank as a disinfectant.
3. For transport of large fishes and spawners suitable anesthentics may be used.
4. For selection of spawners for injection catheter should be used very carefully and only for the final selection.
5. Since the number of the target species is limited, for gaining experience in handling and in breeding and hatching techniques, silver carp which has similar breeding habits as the target species may be tried. In absence of sufficient number of silver carp, *P. gonionotus* may be initially tried keeping adequate controls.
6. For comparison of the effectiveness of crude pituitaries on induced spawning with that of SG-G100, sufficient pituitary glands from mature specimens of *P. gonionotus* may be collected, preserved and extracts made for future use.
7. For conducting induced breeding experiments cool, windy/rainy day may preferably be chosen. Injected spawners may be released in hapa fixed in a pond preferably filled with rainwater or with flow of water in

the pond or in hapa fixed inside the cement cisterns (near the pump house) with continuous shower from overhead pipes and recirculation of water by an electric pump.

8. Early efforts may be made to construct a small glass jar hatchery (Cuttack type) for hatching of eggs. Otherwise, the eggs may be hatched either in double hatching hapa fixed in ponds or in floating trays in laboratory cisterns. Steps should be taken to dechlorinate the water (city water supply) before use. In case of water supply from the underground tank, it would be advisable to treat the water (with U.V.) for eliminating any pathogenic organisms present.
9. Preventive as well as control measures should be taken against parasitic infection such as fin rot, lernaea infection, and bacterial diseases of the spawners.
10. For successful rearing of post-larvae (early fry), the nurseries (H-ponds) should be prepared carefully by eliminating predators, proper fertilization for growth of fish food organisms, control of insects and supplementary feeding for optimum growth and survival of fry.*
11. Spawners may be fed adequately. Supplementary feed may be given in the form of a dough in wooden trays submerged in pond water at one and a half foot under the water surface. Sufficient quantities of *Hydrilla*, *Najas* and duck weeds may be grown in available ponds for feeding grass carp fingerlings and adults.
12. To prevent erosion of ponds and also to provide sufficient food for grass carp, the marginal weed *Enhydra fluctuans*, (Family-Compositae; local name "Chinkro"-Chinese use them as drug and Malays as vegetable) may be grown on the margins of ponds.

[N.B. **A pamphlet on "Technique of Nursery Management" and four other pamphlets on "Induced Breeding", "Breeding of Common Carp", "Intensive Fish Culture" and "Technique of preparation of pituitary extracts in ampoules" have been presented to Research Station Library, Malacca and contents discussed with the spawning group.*]

Conclusion

In conclusion, it may be mentioned that while giving demonstrations and also during practical work in the field, the consultant tried his best to make his service most useful to the research workers of the Malacca Research Station. He kept an open mind and appreciated several local practices and use of quite a few appliances which are very useful and can be introduced at the Research Station at Cuttack. However, from his long, practical experience in the field of fish breeding and fish culture he has found certain well-tried methods and he feels that the introduction of these tried methods at Malacca might give better results.

From the study made by the consultant on the problem of induced breeding of Chinese carps at Malacca, he is of the opinion that breeding these fish should not be a very difficult task. If proper care is taken and adequate number of healthy spawners are raised, at least a good percentage may be expected to mature during the two peak spawning seasons. Contrary to the belief of Hickling that in Malaccan environment female grass carp would not mature, success had already been obtained previously. Recent observations made by the consultant on the maturity of hybrids (silver carp male x bighead female) killed for sale from the farm on 15.2.75, supports the view that fishes do mature fully in this climate. The intergeneric hybrids produced on March 8, 1973 have matured fully even before they were two (2) years old. Six females examined (Wt. 820–1260 g) were either fully mature or resorption of eggs was in progress; the gonado-somatic ratio being as high as 0.22. All the six males (Wt.365–720 g) examined had mature testes under various stages of resorption. This observation revealed that at Malacca the male Chinese carp is capable of maturing at a small size of 365 g and female maturing fully at a small size of 820 g. The spawning group has been instructed to segregate these hybrids and try to breed them during next June-July season.

It is therefore, felt that once fully mature spawners are available, breeding should not be a very difficult task. Only a team of zealous and sincere workers experienced in handling spawners are required who should be prepared to work at any time of the day or night. Such a team should be able to attain success in inducing spawning in these difficult-to-spawn Chinese carps.

C. IDRC Carps (Malaysia) Project (1978) (Second Consultancy)

Editor's Note

This is consultant's second visit to Malaysia having made earlier consultancy in 1975, the details of which have been separately presented. This consultancy covers two phases and the facets of work may be perused from the contents appended herewith. It's a complex project that covers induced spawning, hatching technique, nursery management, nutrition, pituitary bank, seed production by induced breeding, transport of fry, fingerlings and spawns.

The consultant was confident that the project was shaping well and it would help substantially in solving the problem of scarcity of seeds of cultivated species and help increasing the freshwater fish production in Malaysia.

The content reads as follows:

1. Introduction and Background Information
2. Pahse I of the Project
    - 2.1 Objectives
    - 2.2 Programme of Research
        - 2.2.1 Induced spawning
        - 2.2.2 Hatching technique
        - 2.2.3 Nursery management
        - 2.2.4 Nutrition of grass carp
            - 2.2.4.1 Live feed culture
            - 2.2.4.2 Diet formulations
        - 2.2.5 Pituitary bank–fractionation of tuna pituitaries
3. Phase II of the Project
    - 3.1 Objectives
    - 3.2 Programme of Research
        - 3.2.1 Seed production by induced breeding
            - 3.2.1.1 Indigenous species
            - 3.2.1.2 Grass carp (*Ctenopharyngodon idella*)
        - 3.2.2 Pilot hatchery for production of seed

3.2.3 Nursery and rearing pond management

3.2.4 Study on nutritional requirements of carps

3.2.5 Preliminary investigations on some selected natural impoundments

3.2.6 Pituitary bank

3.2.7 Transport of fry, fingerlings and spawners

4. Polyculture Experiments
5. Summary of Observations and Recommendations

Phase I of the Project

Objectives

1. To develop standard procedures for bulk production of fish seed, at all times of the year, by induced breeding of carps using salmon gonadotropin;
2. To develop a pilot hatchery for the production of seed fry and train personnel in fish reproduction, nutrition and hatchery techniques for its practical operation;
3. To investigate local sources of gonadotropin from tuna or other fish;
4. To determine the nutritional requirements of the various stages of the fish being investigated; and
5. To develop suitable feeds from a locally available and inexpensive materials.

Programme of Research

Induced Spawning

At the initial stage the project did not make satisfactory progress mainly because of lack of experience of the young research staff of the station in the technique of fish breeding by neuro-hormonal treatment. The consultant visited the Malacca station previously during Jan-Feb. 1975 as an IDRC consultant, when he demonstrated the technique of pituitary injection and gave instructions for the proper maintenance of the broodstock and on all other aspects of induced breeding including hatchery techniques.* He made several

[N.B. **Activity Report of Hiralal Chaudhuri, Consultant to Carps (Malaysia) Project, 1975.*]

recommendations and also suggested training of the counterpart officer in induced breeding of Chinese carps and nursery management at the Central Inland Fisheries Research Station, Cuttack, India. Accordingly, Mr. Tajuddin (the present Project Leader and Head, FFRS) was sent to Cuttack for training for about six (6) weeks in 1975. This was followed by the visit of another researcher. Mr. Haron to Cuttack, India and Nepal for similar training. The training of these two MARDI officers had greatly helped in the successful implementation of the spawning programme and the first successful spawning in bighead with viable eggs was obtained soon after the return of Mr. Tajuddin from his training.

Although use of salmon gonadotropin was envisaged in the objectives of the project, the spawning team's experiments with SG-G100 were not very encouraging as compared to those with carp pituitary gland extracts. SG-G100 needed the same dose as the carp pituitary for successful spawning, but more consistent results were obtained in the spawning of both bighead and silver carp with carp pituitary glands. Other salmon preparations such as CON A (14), Extract 13, Extract 13 and crude salmon pituitaries were also tried on test fishes (*P. gonionotus*). The first two hormones needed thirty three (33) times higher doses for the precipitation of spawning as compared to carp pituitaries. The dose, appears to be too high. It is not known whether the experiments were conducted simultaneously under similar environmental conditions and keeping controls. It might also be possible that the effectiveness of the hormones deteriorated due to storage. Subsequent experiments using HCG as the primary dose followed by a low dose of carp pituitary gland gave highly successful results and the Project staff was able to breed silver carps and bigheads almost the year round for obtaining their seed. With grass carp, however, the result was less successful. Techniques followed by various institutions were tried and successful results were obtained on several occasions. The technique, however, needs further refinement and standardization. Proper maintenance of large number of broodstock and large number of trials under varying climatic and environmental conditions would help in standardization of the technique.

Hatching Technique

The use of hatchery jars (156 L capacity), made of fiberglass was instrumental in improving the hatching rate. Occasionally premature hatching was observed due to the use of rainwater (softwater) which was greatly remedied by treating the eggs with 0.52 g Tannic acid per litre. The acid is administered to the hatching jar about eight (8) hours after fertilization of the eggs. This, however, delays the incubation period from the normal 15–18 hours to 24–30 hours. This study needs further standardization since delay in hatching (almost double the normal incubation period) involves more risk and more expenditure.

Nursery Management

Nursery management technique has been greatly improved. Trials with various organic manures were given and ground soyaben applied at a rate of 3 kg/10 aq. m. was found to be the best. This manure produces heavy swarms of microcrustacans (Moina). Intermittent low doses of urea may help in maintaining the peak.

The stocking rate of post-larvae used (3.75 m/ha) is quite high. For such an intensive stocking rate the natural feed should be supplemented by artificial feed. Both the growth (2.6–3 cm in one month and survival (30–40 per cent) can be improved by supplementary feeding and stock manipulation (thinning out). The control of aquatic weeds and filamentous algae should not be a very difficult problem in such small nurseries (0.004 ha). These can be eradicated manually. The control of aquatic insects (notonectids) is being done by spraying soap-oil emulsion, a method demonstrated by the consultant during his first consultancy visit.

Nutrition of Grass Carps

Very little progress has been made in Phase I of the project on this important aspect of research, especially on the determination of nutritional requirements of various stages of grass carp and formulation of suitable feed from locally available cheap materials. The reason put forward for this was that the study could not be initiated early due to late arrival of equipment and delay in installation of the nutrition lab.

Live Feed Culture

Preliminary efforts made in producing live feed such as *Brachionus, Moina* and *Daphnia* were very encouraging. The best feed for post-larvae was found to be live feed. The larvae should, however, be fed with minute live feed within forty eight (48) hours after hatching, when the hatchlings start a directive movement. At this stage the mouth is fully formed and the yolk is largely absorbed. Delay in feeding may lead to high mortality.

Diet Formulations

The consultant did not have the opportunity to go through the final report of Dr. Eric Watts, Project Adviser (Nutrition) to learn about the work conducted by him on the nutritional aspect of the project. However, Dr. Watts kindly gave him a copy of the report on "Determination of the Nutrient requirements of the grass carp" prepared by Mr. Allan Castledine, a CIDA scholar. He conducted two preliminary experiments on the 1) determination of an optimum dietary protein level, and 2) determination of optimum protein level and optimum protein to energy ratio for juvenile grass carp. The results of the two-week experiment (first experiment) showed that maximum weight gain was indicated at 40 per cent dietary protein beyond which the growth was depressed. The conclusion made by the worker that "grass carp may be more efficient than other cultivated fish species in utilizing low levels of dietary protein is rather hasty and not very justified since conversion value is less than half of that of 40 per cent and in ad libitum feeding the fish may get substantial quantity of protein for growth since less concentrated diet can meet the requirement if large quantities of it is provided. In the second experiment too, the conclusion arrived at does not appear to have been based on the actual result since the diet containing 26 per cent protein with an M.E. of 3.46 k cal/g appeared to be the best.

Pituitary Bank: Fractionation of *Tuna pituitaries*

The study (grant-in-aid project) undertaken by Dr. Khoo Khay Huat at the University Sains Malaysia, Penang on testing pituitary glands of tunas obtained from the canneries in Penang and comparing that with salmon gonadotropin did not meet with much success because of poor quality of donor fish. Due to the marketing policies of the company good quality fish were frozen as whole-fish and collection of pituitary gland was not allowed. The limited experimental study on testing the crude tuna pituitary extract showed very negligible amount of gonadotropin. As reported by Dr. Huat, attempts to develop simple bioassay on juvenile *Trichogaster, pectoralis* also was not successful.

Phase II of the Project

Objectives

1. To develop standard procedures for induced breeding of some of the local fish of importance, such as *Leptobarbus hoevenii* (Jelawat), *Tor tambroides* (Kelah), *Puntius bulu* and also Chinese grass carp *Ctenopharyngodon idella* (Kam rumput);
2. To develop methods by means of the pilot hatchery for the production, nutrition and transport of the species (such as silver carp and bighead carp) for which fry can now be produced by induced breeding and other species for which breeding is later achieved;
3. To continue the investigations of Phase I of the project after determining the nutritional requirements of the various stages of the fish being investigated and additional fish of local importance;
4. To improve the existing methods of fry culture, including nursery and rearing pond management for the fish being investigated;
5. To make preliminary investigations on methods of stocking in selected natural freshwater bodies and also artificial impoundments in Malaysia;
6. To continue investigations for local gonadotropin sources, including common carp, *Puntius gonionotus* and tuna; and
7. To transfer the improved technologies related to induced breeding, nursery and rearing pond management and transportation of fish fry and fingerlings to the Federal Fisheries Department.

Programme of Research

The consultant had detailed discussion with Mr. Tajuddin, Project Leader and Head, FFRC on the research programmes to be formulated based on the above objectives, priorities to be fixed up and methodology to be followed. The consultant gave his views and suggestions on relevant matters and a tentative schedule of activities prepared.

Seed Production by Induced Breeding

Indigenous Species

In Phase II of the project more stress has been given to develop standard procedures for induced breeding and seed production of a few important indigenous commercial species of carps. Regarding the biology, food and feeding habits, breeding habits and cultural aspects of these local species of carps very little information is available. So before making any suggestion on the programme of research to be taken up on these species the consultant decided to consult literature to collect all available information on these species.

Tor tambroides (Kelah)

The species is available in Indonesia, Thailand (Petchaburi R.) and in Sarawak, East Malaysia (upper Sardong R.). It grows up to 90 cm and 21 kg in weight. 50 cm maximum size is recorded in Thailand. Its supply is very limited. The fish is considered "best among freshwater fishes in Sarawak where successful culture has been carried out in ponds with good flow of water and in the presence of stones and rocks." Practically nothing is known about its feeding and breeding habits. Like Indian Mahseer, it is adapted to rocky bottom in rapid flowing streams. Its growth in stagnant ponds is expected to be very slow.

Immediate attempts should be made to procure fry or fingerlings from Sarawak, carely transported them and stocked in ponds at the Research Station. The gut-contents of the various stages of the species caught from the natural habitat may be studied so as to get some idea of its food habits for facilitating proper rearing of the species. Suitable supplementary feed may be formulated and facilities, if available, for maintaining a mild flow of water in the ponds may be provided.

Puntius bulu (megalan)

The species has distribution in Perak (West Malaysia), Saba and Sarawak (East Malaysia), rivers of Sumatra and Kalimantan (Borneo) and Lakes of peninsular Thailand. It grows up to 3.6 to 4.2 kg. The fish feeds on worms, plant materials, insects and detritus. It is reported to be a suitable fish for pond culture but no information is available as to its rate of growth in ponds and breeding habits.

Procurement of fry, fingerlings and adult fish, if available, has to be made. As in case of *Tor tambroides*, a stock has to be built up by rearing the seed in fertilized ponds and provided with supplementary feed. The growth rate may be studied. In case spawners are available attempts have to be made to spawn them by providing proper facilities and also by hypophysation.

Leptobarbus hoevenii (Jelawat)

Jelawat is available in Indonesia (Sumatra and Borneo), in the central plain of Thailand and in the Pahang river of Malaysia. Maximum length recorded of the species is 50 cm. It is a colourful fish and the young ones (having a black lateral band) are used as an aquarium fish. In Thailand its reputation is not very high as a food fish but in Malaysia it is a delicacy and the fish is sold at a very high price.

Jelawat is omnivorous in food habits and feeds on weeds, insects, molluscs and small fishes.

At Batu Berendam Station, jelawat, culture was initiated in 1974, when 100 fingerlings (3"–4") were brought from Pahang R. The fishes have grown to only 1–2 kg size in 3, 1/2 years. A second batch of 963 (1"–2") was exported from Indonesia through Singapore in 1976 and have grown to about 600 g. These fishes are now regularly fed with trash fish, chicken feed mixed with some weeds and also chicken intestines.

Jelawat cultured in a private fish culturist's (Mr. Khoo Pak Wan) farm, visited by the consultant, have grown to the maximum size of 1.5 kg in one year. The fishes were fed with 80 per cent trash fish mixed with 20 per cent chicken feed. In Singapore, jelawat fed with trash fish reported to grow to 900 g to 1 kg/yr.

For developing the cultural technique of jelawat the fingerlings should be stocked at different stocking rate in fertilized ponds and fed with various types of supplementary feeds.

For broodstock development 40–50 larger specimens of 3, 1/2 yr- old jelawat may be selected and stocked in Pond F12 (0.5 acre) provided with continuous flow of water and fed with 2 per cent of the body weight of feeds rich in animal protein, soyabean meal and some green leaves. The consultant examined the gut contents of a few jelawat which mainly consisted of weeds (Napier grass) and bivalves.

During the visit of the consultant, Jelawat (large size) ponds were netted and the fishes were examined. The weights varied from about 1 to 2 kg. Majority of the fishes appeared to be immature. Twelve fishes selected for detailed study were anesthetized and detailed measurements taken and other studies made. Four specimens were sacrificed for internal study and the rest were released back into the pond. Of the four, one was a slightly oozing male and one female with slightly regressed gonad. The ovary had eggs of dark grey colour having an average size of 1.2 mm. The other two were immature females.

Sexual Dimorphism

Attempts were made to distinguish sex from the external characters of jelawat. There were no distinct roughness in the pectoral fin of the male or any tubercles present on the snout. Cloacal region also did not show much difference. However, the paired fins were observed to be comparatively longer in males than the females. In the mature males the first ray of the dorsal fin is distinctly more elongated than the female. Similarly, the anal fin in males appears to be comparatively longer than that of the female and reaches the base of the cuadal fin. This finding is only tentative and has to be confirmed by examining a good number of fully ripe male and female specimens. Both males and females had coloured fins in males appeared to have deeper orange tinge than those of females.

Induced Breeding

It was suggested that these maturing fishes should be examined occasionally and if found fully ripe, hypophysation may be tried in the same way as in case of *P. gonionotus* or silver cap/bighead. It was also suggested that someone should immediately be sent to Pahang river area where jelawats are available. In case information is received on the availability of mature specimens, a spawning team may go there, fix a breeding hapa in the river and give injections to the spawners. If this is successfully done then a stock of jelawat may be raised and further study on rearing of the larvae, fry and fingerlings can be initiated immediately.

Probarbus jullieni (Temoleh)

The species is available in the Perak and Pahang rivers of Malay Peninsula and also in the Mekong river in Thailand, Laos and

Cambodia. It is the largest carp of Malaysia and is of high eating quality. The population in the local rivers had been greatly depleted. *Probarbus jullieni* was under investigation in a project supported by FAO/UNDP and some useful informations were obtained. The fish has a protactile mouth which suggests its feeding habits from the bottom mud. The food consists of aquatic worms, chironomid and insect larvae, bivalves, snails and plant materials. The fish spawns in rivers. Egg 2 mm in size is light creamy brown in colour. Induced breeding by hypophysation was previously tried but did not meet the complete success.

The consultant feels that considering the importance of the species and recognizing the fact that the population of the fish in natural habitat has been badly depleted, temoleh may be included for the study either in addition to or in place of *Tor tambroides/ Puntius bulu*.

Following the induced breeding technique developed for Chinese carps this species could be attempted to spawn by hypophysation by obtaining ripe spawners from the Perak river above Chendroh Lake. The fry and fingerlings could be collected from the natural habitat and reared in ponds.

Grass Carp (*Ctenopharyngodon idella*)

The Project staff has more or less developed the technique of making, grass carp breed by hormone injections. During the visit of the consultant all the three Chinese carp species were successfully bred by hormone injections. It would be easier to breed the 1976 induced-bred stock of grass carps when they attain full maturity. More breeding experiments may be carried out for the standardization of the technique of hypophysation keeping proper controls. Data on the physico-chemical parameters, rate of fertilization and hatching rate etc. may be regularly recorded.

Hatching Technique

The Institute has been using tannin for hardening the egg shell thus obviating the loss due to premature hatching because of the nature of water (softwater) used for the experiments. However, it has been observed that the hatching time is unusually prolonged to 24–30 hrs. in place of 15–18 hrs. This increases the cost in circulating the water in the hatching jars and also in loss of many hatchlings. Some experiments may be carried out for determining the optimum dose of tannic acid, the duration of the treatment, as well as the optimum time (hours after fertilization) when the treatment of eggs should be initiated.

The spawning team should also further standardize to determine the optimum number of eggs to be introduced in the hatching jar, the percentage of hatching and rearing of the post-larvae in the aquaria before release in the nursery ponds.

Pilot Hatchery for Production of Seed

A pilot fish hatchery has been designed for carrying out experiments on large scale production of the seed of the test species such as bighead and silver carp. Since the Engineers were working on the details, the consultant could not obtain the hatchery design for study. However, the project leader furnished the following information regarding number and size of spawning tanks, hatchery jars and water requirement for the hatchery.

1. Size of the pilot fish hatchery–120′ × 40′ Water requirement
2. Spawning tanks with showers

   [circular with collection chamber as used in People's China]

   (4 nos.) 9′ diameter × 40″ deep — 21,641.76 c ft.

   Collection chamber—4′7″ × 6′ × 40″
3. Smaller spawning tank

   (10 nos.) 6′ × 3′ × 2′ — 13,608.00 c ft.
4. Spawnery (3 nos.) 10′ × 5′ × 2′ — 2,289.60 c ft.
5. Hatching jars (14 nos.) — 8,898.50 c ft.

Total = 47,097.36 c ft.

When the entire hatchery is operational, then total water requirement will be 52,445.35 c. ft. This quantity is 10 per cent of the total volume of A_1 pond (4 1/2 acre) when full.

The water for the hatchery would be pumped in form A_1 pond by the side of which the hatchery is planned to be constructed. According to the plan the water from the hatchery would be returned back to the same pond. The consultant suggested that instead of returning the used water direct to the same pond, it could either be diverted to the adjacent pond or the used water be flown through open channel and released to the distant end of the pond.

With the successful operation of the hatchery, live feed culture system has to be further intensified. More tanks of large size (1/2 to 2 ton) may be required for culturing live feed for post-larvae and fry.

Sufficient supply of quality water for the hatchery and the experimental ponds may be a problem in near future in the Freshwater Station, Batu Berendam. During the period of visit of the consultant the pump station at the intake site was out of action. Since water level in the river had receded further, the intake pipe had to be further extended. The so called Malacca river appeared to be in the process of being converted into a swamp. The whole surface is covered with marshy weeds and there is practically no movement of water in the river. Besides, the water also receives discharge from some factories. Unless an alternative source of water is found out this water might create problems in the efficient working of the hatchery. Alternatively, provision for recycling the water back into the hatchery system by passing through filters should be made.

Nursery and Rearing Pond Management

Management of nurseries and rearing ponds is very important for obtaining better growth and higher survival of fry and fingerlings. FFRS has the advantage that the farm ponds can be drained out completely and refilled.

Weed Control

Weed eradication in small nurseries is not a problem. This can be done manually. But in larger rearing ponds the submerged weeds and filamentous algae can be biologically controlled by grass carp. The floating weeds, excepting *Eichhornia* and *Pistia*, can be efficiently controlled by grass carp. Excessive bloom of blue-green algae is rather a problem in intensively cultured ponds. Microcystis bloom can be suitably controlled by application of Simazine but fishes get killed when depletion of O_2 occurs due to rotting of the dead algae. An impending bloom of *Microcystis* or *Anabaena* can be eliminated by diluting the pond water thereby changing the ecological condition of the pond.

Control of Aquatic Insects

Air breathing aquatic insects (Notonectids etc.) can be controlled by treating the pond surface with soap-oil emulsion (vegetable oil 50 kg/ha mixed with cheap soap 17 kg/ha). However, the expenditure can be reduced if low speed diesel oil is used with a suitable detergent. The larvae of diving beetles are highly predatory. Their occurrence can be prevented by keeping the marginal area of the pond cleared of weeds because the beetles lay their eggs on the stems of the marginal weeds. Small insects like notonectids may not be a big problem if the post-larvae are reared in aquaria up to 13–14 mm size.

Stocking of Nurseries

Stocking rate of post-larvae is very important. When nursery ponds are stocked with high rate of stocking such as 3.75 m/ha, care should be taken to produce sufficient zooplankters in the pond by proper fertilization. In addition, the fry should be supplied with supplementary feeds such as rice bran, mixed with finely powdered peanut meal or soyabean meal or any other suitable substitute. Addition of a very small quantity of growth-promoting substances, such as $CoCl_2$ (micronutrient) with the supplementary feed may help in obtaining better growth and higher survival of fry. One of the most important measures in nursery management is thinning out of fry in appropriate time. The fry should not be kept in crowded condition for a long time. It is advisable to rear post-larvae for 2 to 3 weeks during which the fry is expected to grow to 25–30 mm and then thinned out and stocked in rearing ponds for growing those to fingerlings. Rearing pond management is more or less similar to nursery management.

During the visit of the consultant demonstrations in the ponds could not be given as intended since the ponds if drained out could not be refilled or replenished with water because the intake pump was out of action.

Study on Nutritional Requirements of Carps

This important aspect of study appears to get a further set back in Phase II of the project. As of now, the Project Adviser (Nutrition) has left and the counterpart officer is away on training and he may not be back till the middle of 1979 (as reported). Mr. Hanif is working on this project and is trying his best to formulate some diet from the locally available ingredients. Although he is an enthusiastic researcher, he has no previous experience and training in this field of study. As such, hardly any substantial progress on this important aspect of the research programme could be expected till Mr. Pathmasoothy returns to Malacca after completing his training.

Preliminary Investigations on Some Selected Natural Impoundments for Stocking those with Fry and Fingerlings

One of the objectives in Phase II of the project is to make preliminary investigations on selected natural bodies of water for stocking fry and fingerlings of selected species. Studies on the physical, chemical and biological parameters of the natural water bodies or impoundments may be made initially to determine their suitability for stocking. Study of primary productivity may be initiated by 'light' and 'dark' bottle technique or by C_{14} technique for the determination of carbon assimilation by the planktonic algae, which gives a measure of organic production. Based on the productivity of the waters the suitability of species to be stocked can be determined. Trial stocking of selected species in floating cages and/or pens may be done next. While water rich in zooplankters could be stocked primarily with bigheads, impoundments predominant in algae could be stocked mainly with silver carps. A weed-infested (submerged) water could be ideal for grass carp. However, sufficient precaution should be taken before introduction of any species, and study made on the fish population present in the water bodies both qualitatively and quantitatively and efficiency value of each species determined.

Some preliminary observations have already been initiated by the project staff in a nearby reservoir Durian Tunggal. The consultant visited the reservoir which was filled in Feb. 1977 and has a water area of 480 ha. The average depth of the reservoir is about 15 m with the maximum depth of 27 m. The water above the dam had a light bottle-green colour indicating the water to be fairly productive. This was confirmed by the physico-chemical parameters studied at the time of the consultant's visit. pH of water recorded was 7.3 and dissolved oxygen was optimum. The plankton collected by sieving forty five (45) litres of water was examined by the consultant and was found to be fairly rich. It consisted of *Cyclops, Brachionus, Tetramastix, Ostracods, Phacus, Scenedesmus, Cosmarium, Naviculids* and very minute unicellular green algae. Cage culture in deeper waters and pen culture in shallower fringe areas may be tried. Since both zoo- and phytoplankters are present, both bighead and silver carp can be tried at different proportions and densities. In this connection it may be mentioned that cage culture with bighead and silver carp have given very high production of 25 kg/cu m/10 months in reservoirs in Singapore when stocked at the rate of ten (10) fingerlings (15 cm. size) of bighead/silver carp per cubic meter.

The consultant also visited a 0.75 acre pond belonging to Paya Lang (Dunlop) Rubber Estate and a forty (40) acre reservoir in the Batu Enam Estate (Harrison and Crossfields). The water in both these bodies of water was acidic (6.6–6.7) as was obvious from the rusty colour of water.

Cage culture of *Tilapia mossambica* in the reservoir was suggested. Tilapia in floating cages usually cannot reproduce since nest making is not possible.

Pituitary Bank

The common carp *Cyprinus carpio* is an ideal donor fish because of its comparatively larger size of pituitary gland and is easier to collect and also for the availability of mature common carp for almost throughout the year. Experiments have already been initiated at FFRS in the production of common carp as donor fish in four 0.5 acre 'E' ponds. When sufficient number of donor fish is raised in the farm the cost of pituitary glands would be negligible since the fish could be sold after collection of the glands. Effectiveness of pituitary glands under various methods of preservation will be studied. Pituitary extracts made in glycerine and water (1: 2) and ampouled can remain effective for a longer period. Preliminary work carried out at FFRS gave satisfactory results with pituitary glands when donor: recipient was in the ratio 2: 1. Further work in this line will be carried out such as, correlation of body weight of fresh pituitary gland and correlation of weight of alcohol preserved gland to the weight of fresh pituitary gland. A bioassay technique will be developed with various pituitary materials.

Simultaneously, the grant-in-aid project with University Sains Malaysia would be engaged in trying alternate donor species such as, *Rastralliger* and *Decapterus* because of non-availability of good tuna donors. The bioassay technique would be attempted on common carp as the test animal. Tawas (*P. gonionotus*) may also be a good test animal for such studies. A simple biossay technique is being developed by Dr, Khoo Khay Huat at the University Sains Malaysia.

Transport of Fry, Fingerlings and Spawners

Transport of fry and fingerlings under oxygen packing has been successfully carried out in many countries. If proper care is taken in handling the fish and they are conditioned for several hours with intermittent splashing of water (for emptying the stomach), long distance transport can be made without any mortality. However, experiments have to be conducted to determine the optimum number to be transported depending on the size of the fish and duration of transport. During the consultant's visit a consignment of advanced grass carp fry were conditioned and transported in plastic bag under oxygen packing. For the transport of large fishes or spawners, it may be necessary to use anesthetic in the water. For transport involving tour (4) to six (6) hours duration, anesthetic may not be necessary if the fishes are transported in canvas tank under oxygenation. At Pandan and Tigbauan stations of SEAFDEC 10–12 kg sabalo are regularly transported successfully by this method.

Polyculture Experiments

Mr. Tajuddin, Head FFRS, sought advise of the consultant regarding the stations research project on Polyculture of carps. The expert suggested various combinations of the Surface feeder, Column feeder, Bottom feeder, Omnivorous and Miscellaneous species.

The stocking rate may be 5000/ha and 7500/ha.

Summary of Observations and Recommendations

The Phase I of the Carps (Malaysia) project though initially showed very slow progress, it made good recovery towards the later period. With experience and training received in foreign countries the Project staff gained confidence in the implementation of the project programme.

The project staff have been successful in developing the technique of induced breeding of the experimental fishes and are able to breed the silver carp and bighead with relative ease and virtually every month of the year and raise considerable quantity of seed for supply to the Fisheries Department. The percentage of success in grass carp, however, is much lower, but is expected that in the earlier years of the Phase II project the spawning team would be able to considerably improve their spawning success with grass carp and make further improvement towards refinement of the technique.

The hatching technique has also been improved by the introduction of fiberglass hatchery jars with continuous circulation of water, by use of tannic acid for hardening the egg shell and through the development of a system of live feed culture.

Improved methods in the preparation of nurseries, such as, pond fertilization for production of zooplankton, control of aquatic insects by application of soap-oil emulsion and judicious stocking of fry have resulted in better survival of fry. However, there is still much scope for further improvement on the survival and growth of fry.

While considerable progress has been made in spawning and in hatchery and nursery management, research on the determination of nutritional requirements of the investigated species lagged behind. Two preliminary short term experiments were carried out during Phase I of the project on this aspect of research.

Similarly, studies undertaken in the University Sains Malaysia on the fractionation of tuna pituitaries did not make any progress because of the difficulties in obtaining good quality tuna (donors) from the cannery for gland collection.

In Phase II of the project the programme of research on induced breeding and bulk production of seed have been extended to a few local species of carps. The recommended local species are primarily riverine species and are slow growing as compared to the cultivated Asiatic carps. Several of them have restricted distribution and are not very common in West Malaysia. Very little information is available on the biology and cultural aspects of these species. Growth of *Leptobarbus* (Jelawat) in Batu Berandam ponds had been too slow. However, when jelawat is fed with high animal protein (trash fish and chicken feed), the growth is quite satisfactory but it would be expensive. *Probarbus jullieni* (Temoleh) appears to be a better species and may be considered for inclusion in the project study. Initially attempts have to be made to collect the seed as well as the broodstock of these species, rear those in ponds, study their feeding habits and try inducing breeding at the spot in nature where spawners are available. Jelawat (3, 1/2 years old) are maturing in the farm and so induced spawning experiments may be commenced as and when fully mature specimens are available. The technique followed in spawning *P. gonionotus* and Chinese carps may be tried in this fish too.

Further refinement and standardization of the technique of spawning grass carp may be attempted in Phase II project using HCG in combination with carp pituitary gland and by regulating the frequency of injection and interval between subsequent injections. For synchronising optimum time for stripping, the progress of hydration in the injected female may be closely watched. The synthetic analogue of the nonapept de LH-RH found highly effective in induction of spawning in Chinese carps in People's China may be tried in grass carp.

Hatchery techniques may be further refined with a view to obtaining higher rate of fertilization, hatching and higher survival of post-larvae. The optimum dose of tannic acid, duration of treatment and the stage of development of eggs at the time of treatment may be determined by repeated trials so as to cut down the long extended period of incubation and obtain better hatching.

Sufficient care may be taken in the preparation of nurseries so as to ascertain that no extraneous predatory or weed fishes get into these. Further experiments on the fertilization of the ponds preferably with a mixture of suitable organic and inorganic fertilizers may be tried. Once fry are introduced, use of organic manures for replenishing the depleted live feed is risky. In such case, intermittent low dose of inorganic fertilizer such as urea is advisable.

Soap-oil emulsion is useful for eradicating backswimmers and other smaller air breathing insects. In addition, it is advisable to net the pond several times with a close-meshed net for the eradication of miscellaneous predatory insects. When the stocking rate of fry in fertilized nurseries is very high (> 3 m/ha) supplementary feeding with finely powdered suitable feeds is essential for obtaining high survival. Addition of growth-promoting substances like $CoCl_2$ with feed helps better growth and higher survival. Thinning out of the crowded population is absolutely necessary in appropriate time otherwise heavy mortality may occur. It is suggested that the post-larvae may be grown in a nursery for two (2) to three (3) weeks (25–

30 mm size) after which the fry should be transferred to rearing ponds at a much lower rate of stocking.

A pilot fish hatchery is under construction at FFRS at a cost of M$100,000. The circular spawning tanks with chamber have been designed on the model used in People's China. The source of water is one of the 4, 1/2 acre ponds. Success of the hatchery will greatly depend on the quantity of the water. In future, it may be necessary to provide facilities, for recycling the water after passing through efficient filters so as to maintain the supply of quality water to the hatchery.

The project staff has developed a system of live feed culture. With the establishment of the fish hatchery the facilities for live feed culture have to be expanded considerably. It may be necessary to have pure culture of Chlorella and other suitable algae for supplying feed for the mass culture of rotifers, cladocerans and copepods.

Culture of common carp as a donor fish has recently been initiated. Since common carp is an ideal donor fish, large-scale culture of the species in fertilized ponds with supplementary feeding may have to be developed, keeping in mind the danger of destruction of the pond embankment by the species. Since the pond embankments are protected by packing of car tyres, the common carp might not be able to destroy it. A bioassay technique with the carp gonadotropin may be developed in due course.

Study on the examination of various impoundments and natural bodies of water for stocking of selected fish is being initiated at the beginning of Phase II of the project. Study of the primary productivity of such waters may be made by using either 'dark' and 'light' bottles or C_{14} technique so as to get an idea of the organic production of the water. Floating cage culture in deeper water and pen culture in the fringes with suitable exotic and local fishes, depending on the type of food available, will give an idea of the suitability of the species to be stocked. However, before initiating any stocking programme the existing stock of fish may be ascertained both quantitatively and qualitatively and the efficiency value of each species determined.

With mass production of fish seed and eventual stocking of the water bodies with the fry and fingerlings a well-tried method of transport of the fry and fingerlings has to be developed. Series of experiments to determine the optimum number corresponding to a particular size of fish, correlating with the duration of transport may have to be conducted. Use of anesthetics such as MS 222, 2-phenoxyethanol, quinaldine and others for the transport of adult fish and spawners may be tried.

In conclusion it may be said that the carps (Malaysia) project is shaping quite well and it is expected that the project would be able to help substantially in solving the problem of scarcity of seed of cultivated species and help increasing the freshwater fish production in Malaysia.

D. Grass Carp Project in Fiji

Editor's Note

Two weeks consultancy (in Fiji in June 1974), service was rendered by Dr. Chaudhuri in 1974. He was the Head of Fish Culture of Division at Central Inland Fisheries Research Station at Cuttack, India.

The report that followed from Fiji Government after initiation of Dr. Chaudhuri's suggestion brought immense success and the problems the Govt. faced started showing tremendous success. This saved millions of dollars of the Govt. as import of fish fry was not necessary and the control of weed was successfully implemented.

The extent of work done by the consultant may be visualized from the chapters under contents.

The content reads as follows:

1. Terms of Reference
2. Observations
   - 2.1 Weed infestation in rivers
   - 2.2 Fish population
   - 2.3 Introduction of grass carp in rivers
   - 2.4 Facilities available for rearing of grass carp fry and fingerlings
3. Recommendations
   - 3.1 Rearing of grass carp fingerlings
   - 3.2 Raising donors for pituitary glands

3.3 Preparation for induced breeding of grass carp

3.4 Assignment of a grass carp expert

4. Conclusion

Terms of Reference

The terms of reference were to study the present status of the Grass Carp Project including study of the field conditions and facilities available for rearing and breeding of grass carp and to assist and advise the Government of Fiji on the proper rearing of the grass carp fingerlings recently sent to Fiji for building up of a stock of grass carp for its propagation and their subsequent release in the rivers for the control of submerged weeds.

Observations

Weed Infestation in Rivers

Preliminary observations were made for ascertaining the degree of infestations of the weeds in the rivers. Infestation by *Hydrilla* was very little in the Waimanu river. It was reported that due to cyclone and flood large quantities of the weeds had been washed off. In the Rewa river, it was observed that the marginal fringes had dense growth of *Hydrilla* at several locations. *Potamogeton* was seen further down. The Tongo river especially near its origin from the Rewa river had deep infestation of *Hydrilla*. It could not be however, possible for the expert to make a quantitative evaluation of the weed infestations throughout the length of the Rewa river system during the short period of his stay at Fiji.

Fish Population

Stray observations were made on the fish population of the Rewa river system. Fishing was done by cast netting and seining at certain locations in the Waimanu, Rewa and Tonga river. Hardly any indigenous freshwater fish of commercial importance was encountered in the rivers. *Tilapia mossambica* introduced in the rivers of Fiji during mid-fifties, was encountered in several catches from different locations of the rivers. Besides, a number of brackish water fishes, such as gobies, mullets, *Caranx* sp. and the glass-perch (*Kuhlia rupestris*) were caught in the tidal zone. The glassperch locally called *iko draka* is found in most of the rivers. It is reported that the tidal zone extends as far as 40 km from the Rewa river mouth and that predators like sharks, *Caranx* sp. and *K. rupestris* migrate even up to 50–65 km up. Freshwater eel, a minor predator is also present in the river.

Mrs. Hughes have mentioned in her report of several species of shell-fish from the river. Commonly among those are *Batissa violacea* (Kaibuli and Kai Jamu), *Unio* sp. (Kai ni waidranu), *Melania* sp. (motomoto) and *Neritina* sp. (yamicoko and kadrudru). The mussels are very common and are eaten by the Fijians. Quite a few of these mussels were collected from the river bed.

The only commercial species of prawn available in the river is *Macrobrachium lar* which sometimes get inadvertently introduced in the ponds being brought along with the weeds. This species can be cultured in ponds.

Introduction of Grass Carp in Rivers

The feasibility of introduction of grass carp fry or fingerlings into the river was studied. Since the river harbours a number of highly predatory species, it is not advisable to release the valuable fry and fingerlings directly into the river specially at its lower range. Fishing was attempted at the Tongo river adjacent to the place where the grass carp fingerlings were released in March 1974, but no grass carp was caught. It could not therefore be ascertained as to how many of the released grass carp fingerlings had escaped predation. Besides, since the water of the Tongo river is largely brackish, there is every likelihood of some adverse affect on the fingerlings if released directly from freshwater to brackish water without acclimatization to saline condition. This would, however, depend on the degree of salinity of water at the time of the release. To be on the safe side, it is advisable to release grass carp of about 250 mm or above (10″ size) only and that too into the upper Rewa beyond the tidal range and in places heavily infested with the weeds.

From the observations made by the expert it was felt that instead of making attempts to release a huge number of grass carp in the river at a time, the entire problem should be tackled taking a long range view and grass carp introduced on phased basis. This would also provide opportunity to study the fate as well as performance of grass carp already introduced into the river system. Importation of 0.5 to 1.0 million of grass carp fry from outside may not be a very easy task but even though it is possible, this would involve huge expenditure with considerable risks. Large-scale mortality of fry during transport have occurred on several occasions in the past when attempts were made to import the phytophagous fish. Besides, the imported fry have to be reared to a suitable size before their release in the river and this needs sufficient pond space.

Facilities Available for Rearing of Grass Carp Fry and Fingerlings

As discussed in the foregoing, the introduction of fry and fingerlings of grass carp directly into the river was considered not advisable. Under the circumstances the fry have to be reared upto out if suitable site for construction of ponds could be located at the Nadurulouou Research Station of the Agriculture Department. Naduruloulou Research Station located near the Rewa river, was surveyed. Two probable sites were suggested for the location of ponds pending detailed study of the soil conditions, soil retentivity and availability of adequate water for the ponds. At the first site the soil appeared to be suitable and a perennial small stream was flowing nearby from which water could be tapped. The second location was having a gradual slope surrounded by high lands so that the catchment water during the rainy season could be impounded by constructing a small earthen dam in the narrowest region. Further detailed survey of the sites is necessary before taking up any construction work.

Recommendations

Rearing of Grass Carp Fingerlings

With a view to solving the weed problem of the Rewa river system by introduction of grass carp, great care has to be taken to rear the present stock of fingerlings imported recently from India so that the stock may form the nucleus from which large-scale production of grass carp could be made for stocking the rivers from time to time. It would be essential to have a full-time Fisheries Officer for taking proper care of the present stock. The officer with the occasional help of two to three supporting staff should be able to carry on the work for the present.

The stock of grass carp in the cement cistern had grown to two distinct size-groups. For the normal growth of the smaller fingerlings it would be necessary to segregate the bigger ones and transfer those to Pond 3 after carefully refilling the same with water. Healthy *Hydrilla* should be supplied regularly in excess of their requirement taking care that water does not get polluted by the rotting of leaves. In addition, supplementary feed such as oilcakes, copra meal, wheat short or rice bran may be fed. Care should be taken not to introduce excess of these feeds and supplementary feeding should be stopped in case the water develops a thick algal bloom. Routine analyses of p^H and dissolved oxygen content of water may be made every fortnight. To prevent appearance of thick algal bloom and consequent depletion of oxygen, it is suggested to have an additional constant flow of spring water (as proposed) through a pipe into the Pond 3. The water may be let in, in the form of a spray so that the pond gets properly oxygenated. Plankton sample may be collected fortnightly and examined. Records of various data including growth rate, quantity of feed given etc. may be regularly maintained. However, before release of the yearlings investigation may be made as to the weed infestations, depth of water, water quality and the fish fauna of the area.

The fishes restocked in the ponds may be regularly fed with sufficient quantity of *Hydrilla* supplemented by small quantity of artificial feed. The *Hydrilla* collected from the river may be thoroughly washed and kept for about a day before introduction in the pond. This would prevent entry of *Tilapia* and any other weed fishes in the pond. A rectangular frame of bamboo or wood may be made in the pond inside which the aquatic weeds may be introduced for feeding.

If properly raised, the present stock of grass carp may mature fully by December, 1975 when the fishes would be about 2 1/2 years old. In the tropical waters of India, grass carp male matures even in one year but females take at least two years to mature. At a lower temperature in Fiji it is not certain whether the fishes would mature at this age but it is expected that at least a small percentage of the stock of breeders might mature by the end of 1975. Keeping this aspect in view, raising of as many as 600 probable breeders has been recommended. By the end of 1975, the breeders are expected to attain about 2.0 to 2.5 kg in weight and in the event of their attaining full maturity attempts have to be made to breed them by injections of fish pituitary hormones.

Raising of Donors for Pituitary Glands

Before carrying out any experiments on induced breeding of grass carp it is essential to procure an adequate stock of pituitary gland materials for injecting the breeders. *Puntius gonionotus* was imported from Malaysia with the object of using it as a donor fish for collection of pituitary glands. It is a suitable donor fish and can be reared and bred without much effort. But some pond space have to be provided for breeding and rearing of these donor fishes. About 500–600 specimens of mature *Puntius gonionotus* may have to be raised so as to ensure sufficient stock of pituitary glands. *P. gonionotus* may be reared in smaller ponds initially and then released in the impoundment at the Naduruloulou Research Station as soon as the latter is constructed. As in the case of grass carp, aquatic weeds as well as supplementary feed may be provided for rearing the species. For breeding, fully mature male and female breeders may be selected and released in smaller ponds which had been previously drained out and freshly refilled, preferably with a constant flow of water.

Mature fishes should be selected for gland collection. The pituitary gland, situated vental to the brain proper may be carefully removed by dissecting the head, washed in absolute alcohol and finally preserved in pure absolute alcohol. At the time of injection the glands may be taken out of absolute alcohol, weighed quickly in a chemical balance and the required quantity macerated in distilled water for the preparation of gland extract.

Preparation for Induced Breeding of Grass Carp

As mentioned, the present stock of grass carp if properly raised, may attain full maturity by the end of 1975 when they are 2,1/2 years old. It was observed that a few grass carp brought from Malacca in December 1968 matured in December 1971 when they were little over three years old. It is expected that a few grass carp of the present stock would attain full maturity by December 1975 or at the beginning of 1976.

To initiate a programme on induced breeding some preparatory arrangements have to be made. For large-scale production of grass carp seed and rearing of large number of fry and fingerlings it would be necessary to construct a few more ponds. For this purpose, the barest minimum would be to construct two 0.04 ha nursery ponds, two 0.1 ha rearing ponds and two 0.4 ha stocking ponds. Besides, facilities for obtaining fresh rainwater, if made available, would ensure higher success in the induced breeding programme. Action may be taken to make these ponds available before the breeding season *i.e.* December 1975. These ponds may be constructed at the Naduruloulou Research Station adjacent to the site where two 0.2 ha ponds have been proposed to be constructed. The availability of adequate water for these ponds from the adjoining stream may be ascertained before construction of the ponds.

For conducting induced breeding experiments it may be necessary to get the assistance of a grass carp breeding expert.

Assignment of a Grass Carp Expert

With the provision of proper field facilities for work and raising of an adequate stock of breeders, the services of a grass carp expert may be sought for a period of 1½ years. The expert may arrive a couple of months prior to the breeding season *i.e.* December 1975 for making preliminary arrangements for grass carp breeding. His services thus can be utilized for two breeding seasons.

His terms of reference would be to advise and assist the Government of Fiji in their Grass Carp Project for the effective control of the submerged weeds in the Rewa river system by release of grass carp with special attention to the breeding of grass carp in ponds by administration of fish pituitary hormones, hatchery and nursery management, rearing of fingerlings and raising of breeders; training of local workers and related duties.

Conclusion

In conclusion, it may be mentioned that the Fisheries Division, Ministry of Agriculture, Fisheries and Forests have rightly indicated that the problem of weed infestations in the Rewa river system could be effectively solved by grass carp, the biological control agent. Importing of large number of grass carp fry from foreign countries and releasing those directly into the river is not only a difficult task but is likely to prove very expensive and at the same time may prove ineffective since the river harbours a good population of highly predatory fishes. The only alternative to this important problem is to raise the present stock of grass carp to breeders, breed them successfully and build up a huge stock of grass carp for their systematic release in the river. This no doubt involves quite a bit of expenditure initially for construction of new ponds, but once constructed the same ponds can be suitably utilized for production of fish for food.

In future, a self-contained small fish farm complex of an economically viable 4 ha water area may be planned having six 0.4 ha, four 0.2 ha, four 0.1 ha and four 0.04 ha ponds and some space for a hatchery. By the end of 1975 eight of these ponds would be ready for the grass carp Project. The rest of the construction work may be done on phased basis covering a number of years for its completion. The blue print of the farm may be prepared now but the construction work be taken up as and when funds are available. A good, experienced engineer's service may be sought for the construction. This farm would initially be utilized for producing grass carp for control of weeds in the Rewa river system. Finally, it would form a base and a training centre for the development of aquaculture in Fiji. Polyculture of several quick growing food fishes like the Indian and Chinese carps would give high production of fish in ponds and provide valuable protein to the people of Fiji. These important riverine species when released in the rivers would enrich the riverine fish fauna of the islands. The grass carp would not only control the submerged weeds in the rivers but it would provide thousands of kg of fish by converting the weeds into fish flesh. In addition, the semi-digested fecal matters of grass carp would enrich the water and provide food for other fishes especially the bottom-dwelling ones and also to the shell-fishes thus increasing the production of mussels providing additional food to the Fijians. The expert is convinced that if the grass carp project is continued as suggested, the money that would be spent for the construction of ponds would be realized in no time once success in breeding grass carp is achieved and grass carp successfully released in the river system.

While concluding it may be worthwhile to quote the remarks made by Mrs. Hughes about grass carp in her report "Not only may it be possible to control definite weed areas quickly but a secondary source of income and food may result".

E. Fish Culture Development in Burma (Myanmar) (A UNDP/FAO Project)

Editor's Note

Prof. Choudhuri's unique success in enhancing fish production and improving various research methods in Burma brought him laurels as his work created immence protein food resources to feed millions of poor fishermen and people of Burma.

The extent of work could be ascertained from the contents presented that give various pros and cons of his work.

It is a monumental work and a voluminous work and it will be irrelevant to present in this treatise.

However, Prof. Choudhuri's achievements may be ascertained from the contents and summary of his work as follows:

FAO Report to the Government of Burma of Fish Culture Development based on the work of H. Chaudhuri, Rome, 1971 is worth persuing. (United Nations Development Programme, (Report) No. TA 2954.)

The development of fish culture in Burma from May 1967 to December 1970 is reviewed in this report. Highest priority is given to the solution of the problem of acute shortage of fish seed from cultivated species. Induced breeding of indigenous carps during 1967 by pituitary hormone injections and subsequently the commercial production of fish seed is the first successful achievement. The report shows that it is possible to meet the country's entire present requirements of seed and avoid collection of seed from natural sources.

Other important achievements of this project are demonstrations of the technique of hybridization in fishes for producing improved varieties, improvement in the technique of nursery practices for higher survival of fry, better management of farm ponds, the introduction and acclimatization of quick-growing phytophagous Chinese carps and demonstrations of increased production of fish by intensive fish culture.

The content reads as follows:

2.1.3 Experiments during 1968–69
2.1.4 Experiments during 1969–70
2.2 Experiments on hybridization of carps
2.3 Natural fish seed resources
2.3.1 Spawn collection from rivers
2.3.2 Collection of fry and fingerlings
2.3.3 Identification of carp fry and fingerlings
2.4 Nursery management
2.5 Rearing and stocking ponds
2.6 Demonstration of intensive fish culture
2.6.1 Hlawga fish farm pond
2.6.2 Private fish pond
2.7 Observations on exotic fishes
2.7.1 Chinese carp
2.7.2 Common carp
2.7.3 Other exotic fishes
2.8 Observations on fish farms and other cultivable waters
2.8.1 Government fish farms
(a) Fish farms under the Directorate of Fisheries
(b) Fish farms under the People's Pearl and Fishery Board
2.8.2 Establishment of a composite fish farm
2.8.3 Private fish farms
2.8.4 Potential cultivable waters
2.8.5 Reclamation of swamps
2.9 Present status of fish culture in Burma
2.9.1 Commercial fish culture
2.9.2 Homestead fish culture
3. Training and Research
3.1 Training of personnel
3.2 Government training schemes
3.3 Fellowships

- 3.4 Research
4. Summary of Activities
   - 4.1 Commercial production of fish seed by induced breeding
   - 4.2 Hybridization in carps
   - 4.3 Improvement in the technique of fish pond management
   - 4.4 Increased production of fish by intensive fish culture
   - 4.5 Acclimatization of Chinese carp
   - 4.6 Training of personnel
5. Recommendations
   - 5.1 Commercial production of fish seed
   - 5.2 Fish seed collection from natural sources
   - 5.3 Fish farm management
     - 5.3.1 Nursery practices
     - 5.3.2 Construction of fish farms
     - 5.3.3 Culture of miscellaneous compatible species along with major carp
     - 5.3.4 Studies on pond water and soil
   - 5.4 Culture of exotic fishes
     - 5.4.1 Chinese carp
     - 5.4.2 *Tilapia*
     - 5.4.3 Giant gourami
   - 5.5 Intensive fish culture
   - 5.6 Expansion of area for fish culture
   - 5.7 Survey and reclamation
   - 5.8 Stocking of reservoirs
   - 5.9 Training and extension

Appendix 1: Means of Identifying Characteristics of Fry and Fingerlings of Carps in Burma

Appendix 2: Analytical Data of Soil Samples from Government Fish Farms

Appendix 3: Government Fish Farms: Area Under Culture

Appendix 4: Data on Intensive Fish Culture

a) Hlawga fish pond

b) Private fish pond

Summary of Achievements

Commercial Production of Fish Seed by Induced Breeding

The work on induced breeding of carps by hypophysical injections met with full success. Large-scale production of fish seed of several cultivated species has been achieved in a number of farms, although major production was largely concentrated on the Hlawga fish farm. By this new technique, it was possible to produce spawn in excess of the present requirement and the earlier practice of collection of spawn, fry and fingerlings from rivers has come to an end.

Hybridization in Carps

The technique of hybridization of carps by artificial fertilization of eggs by stripping was demonstrated to the fishery workers so that they could learn the technique for future work in this line. Two hybrids were produced by reciprocal cross of the two important cultivated species, namely, catla and rohu. The growth and survival of these hybrids were quite satisfactory. The hybrids possess some characteristics which are superior to the parent species. Out of the two, catla-rohu (male catla x female rohu) was observed to have grown faster than rohu-catla (male rohu x female catla).

Improvement in the Technique of Fish Pond Management

Loss of fish seed through heavy mortality during the three important operational stages of rearing is very common. Attempts have been made to reduce mortality especially in the nurseries so as to assure optimum production of fish seed. Care was taken at every step of nursery preparations right from the clearance of nurseries of predatory animals, liming and fertilization of ponds, collection and assessment of natural fish food, eradication of predatory insects, judicious stocking rate of fry, supplementary feeding and finally thinning and harvesting of advanced fry. By following improved techniques and proper vigilance, the survival rate could be increased 10–15 per cent to at least 30–40 per cent in several cases.

Increased Production of Fish by Intensive Fish Culture

Although facilities for conducting experiments on intensive fish culture were limited, demonstration arranged in two fish ponds on intensive fish culture gave very encouraging results. Stocking with a number of compatible species and following a programme of pond fertilization and regular feeding with artificial food, the production obtained was very high. One of the ponds gave a total production of 10390 kg/ha/annum, whereas the other pond (0.12 ha) produced 4372.5 kg/ha/nine months. The results demonstrated that by proper management, production of fish in Burma can be increased considerably by intensive culture of compatible species.

Acclimatization of Chinese Carp

Observations made on the survival and growth of the recently introduced phytophagous Chinese grass and silver carps in Burma, have clearly demonstrated that these exotic species have already survived the critical period and have grown to adult size. The stock maintained at present at the Hlawga fish farm is quite adequate for carrying out successful induced breeding experiments by the next breeding season, when these are expected to attain full maturity. It is likely that these two exotic carps would prove to be welcome additions to the indigenous cultivated species of Burma.

Training of Personnel

Throughout his mission, the expert took special care to train as many workers as possible on various aspects of pond fish culture. It is gratifying that the training gave fruitful results as is evidenced by the successful induced breeding and large-scale production of fish seed by the trained staff. Over twenty fishery officers and field staff of the Directorate of Fisheries and PPFB have received training in scientific pisciculture and especially in the technique of breeding carps by neuro-hormonal treatment.

Recommendations

Commercial Production of Fish Seed

Commercial production of fish seed of cultivated carps by the hypophysation technique should be continued and expanded as the area under culture in the country increases. Hlawga fish farm should be converted to a fully fledged fish seed farm for the production of Indian and Chinese carp seeds only. In order to do so, some additions and alterations have to be effected in the farm. More nursery-cum-rearing ponds should be constructed for raising sufficient fingerlings. A few more stocking ponds would be necessary for rearing induced-bred fingerlings and yearlings to prospective breeders. To obtain more efficient breeding and higher survival of hatchlings, it is advisable to make some arrangements for the circulation of water. Hlawga fish farm has the advantage of having a source of water which can be tapped by gravitation flow. At present the Directorate has to pay a rent of K 1.00 per 3 785 l (1000 gal) of water to the Rangoon Corporation since Hlawga lake supplies drinking water to the city of Rangoon. With the completion of the 7 mi^2 (18 km^2) Pugyi reservoir (24 million gallon storage capacity) project, the Hlawga lake would be constantly fed from the new reservoir. Any quantity of water could then be tapped.

It is recommended that the present drain located in between the series of 'A' nursery ponds be converted into a breeding and hatching channel with flowing water fitted with inlet pipes and a sluice provided at the outlet side so that the whole channel could be drained out whenever necessary. The sluice gate should be protected effectively against entrance of weed fishes and predatory fishes. The incoming water from the paddy field could be diverted. The channel with constant flow of water would maintain lower water temperature than the ponds. In the channel, breeding hapa and hatching cloth hapa or circular hatching funnels could be fixed. In future the whole breeding and hatching operations could be carried out in tanks inside the laboratory with a continuous flow of water.

During the coming breeding season special effort should be made to obtain mature specimens of *Cirrhina mrigala*. A number of mrigal breeders at Hlawga fish farm had shown signs of maturity in the last year and it is expected that they will become fully mature during the 1971 breeding season. Once mrigal is bred, the progeny are expected to mature in ponds more easily; the stock of catla breeders should also be increased.

Common carp has been observed to be a satisfactory donor for pituitary glands. At least 1000–2000 mature common carp should be raised for the collection of glands every year. Phalan fish farm, which has been ear-marked for the culture and breeding of the Indonesian strain of common carp, can raise the required number. It is advisable to segregate the males and females and raise them in separate ponds. The segregation would prevent loss of hormone by natural spawning in the pond.

Fish Seed Collection from Natural Sources

With the success in artificial production of sufficient fish seed, the collection of spawn, fry and fingerlings has come to an end. The expert recommends that the collection of mrigal fingerlings continue until the species is bred in large quantities. Mention has already been made of the difficulty in obtaining mature specimens of mrigal for breeding. Yandoon has been observed to be a good centre having a very high percentage of mrigal and other major carp species. Establishment of a good collection centre near Yandoon would be advisable. The Rangoon Yandoon Road which is under construction would definitely help quicker transport of fish seed from Yandoon.

Fish Farm Management

Nursery Practices

Further improvement in the management of nurseries should be made at each operational stage to reduce mortality and ensure a higher survival of fry. Effective filters should be provided in inlets to prevent wild fish entering into the nursery ponds. Treatment with soap and oil emulsion should be given for the control of predatory insects in nurseries before the introduction of fry.

Construction of Fish Farms

As mentioned, it is essential that a civil engineer should be entrusted with the task of construction of fish ponds after making a topographical survey of the fish farm site and preparing a blue print for the whole fish farm. It is recommended that the Directorate of Fisheries should appoint a qualified farm engineer. He should be able to make the necessary modifications of some of the recently constructed farms and properly plan for further extension.

Culture of Miscellaneous Compatible Species along with Major Carp

Possibilities for the culture of the following fishes, along with major carp, should be examined when research facilities are developed:

1. *Osteobrama belangeri*: It forms the bulk of catches in upper andcentral Burma and grows 3–4 kg.
2. *Pangasius pangasius* and *Ompok bimaculatus*: These catfishes are predatory on small fishes and would control minnows in stocking ponds.
3. *Macrobrachium rosenbergii*: Culture and breeding of the giant freshwater prawn may be tried.
4. *Clarius magur, Heteropneustis fossilis* and *Anabas testudineus*: These live fishes are popular food fishes in Burma.
5. *Notopterus chitala*: This species of rapid growth, although believed to be indigenous in Burma, is very rare. It grows to 1.2 m (4 ft), and would control small fishes when stocked with major carps.

Studies on Pond Water and Soil

Information on physico-chemical conditions of water and pond soil is absolutely essential for the proper management of fish farms. The appointment of a hydrologist and a soil chemist in the Fisheries Departments will be necessary to enable such studies.

Culture of Exotic Fishes

Chinese Carp

It is recommended that due attention be given to rear the Chinese carp stocked at Hlawga and Nyaungshwe fish farms. Induced breeding of Chinese grass and silver carps should be taken up as soon as these attain full maturity. Both silver carp and grass carp should be stocked with major carps in composite fish culture. When sufficient stock of grass carp is produced, a programme for the systematic stocking of grass carp in Inle Lake for the control of noxious aquatic weeds and development of fisheries in the lake should be formulated by the Directorate of Fisheries.

Tilapia

Further steps should be taken to extend tilapia culture into the new area of upper and central Burma and a few more tilapia seed distribution centres be opened in suitable places. In deltaic areas in lower Burma, attempts should be made to culture tilapia with piscivorous species such as *Ophicephalus striatus, O. marulius, Lates calcarifer* and *Silonia silondia.*

Giant Gourami

Large-scale breeding of gourami should be taken up at Phalan fish farm. Facilities for breeding and culture of gourami should be provided and a trained technical staff should concentrate on this work.

Intensive Fish Culture

The demonstration on intensive fish culture taken up at the private fish pond near Mingaladon Airport should be continued until the end of August 1971, according to the instructions given by the expert to the full time counterpart assistant. Complete data on management, thinning and final harvest should be collected for the final evaluation of the results.

Repetitions of the experiment, as demonstrated in the said farm, should be made in a few other ponds for confirmation of the results. The production obtained in one year is as high as 10 390 kg/ha and almost all the species stocked have reached about 1.0 kg in size. Since the major carps of comparable size are the prospective foreign exchange earnings in Burma, the observations should be repeated in other fish ponds so as to achieve higher production of marketable fish.

Expansion of Area for Fish Culture

As stated earlier, the initial difficulties of shortage of stocking materials and complete lack of knowledge in fish culture have been overcome. The industry can now develop very rapidly, but the main difficulty still remaining is the availability of sufficient suitable land for fish culture. It is encouraging that the Government has given some consideration to this important problem recently. It is felt however that further relaxation may be necessary for acquisition of suitable land for fish culture. Since cultured fish is being exported and is a foreign exchange earner in addition to its role in human nutrition the fish farmers should be permitted to construct ponds for fish culture more liberally in lands also suitable for paddy culture. This would provide a big incentive towards the development of fish culture in Burma.

Survey and Reclamation

A further survey of potentially cultivable waters should be made and reclamation of a water area suitable for pond fish culture should be taken up. A drive may be given initially to reclaim the swampy areas and weed choked moats in the heart of the townships of Pegu, Toungoo and Mandalay. The cleared water areas should then be stocked with suitable species including grass carp for control of weeds. In addition to supplying fish to the public, these reclaimed areas would eliminate the unhygienic spots and add to the beauty of the towns. Royal Lake and Inya Lake in the City of Rangoon should be stocked with fish for fish production and also for public sport facilities (*i.e.* angling).

Stocking of Reservoirs

With the surplus of stocking materials the Directorate of Fisheries was contemplating stocking of reservoirs. Many of the large and small reservoirs and others to be impounded may not have a good stock of economic and commercial species of fish. Therefore, before finalizing any programme of stocking, it is recommended that a scientific assessment should be made of the composition and extent of natural fish population and their efficiency value in those reservoirs.

Training and Extension

In view of the shortage of trained scientific personnel, more staff should be sent abroad for training after ascertaining the immediate need and giving priority for such training.

Two technical staff with a background of chemistry should be recruited and granted fellowships for training in suitable foreign institutes for specialized training of six months or a year in water and soil chemistry. A few more staff should be sent to a foreign institute to undergo training in Inland Fisheries and fish farm management.

Since the Government intends to promote research on fish culture and has established a fish culture research laboratory at Hlawga Fish Culture Station, it is recommended that qualified staff should be sent abroad to gain experience in the planning and implementation of programmes on research, and to develop specific and technical knowledge, and concepts of fisheries research, so that the expert would be in a position to plan, implement and guide research work in the newly established research laboratory.

With the expansion of fish farming, there will be the necessity for the implementation of extension services. Its importance in playing an important role in the development of fish culture cannot be overemphasized. The extension workers should receive proper training on various aspects of fish farming and periodic refresher courses should be arranged for the extension staff to acquaint them with the latest technique and findings. More demonstration centres should be established in suitable areas. In fact a model farm-cum-demonstration centre should be established in every state.

Appendix 1: Means of Identifying Characteristics of Fry and Fingerlings of Carps in Burma

Catla catla (Nga-thaing-gaung-bwa)

The head is broad and moderately large; no barbels present; the mouth is anterior, lips slightly thick but not fringed. In fry stage (12–35 mm size) the opercular region has a bright reddish tinge due to the red gills seen through the transparent opercular flaps. The dorsal profile of the anterior portion of body is convex and the ventral profile is concave. The first ray and the dorsal edge of the dorsal fin have a blackish tinge. There is a definite circular black spot present at the caudal peduncle which persists until the fry grows to about 150–160 mm in size. In the Indian strain of catla, no such black spot is present. The colour of the fins is either grey or blackish.

Labeo rohita (Nga-myit-chin)

Rohu fry has a stream-lined body, lips are fringed, one pair of barbells (maxillary) is prominent even from 15–20 mm size and is whitish-grey in colour. The caudal peduncle has a dark spot which connects both the edges giving the appearance of a band. In some fry the spot is more circular. The dorsal surface or the back of the fingerlings is slightly greenish. The ventral and anal fins have a reddish tinge and both the lobes of the caudal fin have a reddish tinge with blackish grey marginal areas.

Labeo calbasu (Nga-net-pya)

It is easier to identify the fry and fingerlings of *L. calbasu*. At the very early stage of 7–9 mm size the fry develops an alternate white and dark band on the back. The fins become black at the base and usually the black colour spreads all over as the fry grows. A semi-lunar yellow band appears at the nape and the fry at 15–25 mm size can easily be identified by the alternate yellow and black band along the body. The barbels, maxillary and rostral, appear at an early stage and are prominent and black in colour. The mouth is narrow and the lips are fringed. As the fry grows, the body and fins become blacker and finally at the fingerling stage *L. calbasu* becomes completely blackish-grey.

Labeo gonius (Nga-dane)

The fry of *L. gonius* has a smaller head, narrow mouth, lips are fringed and possesses one pair of maxillary barbels. The rostral barbels appear only in the fingerling stage. An indistinct spot is present at the caudal peduncle. The body has a number of faint white stripes which become more prominent in fingerlings when the black spots in the scales give rise to alternate dark and white longitudinal stripes along the body. The scales are very small. The colour is light green at the back and dull white ventrally.

Labeo nandina (Nga-ohn-done)

Labeo nandina fry has a small ventrally placed narrow mouth with fringed lips and two pairs of prominent barbels which are grey in colour. Like *L. gonius*, the fry develops dark longitudinal stripes. It can be easily distinguished from other Labeos by the elongated dorsal fin. The fry possesses one dark oval spot at the caudal peduncle which becomes fainter in fingerlings. The fringed lower lip, however, develops prominent villi-like protuberances at the fingerling stage.

Labeo stoliczkae (Nga-thaing)

The fry of *L. stoliczkae* has a larger head as compared to other *Labeo* species. There are barbels, the mouth is wide and the lips are very slightly fringed. The caudal peduncle has a dark spot which becomes very faint in the fingerlings. The most important distinguishing characteristic of *L. stoliczkae* is the prominent black band in the body bordering the posterior edge of the operculum. This characteristic persists even in the adult fish.

Labeo pangusia (Nga-loo-me-netha)

The fry and fingerlings of the minor carp *L. pangusia* resemble those of *L. rohita* very much. The fry has a narrow, crescentic mouth with fringed lips and possesses a pair of barbels. The spot at the caudal peduncle also resembles rohu to a certain extent but it persists even in the adult though somewhat indistinct. The colour of the back of the body is not so greenish as in rohu. The main point of distinction between this minor carp and rohu is that the mouth of the former is more ventral, hanging and more crescentic than the latter. Besides, the isthmus of *L. pangusia* meets the lower lip converging to a point and forms a depression on both sides at that point. In the case of *L. rohita* advanced fry and fingerlings, the isthmus instead of converging meets the lower lip at a broad base and no cavities are formed on both sides.

Labeo angra (Nga-loo)

The fry of *L. angra* is pale yellowish in body colour. The mouth is ventral, narrow and the lips are fringed. A pair of barbels are present and the fry is pale whitish in colour. A conspicuous dark band is present at the nape. At the caudal peduncle there is a distinct oval shaped spot. Two other spots are present at the body below the dorsal and anal fins. The caudal spot persists even in the adult. In Indian specimens, *L. angra* fingerlings develop a conspicuous lateral band which is very indistinct in the Burmese specimen. The band also does not persist in the adult in the Burmese specimen as is observed in Indian specimens.

The fry of *L. angra* resembles to a certain extent the fry of *L. calbasu*. The main points of difference between the two are that the former has a yellowish body colour, a conspicuous dark caudal spot and dark pigment is absent in fins. The barbels are dull white in colour in contrast to the black barbels of *L. calbasu*. The dark band at the nape is common for both the species but in *L. calbasu* the band lies in between two yellow bands.

Labeo boga (Kyouk-nga-loo)

The fry of *L. boga* has a terminal mouth and unlike other *Labeo* spp. the thin lips are without any fringes. A black spot is present at the caudal peduncle near the base of the caudal fin, which becomes more conspicuous as the fry grows to fingerling stage. The spot is triangular in shape with its broad base parallel to the base of the caudal fin. It becomes more or less circular in the fingerlings. At 25–30 mm size a pair of barbels is visible. The fry at this stage also develops a spot on either side of the body on the 5th and 6th scales in one or two rows above the lateral line. The spots continue up to 175–180 mm size. At this size the caudal spot disappears. *L. bata* fry and fingerling have similar characteristics.

Cirrhina mrigale (Nga-gyin)

The fry of mrigal has a slender elongated body. The head is narrow and the mouth is wide and terminal. The lips are thin without any fringes. A black spot appears on the caudal peduncle at a very early stage. The spot finally takes the shape of a diamond at the 30–35 mm stage. At this stage a pair of grey barbels is also visible. The fingerlings can more easily be identified by the reddish tinge at the tip of the lower lobe of caudal fin.

Mrigal fry can be distinguished easily from rohu and other major carps but it is rather difficult to distinguish it from those of *Labeo boga* or *L. bata*. The shape of the caudal spot is the most important character by which the two species could be identified. With the appearance of a spot on the lateral line in *L. boga*, and a reddish tinge at the tip of the lower lobe of caudal fin in mrigal they can be easily identified.

Osteochilus chalybeatus (Nga-leh)

The early fry of *O. chalybeatus* could not be studied as specimens were not available. The fingerlings more or less resemble *L. calbasu* because of their uniform dark grey or blackish body and fin colour. Lips are fringed like most of the *Labeo* species. Two pairs of barbells are present which are dark grey in colour. The fingerlings can be distinguished from *L. calbasu* by the elongated dorsal fin.

Barbus sarana (Nga-khon-ma)

The fry of *B. sarana* is characterized by the presence of a black vertical band originating from the base of the anterior position of the dorsal fin and extending to near the lateral line. It has also a conspicuous oval spot on the caudal peduncle. The body is deep and compressed. The mouth is anteriorly placed and the lips are thin and without fringes. Prominent maxillary barbells are present. A pair of rostral barbels-also appears at about 25 mm stage.

Osteobrama belangeri (Nga-hpe-oung)

The fry of *O. belangeri* has a narrow pointed head and a wide and compressed body. The abdomen is trenchant throughout. No barbels are present. There is a conspicuous rectangular spot at the caudal peduncle which disappears in the adult. The spot sometimes extends up to the edges of the caudal peduncle and appears like a band.

In the foregoing a few important characteristics of common fry and fingerlings of Burmese carps have been mentioned. No attempt has been made to describe the salient features of the fry and fingerling at various stages in detail. These are only a few characteristics by means of which each species can be identified in the field so that the economic varieties can be distinguished from the non-economic ones. In doubtful cases, if there is difficulty in distinguishing between *L. rohita* and *L. pangusia* or between *C. mrigala* and *L. boga*, it is advisable to count the number of divided dorsal fin rays. In minor carps the number of divided fin rays in dorsal fins are either eleven or less whereas in major and medium sized carps the number is always more than eleven.

[Source: An abstract from,
Prof. Chaudhuri's work in Burma. (Myanmar)]

Prof. Chaudhuri's Work at Lao P.D.R.

Lao was a communist-ruled and a land-locked country. At that time Lao P.D.R. was one of the poorest countries of the world. The fisherman community did not have modern equipment for fish harvesting although Fish and Rice were their staple food. On the other hand the Thai people used advanced type of net and they caught those fish. So there was only a little amount of fish left for the people of Laos. There were lots of problems for obtaining fish protein for the people of Laos. Firstly, there were no factory for fertilizers or producing fish food. There was shortage of foreign currency so they could not think about importing food from outside. Yet the people needed fish protein very badly. During such days of scarcity Prof. Chaudhuri started his work. Dollar for the project was with the accounts bank, even small instruments had to be brought from Thailand. It was also difficult for him to obtain permission of the Govt. for various petty matters. Hence he was in a dilemma as how he would be able to carry on his job within such limitations.

Prof. Chaudhuri had information that in various communes in China they had a system called "Integrated Fish Crop Livestock, Farming System". He then decided to use such a system in Laos. He also thought to juxtapose the Indian system of Rural Development Programme in order to produce fish at a low cost.

Lao PDR

Rehabilitation of Fish Seed Farm and Fish Farm development in Lao PDR was a Herculean job. Broadly his job was:

- ✰ To increase income and improve food intake of the population by fish culture development.
- ✰ Production of quality fish seed for distribution to the farmers.
- ✰ Development and demonstration of integrated fish/crop/live stock production system.
- ✰ Training of aquaculture technicians, planners and extension workers in improved techniques of fish seed and food fish production.

(This long term assignment from 1970 to 1984 on the Fishery Project was acclaimed as the best successful project of Lao PDR.).

Chapter 13

Some More Notable Works Done by the Professor: Establishment of a Fish Pituitary Bank

The project report for establishment of a Fish Pituitary Bank is a unique contribution of Professor Hiralal Chaudhuri which was submitted as the Project leader in 1993.

Prof. Chaudhuri in his letter dt. 9.4.1993 to Dr. C.P.Madamba, Head Aquatic and Invertebrate Zoology Laboratory at IBS, CAS, UPLB, College Laguna explained that adequate thrust could not be given to the Project due to:

Fund Constraint

Purchase of donar fish could not be made in time. Several essential equipments needed for gland extract could not be purchased which are hamogenizer, centrifuge etc. Surgical instruments for dissection and collection of pituitary glands were not upto satisfaction for which loss of hormones or potency of glands were affected. Pure (analar) acetone was not available for gland preservation. Ampoules of Pituitary gland of 1 ml. was also not available in the market, yet the report was extremely useful for giving a thrust to the effort which was handled later by other aquaculturist as the Professor resigned his post and came back to India.

The Professor drew a background history of the common carps (*Carprinus carpio*), one of the best donar fish as a source of pituitary gland.

Specific Objectives were:

- ✩ To train workers in the Techniques of collection of pituitary gland from nature fishes.
- ✩ To demonstrate the methods for the pituitary glands for longer duration including the method of ampouling the glycerine extract of the pituitary hormone.
- ✩ To demonstrate the methods of preserving the pituitary gland for longer duration including the method of ampouling the glycerine extract of the pituitary hormones. For brevity, the complicated and elaborate methodology adoped by the professor is not presented here.

On the Life History and Bionomics of the Carp Minnows, *Chela phulo* (Ham)

Chela phulo is one of the smaller species of the genus and its distribution according to Day *(op. cit.)* is from Assam, through eastern and central India, to the Thungabhadra and Kistna basins. It is commonly found in ponds at Cuttack. This and other carp minnows generally abound in carp nurseries and stocking ponds. During an investigation of the causes of heavy mortality of carp fry in nursery ponds at Cuttack, attempts were made to elucidate the role of these 'weed' fishes in the economy of pond life. In the present paper the observations made on the breeding, fecundity, development, food and feeding habits of C. *phulo*, are detailed.

The fertilized eggs collected from the pond were identified as those of C. *phulo* by extensive comparison of the range and average size of the ovarian eggs of all the species of *Chela* commonly occurring in the locality, *viz.*, C. *phulo*, C. *bacaila* and C. *gora*. Besides, at the time of collection of the eggs, C. *phulo* predominated in the pond to the almost complete exclusion of other species of the genus. As the very few specimens of C. *bacaila* present in the pond were then immature, they could not have been breeding when C.*phula* was spawning. The characters of the advanced post-larvae rule out the possibility of any doubt in their identity.

Discussion

Like the majority of carp minnows, C. *phulo* breeds in stagnant waters, with a peak period of breeding just before the monsoon rains.

The allied species, *C. argentea* and *C. untrahi* are reported to breed in running waters during the monsoons (Chacko *et al., op, cit.*). The sexual dimorphism in the species, referred to above, occurs in the breeding season in.*C. argentea* and *C. untrahi* also, the female exhibiting a patch of yellow or orange on the ventral portion of the body (Chacko *et al.*). Such pigmentation on the body proper is absent in *C. phulo* in which the males are more brightly coloured than the females. The ovarian eggs of *C. phulo* are less than 0.5 mm. in diameter, comparable in size to those of the larger species *C. gora* (0.65 mm.) and *C. bacaila* (0.6 mm.) (Alikunhi and Chaudhuri, unpublished). The ripe ovarian eggs of *C. untrahi* (diameter 0.75 mm. to 0.95 mm., Chacko, *op. cit.*) and of *C. argentea* (1.20 mm. to 1.50 mm. in diameter– Chacko *et al., op. cit.*) are much bigger than the above. The size of eggs of *C. argentea* is unusually large for such a small-size fish and is almost as big as that of the ovarian eggs of the largest Indian carp, *Catla catla*. In the light of our observations on the eggs and fecundity of *C. phulo, C. bacaila* and *C. gora,* and in view of the similarity in the habits of the different species of the genus, this extraordinary size of the eggs in *C. untrahi* and *C. argentea,* if proved constant and a correct observation, would have to be correlated with the specialized environmental conditions, if any.

The fully swollen fertilized egg in *C. phulo* has a much less spacious perivitelline area than in the major Indian carps. No details of the embryonic and larval development of *C. argentea* and *C. untrahi* are available, but the cleavage and differentiation of the embryo observed in *C. phulo* show that the processes are rather rapid. The period of incubation in *C. phulo* is shorter by atleast 8 hours than in its congeners, *C. argentea* and *C. untrahi.* The hatchling is very small, without a mouth, gill slits or pectoral fins, and the yolk is fully absorbed 1½ days after hatching. According to Chacko *et al. (op. cit.)* the hatchling of *C. argentea* has 4 gill slits at hatching, the mouth and eyes well formed, the yolk fully absorbed and is 8 mm. long in the next 4 hours. Similar conditions seem to prevail in *C. untrahi* also (Chacko, *op. cit.*). As compared to *C. phulo,* the rate of differentiation of structures in the above allied species is unusually rapid. The somewhat longer period of incubation of the egg (20 hours) in these latter may lead to a more advanced stage of differentiation of the embryo at hatching, but even so, it is doubtful whether the yolk would be completely absorbed within 4 hours after hatching.

C. phulo subsists largely on zooplankton, though often feeds on filamentous algae and aquatic insects if zooplankton is scarce, but not on planktonic algae as *C. argentea* does, in which the preference is appreciably less for micro-crustaceans (Cladocerans and Copepods). These comparisons can only give a very general picture of the food habits, when it is seen that the food consumed largely depends on the dominance of items of food present in a given environment.

As *C. phulo* draws largely on the zooplankton resources of the pond, it may be deemed an active competitor to the young fry of major carps which are almost entirely dependent on zooplankton (Alikunhi, 1952). The very careful eradication of this species from carp nurseries thus becomes a necessity to safeguard survival and growth of carp fry. In rearing tanks also the food habits of *C. phulo* will be incompatible, particularly with those of *Catla, catla,* which, even after the early stages, continues to feed largely on zooplankton.

Summary

The common carp minnow, *Chela phula* Ham., breeds in ponds from February to September. Intensive breeding takes place during May-June.

During the breeding season the mature males are easily distinguished from the females by their conspicuous yellow-tinged ventral and anal fins, and by other differences in pigmentation.

The fecundity of the species is relatively much higher than that of the major carps.

Eggs are generally laid in the morning, in the marginal shallows of ponds. The fully swollen fertilized egg is less than 0.75 mm. in diameter.

With the temperature of water ranging from 29.3 to 33.3°C., the early cleavages are completed and the morula stage reached within 1½ hours after fertilization, and the embryo hatches out within 12 hours.

The hatchling, just over 2 mm. long, is extremely frail, small and rather undeveloped, with the mouth and pectoral fins wanting. Absorption of yolk is completed 36 hours after hatching when probably the larva begins feeding. All except the ventral fins are differentiated when the post-larva is 13.0 mm. long. Ventrals are also fully formed in specimens 16.0 to 17.0 mm. long.

C. phulo is essentially a zooplankton feeder, but takes in appreciable quantities of filamentous algae and aquatic insects also. No significant changes in the feeding habits have been observed in the different stages of its life. Food consumed, however, largely depends on the items of food available in the environment.

The feeding habits of *C. phulo* are incompatible with those of the early fry of the major Indian carps.

References

Alikunhi, K. H. "On the food of young carp fry,"*J.Zool.*Soc.India, 1952, 4 (1), 77–84.

Alikunhi, K. H. and Chaudhuri, H... "On the role of 'weed' fishes in the economy of carp nurseries and stocking ponds" (Unpublished).

Chacko, P. I., Venkataram, R. S. and Ranganathan, V. "On the bionomics of the silver fish, *Chela argentea* Day," *Curr. Sci.*, 1946, 15, 167.

Chacko, P. I. "Bionomics of the silver fish *Chela untrahi* (Day)," *Proc.38th Ind. Sci. Congr.*, Kolkata, Part 3, Abstracts, p. 226, 1951.

Day, F. *Fishes, Fauna of British India, London,* 1, 1889.

Khan, M. Hamid. "Development of fishes in the Punjab–III, Culture and propagation of indigenous species of fish," *Ind. Farm.*, 1947, 8 (9), 447–53.

The Role of Plankton in the Survival and Growth of Carp Fry in Nursery Pond

As in China, the collection of millions of early fry of major carps from rivers and stocking them in nursery ponds for rearing is a common practice in the fish cultural methods followed in India, particularly in Bengal, Bihar and Orissa. There is, however, very little reliable information about these practices, most of which are known only to a few professional fish farmers who keep the information as family secrets, handed down from generation to generation. The results of rearing are, however, generally poor, and even in ponds managed by expert fish farmers not only are the rates of survival of fry often low but even reports of total failure of the crop are not uncommon. Exceedingly high rates of mortality of fry in nursery ponds (9° to 98.7 per cent) were reported from Orissa State and a detailed investigation on this important problem was undertaken at Cuttack in 1949 with a view to ascertain the probable causes thereof and to find out methods by which the percentage of survival of the fry stocked could be enhanced for increased fish production. Some of the experiments carried out and observations made during 1949–52 are summarized in this communication.

Discussion

Successful rearing of the tender baby carps that are released in nursery ponds is one of the fundamental problems of carp culture in India. Extreme care should, therefore, be taken at all stages of the operation, right from the collection of fry, through conditioning and transport, to preparation of nurseries and stocking and rearing the fry in them.

Elimination of predators other than insects is not very difficult but the control of aquatic insects has become the most important problem and until this is satisfactorily achieved the survival of fry is bound to be considerably low. Occasional high survival is to a large extent due to the coincidence of minimum insect population with rich plankton at the time of stocking. Netting the pond prior to stocking in order to remove predatory insects is not adequate to ensure satisfactory results.

Zooplankters are preferred by the tender fry and are easily digested. Plankton algae are not only not utilized as food but they are also difficult to digest at that early stage. Besides, available information on the chemical composition of certain phyto-and zoo-plankton organisms (Welch, 1935) also tends to show that the nutrient value of phytoplankters is generally less than that of zooplankters. The fact that all the ponds which have given high rate of survival of fry had an abundance of zooplankton at the time of stocking and a minimum of phytoplankton during the period of rearing, partly substantiates the laboratory findings, though complete confirmation cannot be expected in view of the depredations caused by the aquatic insects.

In spite of systematic manuring, production of a swarm of zooplankters in the pond at the desired time is still a matter of chance only. The indication that application of heavy doses of cowdung in cisterns usually results in the early production of zooplankton swarms, might well be worth trying in natural nursery ponds.

Though only a few of the important factors causing mortality of fry were eliminated in the field experiments in 1950, on an average 46 per cent of the fry stocked survived in the nursery ponds, indicating approximately 500 per cent increase over the previous year's average.

References

Alikunhi, K.H. (1952). On the food of Young Carp fry. *J. Zool. Soc. India*. IV (r): 77–84.

Alikunhi, K.H. *et al.* (1951). Mortality of carp fry under supersaturation of oxygen in water. *Proc. Nat. Inst. Sci. India* 12(4): 261–264.

Welch, Paul S. (1935). *Limnology*. McGraw Hill Book Co., New York.

(It is a joint paper, Alikunki, Chaudhuri and Ramachandran).

Experiments on the Effects of Organochlor Insecticides Aldrin, Dieldrin and Endrin on Fish and Other Pond Organisms

Editor's Note

Complete kills of miscellaneous fishes, prawns and insects (back-swimmers) were obtained in the laboratory at a concentration of 0.3, 0.01 and 0.001 ppm of aldrin, dieldrin and endrin respectively. Zooplankters were not affected at any of these doses. A few zooplankters were found alive even at 0.5 ppm of endrin. In the field-conditions, endrin killed all fishes and common pond organisms at 0.01 ppm excepting molluscs and zooplankters. It could be used economically in cleaning ponds of miscellaneous fishes and other unwanted organisms but its regular use is not recommended because of the prolonged high toxicity, probable harmful cumulative effect on pond productivity and the risk involved in improper handling and use of the chemical.

Introduction

The present study was undertaken in 1956 to obtain firsthand knowledge on the degree of toxicity of the organochlor insecticides on fish and other pond organisms. The study was also aimed at determining their suitability and economics for clearing nursery ponds of miscellaneous predatory and weed fishes and other harmful organisms like predatory insects, tadpoles, prawns, crabs, etc., the presence of which is highly undesirable in nursery ponds (Alikunhi *et al.*, 1955). The details of the experiments conducted and results obtained are summarized in this paper.

Discussion

The results of the experiments indicate that the organochlor insecticides are highly toxic to fish, prawns and insects. Even lower doses of the chemicals affect majority of the pond organisms and the toxicity lasts for a long time at slightly higher doses. Among the three insecticides, endrin has been found to be the most effective and aldrin the least. When compared with other more common insecticides, endrin and dieldrin seem to be much more effective. Experiments on bluegill and bass showed that the test fishes were completely killed at 0.05 ppm of toxaphene, 0.15 ppm of DDT, 0.2 ppm of Methoxychlor, 0.2 ppm of Chlordane, but BSC at a concentration of 0.1 ppm could kill only 50 per cent of these fishes. In earthen ponds, Toxaphene could kill bluegill and bass at a high concentration of 0.2 ppm only and the toxicity lasted for more than 10 weeks (Lawrence, 1950). According to Surber (1948), the effective dose for eradication of fish population by Toxaphene is 0.25 ppm. While Chlordane was more toxic to fish, an application of 0.45 ppm of gammaisomer of Benzene hexachloride failed to kill bluegill population. Parathion (organic phosphate insecticide) at.063 ppm was observed by him to kill a few species of fish. Hemphill (1954) got good results in eradicating rough fish from a lake by applying 0.1 ppm of Toxaphene. Tarzwell and Henderson (1957) determined the median tolerance limit of fathead minnow (*Pimephales promelas*) at 96 hours time interval as 0.023 to 0.036 ppm of dieldrin. Dieldrin was found by them to be more toxic in hard than in soft water. Harrington and Bildingmayer (1958) observed complete kill of fish and crustacean population at a concentration of 0.13–0.4 ppm. A lower dose, however, was not tried. Molluscs were not affected at that concentration. Preliminary experiments carried out by Hogan and Gray (1950) to determine the toxicity of aldrin and dieldrin using goldfish and bluegill as test animals indicated that aldrin is more powerful with a killing threshold of 1 part in 4000 million when dieldrin killed at 1 part in 32 million of water. Iyatomi *et al.* (1950) while studying the effects of endrin on fish observed that the toxicity of endrin to fish increases with the increase of temperature in water and that eggs and larvae of fish are more resistant to endrin than adults. The median tolerance limit of carp and goldfish determined by them was 0.005 and 0.002 respectively at an interval of 48 hours and temperature range of 27–28°C.

The relative efficacy of the organochlor insecticides can be made out from the observations of several workers referred to in the preceding paragraph. Endrin is undoubtedly the most toxic chemical that has been tried on fish so far. The observations made by the author on the toxic affects of aldrin on fish differ from those of Hogan and Gray (1950). According to them aldrin is more effective than dieldrin and kills fish at an extremely low dose. On the contrary, the present author found dieldrin more toxic than aldrin which needs a comparatively higher concentration to kill fish and other organisms.

Endrin was observed to be the most suitable of the chemicals tried for the eradication of miscellaneous fish and other predatory organisms from ponds when a complete kill is necessary for cleaning a pond. The use of endrin is much more easier and cheaper than draining the ponds. Fish food organisms, especially zooplankters, are not killed at a dose which kills fishes and most of the other pond organisms. The toxicity of water at the lowest dose effective on fishes is lost within two days. In the field, however, when 0.01 ppm of endrin is applied, toxicity continues even up to a month or longer. There is, however, a great risk in the use of endrin in nursery ponds because of the residual toxicity of the chemical, if given in overdose or applied repeatedly.

Role of Chinese Grass Carp *Ctenopharyngodon idella* (Val.) in Biological Control of Noxious Aquatic Weeds in India: A Review*

Introduction

In India, according to recent estimates (Mitra, 1964; Philipose, 1968), about 320,000 ha have to be cleared of weeds annually for fish culture out of nearly 800, 000 ha of the available cultivable waters. The problem is severe in the eastern states where 40–70 per cent of water area is infested by weeds (Philipose *et al.*, 1962). Weed infestations send to convert valuable low lying inland waters into swamps. The adverse effects of excess vegetation in fishery waters are summarized by Ling (1960).

The Central Inland Fisheries Research Institute, besides other state and private organizations, has been engaged in the development of methods of weed control suitable in Indian conditions. A major step taken by the Institute in this direction was the introduction of the Chinese grass carp, which has gained popularity throughout the world as an effective biological agent for aquatic weed control.

Introduction of Grass Carp into India

Experimental consignments of grass carp from Hongkong were introduced at its Pond. Culture Sub-station, Cuttack, during December, 1959 and January, 1962 with the object of assessing its response to the new environment. The cultural qualities of grass carp, vis-a-vis indigenous carps have been carefully studied and the species has been found to be invaluable addition to indigenous culturable species (Alikunhi and Sukumaran, 1964).

The food selection depends, among other factors, on the size of the fish and its capacity to consume. According to Singh *et al.* (1967) grass carp has almost equal preference for floating weeds such as common duck weeds, *Wolffia, Lemna, Spirodela* and *Azolla. Eichhornia* and *Pistia* were not consumed even when pieces were fed: however, Ling (1960, 1967) reported the use of *Salvinia, Pistia* and young leaves of *Eichhornia* as feed for grass carp. Control of *Eichhornia* under experimental conditions has been observed by Avault Jr. (1965).

A tentative order of preference of the fish is *Hydrilla, Najas, Ceratophyllum, Ottelia, Nechamandra* and *Vallisneria* (Singh *et al.*, 1967). *Potamogeton* and *Halophyla* are also consumed by the fish. Grass carp of about 0.5 kg size and above consumes *Nechamandra*. Though the consumption of all these weeds by grass carp has been reported by other authors, opinions differ regarding the order of preference.

Among the common emergent weeds, *Trapa, Myriophyllum, Limnophila* and leaves of *Monochoria* and *Ipomoea* are also preferred by grass carp, almost in the same order. *Nymphaea, Nelumbo* and *Nymphoides* are not, however, consumed by the fish. It has often been observed that in ponds where marginal weeds like *Marsilea quadrifolia*, *Ceratopteris thalichoides*, *Panicum fluitans*, *Cyperus* sp., *Scirpus* sp., *Jussiaea repens* etc., were present, they were cut and tender parts eaten by grass carp keeping the pond margins trimmed.

Performance of Grass Carp in Composite Culture vis-a-vis Utilization of Weeds for Fish Production

Grass carp is extensively used as a compatible species along with indigenous fast growing species not only in China but also in many South East Asian countries. (Lin, 1954; Hickling, 1962; Hora and Pillay, 1962).

The advantage in utilizing grass carp as compatible species in composite culture is that it occupies an ecological niche which otherwise remains unfilled by indigenous carps. The herbivorous feeding habit of this species brings in the problem of feeding the fish with desired aquatic weeds from external sources if they are not naturally available in the pond to meet the food requirement of the stock. The weeds are thus converted into more useful fish flesh by the species.

Due to the low assimilation of food, the activity of grass carp results in a quick passage of digested and semi-digested plant mass into the medium thereby manuring the ponds. This accelerates the growth of phytoplankton.

The role of grass carp in enhancing fish production from fish ponds in India is encouraging. Lakshmanan *et al.* (1971) observed that grass carp representing 7.5 per cent to 17 per cent of the initial total stock density have attained an average weight of 501 to 1307 g in one year. The total weight of aquatic weeds supplied ranged from 4375–14662 kg/ha/yr. Singh *et al.* (1972) report that grass carp representing 16.67 per cent to 22.22 per cent of the initial stock density have attained an average 639 to 1250 g in one year. The quantity of aquatic weeds supplied ranged from 10,070 kg to 17,800 kg/ha/yr.

Summary

The Chinese grass carp introduced in India, in recent years, is gaining popularity not only as an important species in composite fish farming but also as an efficient biological agent for controlling weeds. In water supply tanks of heavy industries where the use of chemical weedicides is prohibited, eradication of submerged weeds by grass carps is the most effective and cheap control method.

The efficacy of grass carp in controlling aquatic weeds depends on several factors such as water temperature, species composition of weeds and the size and stock density of grass carp.

The technique of hypophysation adopted for inducing grass carp to spawn is almost the same as that adopted for indigenous carps. Rearing of fry and fingerlings and maintenance of broodstock of the species are also easy. The progress of large-scale production and utilization of this species for biological weed control on an intensive scale depends mainly on the concerted efforts by the various state fisheries departments.

References

D. S. (1963). Problems of the fisheries exploitation of plant eating fishes in the Water Bodies of the U.S.S.R. Experience in the use of white amur in the struggle against overgrowth of water bodies. Trudy Inst. Zool. Parazit, Ashkhabad, 6 p.

Alikunhi, K.H. and Sukumaran, K.K. (1964). Preliminary observations on Chinese carps in India, Proc. Indian Acad. Sci. (B), 60 (3): 171–188.

Alikunhi, K.H. and Sukumaran, K.K. (1965). Observations on growth maturity and breeding of induced–bred pond–reared silver carp, *Hypophthalmichthys molitrix* and grass carp, *Ctenopharyngodon idellus* in India during July, 1962 to August, 1963. Bull. Cent. Inst. Fish. Educ. Mumbai, (2): 19.

Avault, J.W. Jr. (1965). Preliminary studies with grass carp for aquatic weed control. Prog. Fish–Cult., 27(4); 207–209.

Blackburn, R.D., Sutton, D.L. and Taylor, T. (1971). Biological control of aquatic weeds. Journal of the Irrigation and Drainage Division. ASCE 97: (IRS). Proc. paper 8362, 421–432.

Chaudhuri, H., Singh, S.B. and Sukumaran, K.K. (1966). Experiments on large–scale production of fish seed of the Chinese grass carp, *Ctenopharyngodon idellus* (C and W) and the silver carp, *Hypophthalmichthys molitrix* (C and V) by induced breeding in ponds in India. Proc. Indian Acad. Sci., 63 (B) (2): 80–95.

Chaudhuri, H. *et al.* (1967). Note on natural spawning of grass carp and silver carp in induced breeding experiments. Sci. Cult. 33(2): 493–494.

Singh, S.B. *et al.* (1967). Observations on the efficacy of grass carp in controlling and utilizing aquatic weeds in India. Proc. IPFC, 12(2): 220–34 (1966).

Singh, S.B. *et al.* (1972). Observations on composite culture of exotic carps. J. Inland Fish. Soc. India, 4: 38–50.

Stevensen, J.H. (1965). Observations on grass carp in Arkansas. Prog. Fish Cult., 27(4): 203–206.

Stott, B. and Orr, L.D. (1970). Estimating the amount of aquatic weed cons–med by grass carp. Prog. Fish Cult. 31(1): 51–54.

Verigin, B.V., Nguyen Viet and Nguyen Dong (1963). Data on the food selectivity and the daily ration of white amur. Ru. Publ. Problems of the Fisheries Exploitation of plant eating Fishes in the Water Bodies of the U.S.S.R. Askhabad 192–195.

Chapter 14

Biotechnology to Enhance Aquaculture Production

Two writeups of Dr. Chaudhuri are presented. He stresses on "Hormonal Manipulation" and "Genetic Manipulation" and application of biotechnology in enhancing improved seeds. Also application by advanced technologies in fish food production.

In this connection it is worthwhile to peruse the paper of Dr. P. Das entitled "Genetics in Indian Aquaculture."

Biotechnology to Enhance Aquaculture Production

Potential for Using Biotechnology to Aquaculture Production

The main approaches are:

Genetic Manipulation

Genetic manipulation in broad sense means changes in the genetic make up of any organism by application of improved techniques. Fishes and other aquaculture species can tolerate a wide range of genetic manipulations than terrestrial farm animals. Fishes have additional advantage that they produce large number of gametes and because of *in vitro* fertilization of eggs, genetic manipulation becomes easier than the former group.

Manipulation of Chromosomes Sets

By manipulation of chromosome sets, the phenomena like Polyploidy, Gynogenesis and Androgenesis could be induced which have definite advantages in enhancing aquaculture production.

Polyloidy

It means production of individuals having extra sets of chromosomes than normal fish species. It is produced by subjecting fertilized eggs of fishes to heat, cold, pressure or chemical shocks. The advantages of polyploidy are many. It produces (*i*) sterile individuals, (*ii*) improved quality of flesh in some, (*iii*) enhances growth and survival, (*iv*) more viable and ideal for producing new hybrids. Induced triploidy, commercially applied to rainbow trout and Pacific oyster, appears to be well established as a biotechnology tool in aquaculture. Triploidy leads to sterility which in turn leads to better growth and survival.

Gynogenesis (All Maternal Inheritance)

Gynogenesis is induced in fishes by stimulating parthenogenetic development of fish eggs by artificially inactivated spermatozoa. Sperms are irradiated by X-rays or UV-rays and their genetic contents (DNA) are destroyed. Fertilization occurs without any genetic contribution from the sperm. The production of diploid parthenogenetic individuals is further induced by cold, hot or pressure shock at an appropriate time after fertilization. Example: Grass carp, Salmon, Rainbow trout.

Androgenesis (All Paternal Inheritance)

Androgenesis is generally induced in homogametic males. In the process, ovum is irradiated with X-rays or UV-rays and its genome is destroyed. Fertilization occurs but without any genetic contribution of ovum and the offspring is a male having paternal inheritance only, Example: Common carp.

Advantages of Andro-or Gyno-genetic offsprings:

1. Useful in the generation of inbred line, progeny testing and selective breedings,
2. Production of uniparent for monosex culture for enhancing production.

Transgenesis or Gene Transfer

Transgenesis is the process of *in vitro* transfer of exogenous and usually heterogenous genes into the genome of a fish or any organism. Such a fish is termed as a "Transgenic fish". The main techniques involved in transgenesis are, microinjection and electroporation. Genes are introduced into one-celled embryo as well as oocyte. The aims of gene transfer technology in aquaculture are mainly genetic selection and production of the most efficient foodfish suitable for a particular environment, as transgenetic work offers great promise to produce "lsuper fish".

A few of the general potentialities of transgenesis are:

1. To accelerate growth by using hormone genes.
2. Tolerance to physical factors, such as cold using antifreeze gene.
3. Behavioural modification through regulation of endocrine function with regard to reproduction, maturation and sex control etc.
4. More efficient use of food by fish by manipulation of biochemical pathways.
5. To increase disease resistance: Resistance to parasites; pathogens or predators.

The application of transgenic technology for increased growth in fish and their cold tolerance are the two main areas of research at present, which have been successfully done. Growth hormone (GH) genes were transferred into eggs of a number of fish species and showed integration, expression and also inheritance. Therefore, the production of transgenic fish with increased growth rate is technically feasible but their commercial culture requires consideration of several factors of which the most important one is the risk to the natural populations in case of transgenic fish escapes to wild waters. Another area of great benefit to fish culture is disease resistance using specific disease resistant genes *viz.*, T-cell receptor, immunoglobulin, lymphokines etc.

The above studies on trangenesis of a aquatic organisms show great promise in their application for the benefit of aquaculture.

However, a number of technical problems need still be solved before gene transfer is widely accepted in aquaculture.

Application of Cell Culture Biotechnology to Marine Macro Algae (Seaweeds)

In recent years, millions of tons of seaweeds are produced through mariculture in many maritime countries and are used as human food (mainly green seaweeds) and also for commercial production of phyco-colloids for industrial purposes, such as, (*i*) Agar (from *Gracilaria*), ii) Algins (from *Sargassm*), (*iii*) Carrageenans (from *Eucheuma*). Seaweeds could be easily genetically engineered. More consistent cell cultures of certain red seaweeds was obtained as compared to those from natural thalli. In several cases entire plants could be regenerated from isolated cells and protoplasts. It was also possible to select disease resistant strains of seaweeds using selection in cell culture. Also, new strains of seaweeds have been developed through hybridization of gametes from separate populations. Increased production of red seaweed *Porphyra* was obtained by the use of biotechnology.

Biotechnology in Single-Celled Algae Culture

Principles of biotechnology can be readily applicable to single-celled algae which are generally used as feed in aquaculture and also have the potential in the production for human food as well. The single-celled green alga *Chlorella* and the Cyano-bacterium *Spirulina* can be used as a valuable human food source. These algae could be easily manipulated genetically. Hybrid strains of single celled algae could be produced by protoplast fusion techniques and also methods to introduce foreign genes are being investigated.

Cryopreservation Technology

Cryopreservation means preservation of gametes and embryo usually at below freezing temperature for ready availability of genetic material for breeding and research purposes. Cryopreservation of milt in certain cultivated fishes is very important since the males of some fish species mature earlier than the females and there is dearth of oozing males when fully mature female fishes are available. While cryopreservation of fish milt has been very successful, cryopreservation of eggs of fishes have met with very limited success.

Hormonal Manipulation for Sex Control

(Masculisation/Feminisation and sterilization in fishes)

Sex Reversal

Use of sex hormones (steroids) to produce monosex for culture are being used in aquacultural practices for over two decades, especially in tilapia culture. The problem in tilapia culture is because of their prolific breeding in ponds resulting in over-crowding with small size fishes and decrease in production. Monosex culture controls reproduction and, therefore, to obtain only males by sex reversal method the tilapia. O. niloticus or O. mossambicus early fry are fed with the androgen 17-of methyl testosterone for induction of sex reversal from genetic females to phenotypic males. Since males in these tilapia species are larger than the females and grow faster, culture of only male tilapia gives much higher production. This practice has been successfully tried and practiced in a few other cultivated species. In fish species where females are bigger and grow faster than the males, then only females are produced for monosex culture by feeding fry with estrogen 17P- Estradiol benzoate or diehyl stibestrol.

Sterilization

Hormonal sex manipulation is also used in the production of sterile individuals by administering steroids and profitably used in aquaculture.

Application of Biotechnology to Produce Improved Feeds

Aquaculture production can also be increased by developing improved feeds, such as:

1. Development of new artificial dies, especially for the filter-feeders (oysters and mussels) through micro-encapsulation of dietary compounds.
2. Development of food additives, such as, anabolic steroids and thyroid hormones, have led to enhance growth of salmonid fishes.
3. Improved culture of single-celled feed organisms, such as, algae, yeast and bacteria also helps increase in aquaculture production.

Use of Biotechnology for Stock Assessment in Marine Environment

Molecular technique or DNA finger printing is an useful technique of sock assessment of migratory fishes and also for species identification in the sea. For comparison of mitochondrial DNA from different individuals, a method has recently been developed in mapping the mt DNA genome. Analysis of mt. DNA offers a valuable mechanism in assessing the population structure in marine fishes and invertebrates.

In the foregoing, an attempt has been made to present briefly the impact of various advanced technologies developed during the last five decades of the century in enhancing fish production in India through aquaculture. It may now be concluded that the aquaculture industry has reached such a stage that it is now capable of contributing an appreciable increase in animal protein content in the diet of poorer masses in India.

The contribution of biotechnology, towards increase in aquaculture production, on the other hand has not been appreciably realized, so far, as its application in India is still in the preliminary stages. Also, there are some genuine concerns regarding capatibility of biotechnological approaches with the existing ones and the potential negative impacts of genetically improved domestic stocks on wild populations and the environment, in case the former escape into wild waters.

Biotechnology, however, has tremendous potentialities, and it is hoped that with judicious application, biotechnology will play a crucial role in future in substantially increasing fish production in India.

Application of Advanced Technologies and Potential for Biotechnology in Enhancing Fish Production in India

Application of Advanced Technologies in Food Fish Production

Induced Breeding Techniques for Seed Production

For centuries, the stumbling block in the development and expansion of carp culture was the acute shortage of seed for culture since he cultured carp species failed to reproduce in ponds. The major breakthrough was achieved in the year 1957, when the carps could be bred successfully for the first time in the world, in India, by injection of fish pituitary hormones and a vary useful technique was developed to produce millions of quality seed for culture. The success revolutionized prevailing aquaculture practices and the technique was effectively introduced all over India and also to many other countries in the world. Following this success, in subsequent decades it was possible to induce breeding in several other important species, such as, catfishes, gray mullet, milkfish and others.

During the last four decades, the induced breeding technique was modified and in addition to fish pituitary hormones, hormones of mammalian origin (both natural and synthetic) as well as several other hormones are progressively being used. A few important hormones used in seed production programs are mentioned below:-

1. HCG (Human Chorionic Gonadotropin obtained from human pregnancy urine) alone of in combination with FPE (Fish Pituitary Extract) gave various successes in breeding fishes.
2. Luteinizing Hormone-Releasing Hormone (LH-RH) and its analogues. Here attempts have been made to intervene at the Hypothalamic-hypophysis interface in the endocrine chain instead of injecting exogenous gonadotropins to stimulate hypophysis directly.
3. The mammalian (mGnRH) and/or piscine (sGnRH) gonadotropin releasing hormone analogue combined with Dopamine antagonists, is a further improvement over the previous ones.
4. Salmon gonadotropin–releasing hormone analogue (sGnRG-A) in combination with Domperidone, known as 'Linpe Method' is largely used in China. "Ovaprim" its trade name, prepared by Syndal Laboratory, Canada, is used for induced breeding in several countries.

However, the poor fish farmers of India and many developing countries of the world still prefer to use the whole pituitary gland extract mainly because of lower cost, simpler technique and ready availability of hormones.

Thus, availability of adequate quantity of seed at will has

extended carp culture to every nook and corner of India that has substantially enhanced fish production in the country.

Selective Breeding and Hybridization of Cultured Species

Carps which could not be hybridized for centuries, were successfully hybridized in India for the first time in the world. Selective hybridization is an approach by which strains are produced superior to the parent species. Several of the interspecific and intergeneric carp hybrids showed improved characters and faster growth.

Many fish hybrids are sterile and are utilized for getting increased production. Cross between two Tilapia species *i.e.* *Oreochromis mossambicus* ♀ *Tilapia hornorum* ♂ produced 100 per cent male hybrids. The male hybrids are used in monosex culture (for prevention of reproduction) and since Tilapia males grow faster and attain size, higher productions obtained.

Further by genetically research and intensive farming technology, a red hybrid strain (*Oreochromis niloticus* x *Oreochromis mossambicus*) of tilapia has been produced (termed as aquatic chicken) which is capable of giving production as high as from 600-800 MT/ ha/yr.

Polyculture of Compatible Species for Increased Production

Development of the system known as polyculture or mixed farming of a number of quick growing fish such as Indian carps (catla, rohu and mrigal) and introduced exotic Chinese carps (bighead, silver carp and grass carp) and a few other compatible species of fishes having different feeding habits and feeding at different zones, so that plenty of food that are produced and are available in the pond are fully utilized and converted to fish flesh. By mixed farming, fish production could be increased to 8-10 tons and as high as 20 mt/ha/yr.

Integrated Fish-Crop-Livestock Farming Systems

Integration of aquaculture with agriculture and livestock production systems is yet another milestone in increasing fish production. In this system there is mutual benefits among the different commodities and the wastes and bye-products of one is utilized by another and practically nothing is wasted. Fishes get the maximum benefits since they utilize the wastes and bye-products of livestock and crops, thereby saving expenditures on feeds and fertilizers. Also recycling of animal wastes is very important since it converts the animal wastes to fish flesh. By simple recycling of wastes/5-10 mt/ ha/yr. fish can be produced without using inorganic fertilizers or supplementary feeding.

Running Water Culture

The principle behind running water fish culture is that the flowing water constantly removes the nitrogenous and other metabolites from the culture environment, thereby eliminating these toxic products which when accumulated inhibit fish growth. Running water culture can give unbelievable production, such as over 4000 MT/ha/yr (estimated) of common carps in Japan and over 2000 MT/ha/yr of rainbow trouts in Idaho, USA. Such a high production is possible because of intensive stocking, elimination of metabolites and feeding ad libitum in oxygen-rich flowing water.

Intensification of Monoculture of the Tiger Shrimp (*Penaeus monodon*) for High Froduction

Shrimp culture during last two decades have reached the peak of very high production by intensification and improvement of culture techniques, supplemented by sophisticated water exchange system, and supplementary feeding replaced by complete (high protein) balanced diet. The production may even go up to 10-40 tons/ha/yr. by sophisticated intensive culture. However, such highly intensive culture is very risky since very often it leads to disaster due to disease. Such intensification has also adverse effects on the environment and may cause socio-economic conflicts.

Application of Biotechnology in Aquaculture

Biotechnology

The term Biotechnology, coined in recent years has several definitions. One of the simplest one is, "*Controlled use of biological agents such as, micro-organisms*" or cellular components, for beneficial use. Also, Biotechnology is defined as "*The collection of industrial processes that involve the use of genetically engineered micro-organisms*". Yet there is the third definition "*The application of scientific and engineering principles to the processing of biological agents to produce goods and services*".

Biotechnology offers great promise in finding novel solutions to biological problems in terms of food production. As mentioned earlier that while biotechnological approaches have been very successfully applied to Agriculture and Animal husbandry by transfer of desirable traits into economically important plants and animals using cloned genes, very little progress has been made in such studies when applied to aquaculture species.

Genetics and Aquaculture

Prof. Chaudhuri's Work inspired Dr. P. Das, Ex-Director NBFGR, to observe the following:

The green and white revolutions in India, as is known, became possible on account of deep understanding of the plant and animal genetic principles and through application of specific genetic manipulations. Similarly, a sustainable blue revolution will be possible through application of genetics for improvement in the quality of fishes used for aquaculture, in addition of genetics for improvement in the quality of fishes used for aquaculture, in addition to the effort of fish husbandry improvement. It is heartening to note that the work in this direction has already been initiated in the country in several Universities and Research Institutions.

Genetic Characterisation

Genetic characterisation is being done through cytogenetic, biochemical-genetic and the most sophisticated molecular-genetic techniques.

When the DNA in chromosomes is coiled and condensed, the chromosomes in different species take characteristic spahes which are studied under cytogenetics. In addition to Karyotyping, Gycogenetic study is now done by C,G,R,Q, and NOR banding techniques. BRDU and Restriction Endonucleases are also being used.

Biochemical-genetic investigations, use electrophoretic techniques which enable identification of genetically variable polymorphic enzymes, sophisticated devices like ultrathin isoelectric focussing of eye lens proteins and gradient SDS polyacrylamide gel electrophoresis are used in the investigations.

All the genetic information of an individual are contained in DNA. With the rest of the world, in molecular genetics, we also use genomic and mt DNA fingerprint techniques.

Hybridisation

Hybridisation aims at flow of genes resulting in heterosis introgression due to overdominance and heterozygosity at many loci. Though crosses between different species and genera were successful, so far no viable superior hybrid with the exception of the catla-rohu hybrid with intermediate characters, could be produced in India.

Some intraspecies trout hybrids abroad are disease resistant but show lower percentage of survival. In India cross breeding of trouts of north and southern parts could yield some results in a collaborative research by NBFGR and Himachal Pradesh Fisheries Department.

Once identification of strains of our fishes is complete, cross breeding of different strains and inbred lines may produce hybrid vigour in due course, thereby helping in increasing production.

Sex Reversal

Sex reversal aims at aquaculture upgradation using only a single six which grows faster or with sterile fish in case of profile breeders. Artificial reversal of phenotypic sex from male to female and vice versa without changing the genotype is possible in most of fishes owing to lack of distinctive heteromorphic sex chromosome pair. It has been possible to reverse the sex in some Indian species also with not much of genetic gain so far.

A report indicates a production increase of a specific fish by 100-200 per cent through mono-sex culture of sex reversed seed in Malaysia.

In view of about 100 per cent higher growth rate of male *M. rosenbergii* than the female, its all male culture in India can automatically increase the yield and also avoid cannibalism in the pond. Sex reversal of this species when standardized could be expected to boos cultured freshwater prawn production in the country first decade of 21st century.

Gene Banking

Successful freezing of fish gametes and storing them for indefinite periods without deterioration and loss of genetic properties are of considerable importance in aquaculture. While the work in cryopreservation of eggs and embryos are pursued world over, it has been successful only in respect of sperms/milt in India and abroad. There is a milt gene bank at the NBFGR at Lucknow. Some stock is availabe at CIFA, Dhauli as well.

Milt gene banking facilities availability of male gametes of good quality all the year round for early or late breeding if females are available. It can also cut down the need for brood fish transportation, and help in selective breeding and hybridization programme, maintenance of genetic diversity in hatcheries and in the conservation of fish biodiversity in the country in years to come.

Selection for Faster Growth

One of the techniques for improving the genetic worth of stocks is through the modern concept of selection. One fundamental concept is 'like begets like.' But there is always some variation within a group of individuals from which preferred progenitors can be selected to derive the domesticated line with desirable characters.

There are some traits such as growth rate, food conversion, disease resistance, fecundity etc., which are influenced by unknown but sizeable number of genes which are additive in nature. Selection of such additive genes will have the effect of increasing frequencies of favourable genes for the characters in question for stepping up heterozygosity and the number of favourable gene combinations which can be utilized through cross breeding after studying heritability.

High yielding wheat, rice, broiler chicken, milch cow etc., are the benefits of adopting such a selection. Response to selection in the form of genetic gain in fishes has been at 5 per cent in salmonids in 3 generations, 33 per cent in channel catfish in 3 generations and 30 per cent in Coho salmon in 4 generations. In Norway, the Atlantic salmon (*Salmo salar*) stock used in aquaculture through selective breeding, improving about 2 per cent per year in terms of production. In India, predicted response was 8 per cent in *Etroplus suratensis* in one generation. Likewise there was 34 per cent gain in 3 generation selection in *Labio rohita*. These can be taken as bright indications of possible genetic gain through selection.

With intensified and systematic mass scale commercial selection programme of cultured species in general and rohu in particular, tested genetically superior higher yielding fish seed is likely to be available in India for aquaculture with promise of increased production at the beginning of the next century.

Genetic Engineering

Chromosome in a normal individual is in a pair of two. During cell division, manipulation in chromosome number has become feasible. Changes in the number of chromosome sets may be brought about by the simple destruction of one set or by disruption of metaphase spindle.

The produciton of viable progency with all maternal or paternal inheritance alone and increasing the ploidy of individuals to triploidy (3n) or tetraploidy (4n) are results of genetic engineering at different stages, as explained hereunder.

Gynogenesis

It involves stimulation of parthenogenetic development of an egg by a genetically inactivated spermatozoa. The diploidy of the haploid egg is restored by the retention of second polar body normally through through a temperature shock at an appropirate time after fertilization. In other words, the DNA in the progency is constituted only by the maternal genome.

The high level of inbreeding achieved in the process may help in the genetic studies and also in the hybridization programme. The gynogenetic offsprings, being all females, provide scope for monosex culture too.

Androgenesis

It is based on the same principle as gynogenesis. The androgenetic progeny is produced with the paternal genorae alone avoiding maternal genetic contribution altogether.

Androgenesis, when successful in India, would help in the conservation of fish biodiversity since it would be possible to reconstitute the endangered or extinct species from its cryopreserved milt, if present in the Genebank.

Polyploidy

This is again a manipulation of chromosome number after normal fertilization of both sexes. Triploids are produced by suppression, it, normally done by heat or pressure shock shortly after fertilization. In such fishes there is a set of 3 chromosomes instead of normal 2. These fishes being sterile, generally grow faster and bigger due to avoidance of gametogenesis. Production of tetraploid fishes by disruption of the rifrst cleavage mitosis is possible. The tetraploids also may grow faster.

Sterile triploids with higher growth rate when standardised shall provide scope for higher production in aquaculture in time to come.

Transgenetic Fish

The most sophisticated approach to genetic manipulation is production of transgenics through gene cloning. The successful injection of a novel gene into the host organism produces a transgenic individual with better growth/disease resistance/other desired characters as the case may be.

Although human insulin gene was the first gene that was successfully used to produce transgenic loach in 1985, subsequently, growth hormone gene of cattle, rat and chicken were used to produce transgenic pike, trout and medaka respectively. Recently, genes of rainbow trout, common carp, salmon, flounder and tilapia have also been used for the production of transgenics. Such an usage has been successful in prawn also. Successful transgenics grow significantly larger than their normal ones.

With the recent successful experimental production of transgenic rohu and mrigal and initiation of a collaborative programme, India has entered the race with rest of the world.

The commercial potential of transgenic fish, however, would lie in developing transgenic broodstock lines. This may require combining transgenic technology with selective breeding for producing superior strains of fish. The ecological and psychological aspects are also required to be studied side by side.

In view of the above progress, we may hopefully expect in achieve a breakthrough in the application of genetics of augment aquaculture production in the country at the beginning of the 21st century.

Chapter 15

Professor Chaudhuri's Very Short Deliberations as Chief Guest or as an Invitee Dignitary at Various Functions

Only a few such deliberations could be compiled from his innumerable such as no detailed documents were available. These were on a wide range of subjects and only a few were chosen which are presented herein after:

1. Biodiversity and Environment
2. Training Course on Prawn Farming (Pond culture)
3. Intensive Carp culture.
4. Eco-friendly management of resources.
5. International Shrimp Hatcheries operations and Management.
6. Fisheries development and impact on environment.
7. Waste water aquaculture.

On Biodiversity and Environment

Mr. Chairman, distinguished dignitaries in the Dias, Learned Professors and Scientists and my young friends.

I am grateful to Principal Prof. Mukherjee for inviting me as one of the Guests of honour in to-day's gathering. When Prof.

Chakrabarty and his colleagues approached me and also requested me to say a few words to the delegates attending the seminar, I told them I myself am not very clear about the modern concept of the new coinage. By biological diversity we ordinarily mean diversities or varieties in the living world which includes microbes, biota, flora and fauna. However I searched for a modern definition of the term and here it is:

"Bio-diversity refers to the variety and variability among living organisms and the ecological complexes in which they occur. Diversity may be defined as no. of different items and their relative frequencies. For bio-diversity these items, organized at many levels ranging from complete ecosystem to the chemical structure that are the molecular basis of heredity. Thus the term encompasses different ecosystems, species, and their relative abundance. (Source: US Congressional Office of Technology assessment (OAT) 1987).

This definition suggests the necessity of mathematical approach in determining relative abundance and evaluating each community on the basis of diversity index. The term also suggests status besides some other parameters.

Some of my young friends might be finding this definition too complicated then there is a similar one for them: "Bio-diversity is an epi-phenomenon an expression of the interaction of the variety of biotic and abiotic factors (Henry Gee 1992).

Bio-diversity has become a global issue these days since it is being threatened by degradation of the environment all over the world. It is Nature which promotes bio-diversity keeping in view to maintain a perfect harmony among living organisms and the ecological complexes surrounding them. Trouble begins only when the harmony is disturbed or broken. Many of the current ecological problems of environment are really based on ecological principles. To be more explict, the balance of Nature remains a balance only if there exists a proper give and take relationship among the following 4 groups, namely, (1) The Produces *i.e.* Autotrophs, the photosynthetic algae that produce food staff for all forms of life. (2) The Consumers *i.e.* Heterotrophs: organisms living directly or indirectly for food on autotrophs. (3) The Decomposers, that help dead organisms to decay and contribute to dissolved substances and detritus. These are bacteria and some Fungi, and (4) Inorganic dissolved substances which are needed for nutrition of both producers and consumers. And the Cycle goes on and a balance is maintained, till there is interference from some quarters-and that is MAN, who has been threatening the delicate balance existing among these 4 groups. He is increasing his threats to the point that he not only threatens the existence of non-human organisms but also his own (*e.g.* Depletion of Ozone layers, global warming etc.); for Man is the only organism that pollutes the environment, that mercilessly destroys the green forests and so on and he is the only organism that manipulates his environment to serve his own needs and greedy desires.

Due to shortage of time it is not possible to cite scores of examples. I am just mentioning a few where unauthorized indiscriminate introduction of exotic plants, animals and especially fish without proper studies on their effect on the indigenous ones, by man, have done irreparable harm to the indigenous species even to the extent of bringing them down to the status of endangered species. To cite a few examples, these *Eichhornia crassipes* the water hyacinth in tropical Asia and Africa; introduction of rabbits in Australia, piscivorous sportfish Bass to Porto Rico, *Tilapia mossambica* in India and SE Asia, *Clarias batrachus* in Philippines and so on. We should be very careful about the recent unauthorized introduction of the African catfish *Clarias gariepinus* and also there should be thorough screening of the imports of aquaria fishes. Recently, I happened to visit a district Aquarium and was shocked to find that the deadly carnivorous PRRANHAS of the Amazon river of Brazil were being exhibited. If by chance it gets into the open waters and gets established then they may create havoc. Appropriate Govt. authorities should be very strict and see that no such harmful animals or plants get their entry into the country without proper investigation.

In closing, I thank the Organizers for selecting such an important national as well as an International problem for discussion in the seminar. I wish the seminar a grand success. Thank you all.

On Pond Culture

I am thankful to Ramkrishna Ashram Krishi Vigyan Kendra, Nimpith and the Zoological Society, Calcutta, for inviting me to deliver the key note address at this inaugural programme of a ten-day residential training course in 'Prawn Farming'.

Prawn farming has shown tremendous interest and development in recent years, and has established itself as a leading export industry in many developing countries of the world. The dramatic increase in the production of prawn in these countries, primarily for the earning of foreign exchange has led to the proliferation of the industry, spurred to the adoption of high intensive type of culture technology. This technology, however, has also its negative aspects and draw-backs which I would be mentioning very briefly later on.

The training course in 'Prawn farming', as I understand, includes both brackishwater shrimp culture as well as freshwater prawn culture. Here, if I am permitted, I would like to discuss a little on the terms 'Shrimp' and 'Prawn' in English. In Bengali we have to distinct terms "Bagda Chingri" and "Golda Chingri". But many shell-fish workers do not distinguish between the terms 'Shrimp' and 'Prawn'. To be little precise, the term 'Shrimp' is ordinarily used for the brackishwater or marine penaeids including all the species under the related genera of *Penaeus Metapenaeus, Parapenaeopsis* etc. whereas term 'Prawn' is generally restricted to the freshwater prawn species of the genera *Macrobrachium, Palaemon, Leander* etc. Of course, there is no hard and fast rule restricting the use of these two terms. Of the two groups, the shrimps have greater demands in the world market, especially, in the major economically developed countries like USA, Japan, France and U.K. I welcome this training programme and commend the organizers for initiating such a training programme in the culture of two very important commodities in view of the rapid expansion of the prawn farming industry in India and at the same time because of the extreme dearth of qualified trained personnel on various aspects of prawn culture in the country. I wish many more programmes similar or of longer duration would follow in the coming years so that our young fish farmers may get proper training in this highly profitable business so as to earn their livelihood. However, at the very outset I would like to give a note of caution to these young potential entrepreneur that they should not get lured to venture straightway for the intensive type of shrimp culture.

As I mention about the intensive type of culture', I think it would be worthwhile to give the participants some idea about the limitations of various intensities of prawn culture, namely, the extensive (improved traditional), semi-intensive (much improved traditional) and intensive (sophisticated) or super-intensive type of cultures. Each category has its special requirements with regard to the pond-size, water-depth, fertilizers to be used, quality of supplemental feeds, water exchange and management, stocking rate of fry etc. The more intensive the culture is the more sophisticated, the more expensive it becomes and the more risk it involves. To give an idea, the stocking rate of fry starts from less than 1-2/m^2 in extensive, 5-15/m^2 in semi-intensive, and 50-250/m^2 in intensive and may be as high as 500–600 fry/m^2 in high super intensive culture. Indeed, the shrimp production is also correspondingly enhanced from a meager 1 mt less in traditional culture to over 40 mt/ha/crop in super intensive–farming. Such high production, naturally appears very tempting to the elite farmers and industrialists-turned shrimp culturists who get attracted towards the intensive or super intensive types of culture. But by doing so, they pollute the environment bringing disaster to the flourishing industry very soon.

I am tempted here, to share with you some of my personal experience in the countries of Southeast Asia, where I have seen the industry growing almost overnight like mushrooms, especially, in the Philippines and also in Thailand and to some extent in Taiwan. I have also got first hand knowledge about the collapse of the flourishing shrimp industry in the Philippines. The huge profit enjoyed by shrimp producers, proved to be short-lived. The same story of rapid rise and equally rapid fall of the industry was experienced by the countries like Taiwan, China and Ecuador. In India, we had and rather still have the advantage of learning from the failures of the above mentioned countries.

In 1993, when I returned back to my country, I was approached by a couple of industrialists who were engaged in shrimp farming. I tried to explain to them the danger of rapidly expanding intensive type of culture since over–expansion and excessive intensification are sure to result in environmental pollution with serious impact on water quality which would ultimately lead to various types of disease syndromes as has happened in the countries mentioned earlier. Taura syndrome in Ecuador, which baffled the shrimp farmers for a long time, regarding the cause of mortality. The mystery was finally solved by the recent discovery of a very small virus-particle supposed to be the infectious agent. Taura type syndrome in the Philippines also probably be due to some such micro-virus. The yellow head virus which was responsible for the collapse of the shrimp industry in Taiwan was also reported from Thailand; but fortunately in Thailand the collapse was averted by strong research effort by pathologists followed by successful management and massive educational effort instituted by the Thai government. In China, with the discovery of a non-occluded baculo-like virus causing the massive mortality of shrimp, management counter measures are being developed to minimize mortality problems. Here, I would like to mention that China was the World's largest producer of farmed shrimp, producing 200,000 in 1991, which was expected to climb up to 400,000 mt/yr. However, because of the new shrimp virus, the production has crawled down to 50,000 mt in 1993 and 1994.

Sometime back, I was shocked at the newspaper report that in West Bengal, already the disaster has started and mass mortality of shrimp have been reported from many farmers. I didn't apprehend this to happen so soon. So far my information goes, the cause of mortality or the pathogen causing it has not been identified as yet. I hope by now appropriate action has been devised to minimize mortality and remedial measures taken.

Lastly, I would like to give a bit of advice to the participants of the training programme who are the potential shrimp farmers that they should try to restrict themselves to semi-intensive type of culture since it has several plus points:

1. It has less adverse impact on the environment.
2. Stocking rate being moderate there is less chance of disease due to overcongestion.
3. It has several other advantages such as its internal rate of return is substantially higher than all other culture systems including the intensive ones.

With these words I conclude my address, thanking the organizers again and wishing the participants of this very useful training in this beautiful surroundings of Ramkrishna Ashram Krishi Vihyan Kendra, Nimpith.

[N.B.: Key Note address by Dr. Hiralal Chaudhuri at the inaugural programme of Training Course on 'Prawn Farming' at Nimpith on June 19, 1995]

Intensive Carp Culture

I am thankful to Dr. Ayyappan for inviting me to visit the beautiful Institute CIFA–once my dream Institute of Dhauli during mid-70's and also to inaugurate the Training programme on "Intensive carp culture". I am happy to see before me so many enthusiastic young men representing various organizations from a number of states, who have come to participate and make this training programme a grand success. I join Dr. Ayyappan and other organizers of the training course, in extending you all a warm welcome and sincerely hope that the intensive training which you would be receiving here would help you to enhance fish production in the country to a great extent.

I am glad that the training course is on a very important subject that is carp culture. The carps belonging to the Family Cyprinidae are being cultured in fresh-water ponds in China for over 3,000 yrs and in the eastern region of India for at least 1,500 yrs. The ancient Chinese and Indian farmers had selected this group of fishes because of their many cultural qualities. They are quick-growing, attain quite large size, tasty and non-predatory in food habits. Over the above, the greatest quality of carps lies due to the fact that they feed very low in the food chain. They either feed on algae/vegetative matters which are the primary producers or feed on zooplankters and other smaller organisms which are the primary consumers. As you all know that according to the basic principles of fish production their feeding habit itself is supposed to give very high production as compared to the predatory/piscivorous fishes having longer food-chain. Both India and China belong to the group called developing countries having fish and rice as their staple food. Fish culture is not a luxury for them but a means to produce cheap fish at a very low cost. The very selection of carps as the cultured commodity may mainly be because has assumed the farmers of a low-cost farming technology suitable for their production.

In recent years, there had been a craze in many developing countries of the world for shrimp culture primarily for exporting shrimp/prawn and earn foreign exchange in general. Very often shrimp culture gets priority and flourish at the expense of fish culture. I am scared at this trend and would like to give a note of warning to fish culturists that carp and popular brackish water fish culture of aquaculture in general may get a set back because of this unusual enthusiasm shown towards shrimp culture. Here I wish to share with you some of my personal experiences in SE Asia during the last two decades. Please excuse me for digressing a little bit. I have seen the shrimp culture industry growing like mushrooms almost overnight in the Philippines and in Taiwan and also to certain extent in Thailand. I have also the first hand knowledge of the sudden collapse of the flourishing industry in these countries. Although India started quite late but we didn't learn from others failure and we had the same fate. This happened only because of the over-greediness of the elite shrimp farmers, entrepreneurs and multinational industrialists-turned shrimp culturists. To become rich overnight they straightway go for highly sophisticated intensive or superintensive type of cultures thereby bringing in mass mortality due to diseases and polluting the aquatic environment. These have also brought in ecological imbalance giving a negative image of modern aquaculture among the public. The result is the 'Moratorium on new fish farm'. On May 1995, the supreme court issued interim orders to Tamil Nadu, A.P. and Govts. prohibiting construction of shrimp farms on agricultural lands. Also, there was another order prohibiting construction of fish farms within 500 m. from the coast. These orders have definitely given an impression detrimental to aquaculture in the public mind. A few months back I was invited to participate in a seminar jointly called by the Environmentalist and Shrimp farmers to discuss about the pollutional hazard due to highly intensive culture. In that meeting one Law officer of the W.B. Govt. held aquaculture responsible for all pollutions in aquatic environment. He was of opinion that this type of culture should be prevented by suitable legislatures. Also, being an irreversible process, conversion of fish farms to shrimp farms might bring disaster to aquaculture.

With carps however, intensive type of culture can be carried out without any adverse effect since the carps also utilize natural food produced in the ponds supplemented by artificial feeding.

Unless the stocking rate is very high which need complete balanced feed, there is not much fear of pollution of the environment in scientific carp culture. However, semi-intensive type of culture is generally more suitable for small fish farmers, especially, because of moderate rate of stocking and consequently there is less congestion and less chance of disease and chances of having any adverse effect on the environment is minimum. Besides, the internal rate of return is much higher in semi-intensive than any other type of culture system. However, highly sophisticated intensive or super intensive type of carp culture should be avoided to keep the aquatic environment free from pollution as far as practical. In rich countries like USA, Japan and Europe, the highly productive vertically integrated aquafarming problem effectively.

I have taken lot of your time and I should stop now. With these words, I conclude my inaugural address and welcome you all again to this training programme wishing you a very profitable time.

Ecofriendly Management of Resources

I am thankful to Dr. Maniranjan Sinha, CIFRI for inviting me to participate in this National Seminar on "Eco-friendly Management of Resources" for doubling fish production–strategies for 21st Century. Eco-friendly or Environment friendly management of resources is of great importance in substantially increasing fish production in India, since protection of environment has become a global issue these days, and any attempt to disrupt the balance of nature could be disastrous.

The future of Inland Fisheries and Aquaculture in the decades ahead, is viewed against the background of increasing world population and the deteriorating aquatic environment. World population is expected to reach 6 billion and in India alone population may be touching one billion mark at the onset of the 2nd millennium. While fish production have to be increased manifolds to meet the increasing demand of animal protein, there is also the great danger of adversely polluting the environment. It is well-known that super-intensive culture could give very high production. However, very often this results in bringing ecological disbalance and adverse impact on the environment due to discharge of metabolites in the aquatic environment. The aquaculturists are therefore, in a dillema in fixing their production target which should also be eco-friendly. Hence achievement of a real balance between maximizing of aquafarming production and conservation of nature, would be the ideal solution. Also, globally attempts are being made to bring about a compromise between these two extremes.

As for the strategies to be developed to achieve the targeted production of doubling fish production by the year 2020 (which would be discussed in this seminar), biotechnology offers great promise in finding novel solutions to biological problems in terms of fish production. Of the main approaches in biotechnology, Genetical Engineering which includes genetic manipulations such as manipulation of chromosome sets, involving phenomena of polyploids, gyno and androgenesis etc. transgenesis or gene transfer technology, application of cell-culture technology and others have tremendous potentialities. It is expected that in early 21st Century, judicious application of biotechnology would play a crucial role in substantially increasing fish production in India.

I wish the National Seminar a grand success.

An Inspiration Talk: Shrimp Hatcheries

Mr. Fukumoto, Deputy Chief, SEAFDEC, the earned participants of the International Shrimp Hatcheries Operations and Management, my colleagues, Ladies and Gentlemen.

At the outset I wish to thank the organizers of the training course for inviting me to participate in this closing ceremony and to deliver the inspirational talk to the participants. This International Training Course of SEAFDEC is also supported by FAO and a few other International Organizations. The training course was initiated and concluded when I was away from the Philippines and so I missed the chance to meet the participants personally. This batch seems to be the biggest group that we had in recent years and I am happy to learn that as many as 39 young men and young women have attended the course, representing 15 countries from various inmates of the old as well as the New World. From the profile of the participants, I observe that a large contingent have come from the private sector many venturing into this lucrative newly developing industry. Shrimp culture, especially the culture of the Tiger shrimp *Penaeus monodon* has attracted world-wide attention and is considered at present to be the most important aquaculture commodity mainly because of its value as a Foreign exchange earner has been offering this particular course from 1977 and so far 17 short term courses have been conducted, of which NACA has participated in two only including the present one. When we consider its impact in the Philippines, we find that there has been a significant increase in the number of hatcheries spread across the nation and I think I won't be wrong to attribute this modest contribution to the industry as a direct result of our training efforts. The Government is encouraging shrimp culture and to exemplify its commitments, the(ADB)has earmarked $36.4 million on loan to the Philippines to support shrimp culture development here in Panay island. In world-wide picture we observe that shrimp culture has made a tremendous impact during the last ten years because of the ever increasing in the demand and limited supply of shrimps in the world market. To give you an example in late sixties, out of 1.7 million tons of shrimps production the contribution from culture was a meager 1,000 tons which increased to around 58,000 tons (3 per cent) in 1983 and the estimates for 1990 are over 400,000 tons (18 per cent). In the course of a short period of time many research and development projects have emerged in different parts of the world creating new industries with brilliant prospects.

The training course which you have just completed is the most crucial and I should say the most important aspect of shrimp culture, that is, the seed production. Its success depends on each and every step you take in your hatchery, operations and management. Of these, I may particularly mention of 'water quality control' to be the most important. To substantive this I must tell you a small incident. I went to Hawaii in 1976 and visited the *Macrobrachium* (giant few prawn) hatchery of the famous pioneer worker Fujimura. I asked him what was the secret of his success. At the beginning he was hesitate to say but when in the course of our discussion he observed that he could also act lot of information from me then he dinelged the secret that it is "water". So I would like to advise you that you should take utmost care in selecting the site and source of water for your hatcheries.

I have no doubt that after finishing this intensive practical training course when you go back to your respective countries, most of you if not all, will try to put to practice what all you have learnt here with the zeal of a missionary for the development of shrimp culture of your country at large.

Being relatively new, shrimp industry has a place for each and every one of you, not like other overcrowded business industries. Here you can be bold and innovative but don't restrict yourself to only what you have learnt in the classrooms–the challenge is to go beyond it–and once you have done it successfully be sure to share it with us. Also, in case by chance you fail to attain the desired success, don't hesitate to consult us at SEAFDEC, we would surely try to help you in your enterprise.

In closing, may I congratulate each and everyone of you tonight for making a bold career decision, investing your valuable time and money in a new but a promising career in aquaculture.

Good luck to you all.

On Fisheries Development and its Impact on Environment

I am thankful to the organizers of the Seminar on "Fisheries Development and its Impact of Environment" for inviting me to deliver the keynote address. Since aquaculture has played a major role in the development of Fisheries in India, I would largely confine my talk on 'Aquaculture Development and its probable impact on the environment'. Fish culture in India, that is mainly culture of carps Catla, Rohu and Mrigal, is an age-old practice, particularly, in the Northeastern part comprising mainly Bengal and to certain extent its adjoining provinces of British India. However, till the independence of India, only traditional carp culture was practiced by the farmers using the empirical knowledge generated from experience gained through generations. Scientific research for the development of new technology on fish culture was actually initiated with the establishment of several Research Institutions by the Govt. of India immediately after independence, although some useful research on fish and fisheries were carried out earlier by our revered Late Professor H.K. Mookerjee at the Fisheries Lab. in the Calcutta University and some work initiated in Punjab.

For centuries, the main stumbling block in the progress and development of carp culture was the acute shortage of seed for culture, since these commercial species of cultured carps failed to reproduce in ponds. The only source of their seed were the big rivers like the Ganges and its tributaries, where the carps breed during monsoon months. The seed from the rivers are then collected by seed collectors and are sold to the fish farmers at Sealdah and Howrah fry markets in Calcutta. Since the carps could not be bred under captivity, the riverine source alone had to be tapped for seed for thousands of years both in India and China.

The major breakthrough, however, was achieved only in the year 1957 when the carps could be bred successfully for the first time in the world in India by injections of fish pituitary hormones and a field technique was developed to produce millions of quality seed for culture. The success revolutionized aquaculture practices and the technique was effectively introduced not only all over India, but also to many countries of the world for production of seed of carps and other cultivated species. The new technique solved the problem of shortage of carp seed and made carp culture popular throughout India.

Thus with the expansion and extension of cultured area, aquaculture was also intensified from the traditional and extensive types to semi-intensive and intensive types. Also, new prized fishes of superior qualities were introduced from outside India after proper screening, so as to enrich our indigenous species. The average production was thus increased from a meager 200 kg/ha/yr to over 1,000 kg/ha/yr.

The next remarkable success achieved in increasing further production of fish in ponds, was development of a system known as Polyculture or mixed-farming of a number of quick-growing fishes, such as, Indian and Chinese Carps, having different feeding habits so that all kinds of available food in a pond could be utilized and converted into fishflesh giving very high production upto 10 metric tons/ha/yr and above with proper pond management practices.

Integration of Aquaculture with Agriculture and Livestock production system is yet another milestone in increasing fish production. The system is likely to play a major role in Govt. of India's rural development programme. Such a farming system offers greater efficiency in resource utilization, reduces risk by diversifying crops and provide additional food and income to the farmers. In this system, there is mutual benefit among various commodities and fish gets the maximum benefit as it utilizes the wastes and by-products of others, thus saving expenditures on feeds and fertilizers.

So far in India, researches carried out for increasing production of fish/ha were mainly aimed at producing more animal protein to supplement the protein-deficient diet of the poorer masses. The very selection of carps as the cultured commodity, was largely due to the fact that low-cost farming technology could be suitably applied for their production. Even in intensive carp culture, pond fertilization is generally done by recycling of animal wastes and supplementary feeding done by cheap feeds only. Unless the stocking rate is very high, that needs complete balanced feeds in large quantities, there is not much fear of environmental pollution in scientific carp culture. However, highly intensive or super-intensive types of fish culture should be avoided so as to keep the environment free from pollution since too much intensification may cause innumerable problem. In the rich developed countries like USA, Japan and Europe, the most sophisticated verti-integrated Aquaculture Industry, an extremely high production system as well as a high risk industry needing high capital investment did not succeed as it couldn't effectively solve the serious environmental pollution problem.

Although carp culture in India is yet to reach the stage where there is sufficient risk of environmental pollution, the same cannot be said about present day shrimp culture in India, shrimp being a dollar earner is being cultured mainly for export. Over enthusiasm together with the desire to get very rich overnight have tempted many elite shrimp farmers, entrepreneurs and multi-national industrialist turned shrimp culturists to start straightway highly sophisticated intensive or super-intensive types of shrimp culture very often leading to mass mortality due to congestion and diseases and polluting the aquatic environment. This has brought in ecological imbalance accompany by various adverse impacts on the environment; to name a few, such as, overloading of environment or ecosystem leading to hypernutrification, destruction of mangrove wetlands, conversion of fish farms–an irreversible process etc. and lastly giving a negative image of modern aquaculture in the minds of public. The result is the 'Moratorium on new shrimp culture farms'. On May 9, 1995, the Supreme Court of India had issued an interim order to Tamil Nadu, A.P. and Pondicherry Govts. prohibiting construction of shrimp farms on agricultural lands. Also, there was another court order prohibiting fish farms within 500m from the coast. These court orders quite likely have given an impression in public minds which is detrimental to aquaculture in general and shrimp in particular. Just to cite an example, a few months back, I was invited to participate in a seminar conducted by the Environment Protection Council in collaboration with some other Institutions, to discuss about the pollutional hazard due to highly intensive shrimp culture among others. In that meeting a speaker, a Law Officer in his deliberation held aquaculture responsible for all pollutions in aquatic environment. He probably forgot about the extent of industrial, agricultural and domestic pollutions in aquatic environment. He also said that aquaculture should be prevented by legislations. The word 'prevented' appeared to be to biased; instead suitable legislations guide and regulate excesses in shrimp culture go as to protect the environment. I am afraid, that if such biased steps carp culture or brackish water fish culture may be get the axe and progress of fish culture in general get hampered because of the over greediness of a few elite shrimp farmers whose benefit only the consumers in USA and Japan.

About two decades ago, Mr. Ray Jackson, ADG (Fisheries) of FAO in one of his lectures on the future of World Fisheries in the decades ahead, gave a very gloomy picture against the background of increasing world population and deteriorating aquatic environment. The present day fish culturists are at a fix to determine how much high production they should aim in view of the risk of polluting the environment? A compromising solution has to be found out limiting high production as well as preventing adverse effect on the environment. In this connection I must place before you the example of Thailand the only major shrimp producing country among a host of other Asian countries, which alone could avert total collapse of the industry. Thailand has also taken bold steps to protect the environment and at the same time guaranteed the sustainability of shrimp farming by framing new legislations guiding shrimp farms and controlling their haphazard construction. Incidentally I may inform you that Thailand's farmed shrimp production of 220,000 mt in 1995 is the highest in the world.

The new legislations are as follows:

1. All farms must register with the Government.
2. Farms larger than 8 ha must get Govt. approval before construction.
3. Farms must use sediment ponds that contain molluscs and seaweeds.

4. Farms must have reservoirs occupying at least 30 per cent of the total area.
5. Farms must dispose sediments on land and not in public waterways.

Also, FAO (1991) have formulated strategies to reduce impact of shrimp culture on the coastal environment. Every shrimp producing country should try to adhere to the strategies as formulated by FAO.

In conclusion my humble suggestions to the aquaculturists is that they should keep in mind that environmental integrity should be the primary concern for attaining sustainability of aquaculture.

With these words I conclude my address wishing all participants attending the seminar a very useful and profitable time.

On Wastewater Aquaculture

At the outset I want to thank Mr. Trpallic and the Organizers of the training course for inviting me to deliver the valedictory address and also participate in the closing ceremony of the training course in Waste water aquaculture for the State Govt. officials. With the rapid increase in world population, the concern for environmental upkeep has assumed as much importance as the basic needs of food, clothing and shelter for the mankind. When United Nations declared the "decade 1981-90" as the International Drinking Water Supply and Sanitation Decade, one of the goals was to provide all the world's population with adequate excreta disposal facilities that indicates the magnitude of the problem of waste disposal. While effluent treatment was the major concern a few decades ago, the concept that 'wastes are resources out of place' has changed the approach in dealing with them. The thrust is gradually shifting to reuse of the wastes considering their high nutrient levels and the great possibilities of utilizing them for fish production. Twin objectives of waste water treatment and food production are amply achieved in case of aquaculture and potentials of the same could also be utilized for horticulture and agriculture.

Raw and treated domestic and municipal sewage have often been used to feed fish ponds for increasing the productivity. This is done on the basis that nutrients contained in the sewage encourage production of fish-food organization. For successful pisciculture, compatibility between fish, then food and the aquatic environment is highly desirable.

Reuse of treated effluent from waste stabilization tanks or from conventional sewage treatment system (activated sludge process or trickling filter) may be carried out to fertilize fish ponds in order to increase production.

India has made great strides in polyculture of various Asiatic carps and a few other species. The idea is mainly based on utilization of all niches of a water body in the most useful way. Water management and supplementary feeding are the two major factors for successful implementation of the technology. Around 60 per cent of the running expenditure however is towards feed and fertilizers resulting in high cost of production.

Sewage-fed aquaculture offers a low-cost technology eliminating the costly inputs like feed and fertilizers. Fish production in sewage-fed system is quite high varying from 4 to 6 t/ha. In some selected well managed wastewater fisheries the production is even much higher. Wastewater aquaculture is a modern concept of reuse of waste for nutrient recycling and biological treatment of wastes before discharging to natural water. This gives high yield of fish on one hand and maintain ecological balance on the other.

Wastewater may however, contain some toxic elements. Measures must be taken to study the accumulation of such materials in the biotic community.

Wastewater may also contain some pathogenic organisms which may pose problems to the community if proper management is not followed.

It is presumed that the training programme organized by CIFP has taken good care of all the aspects of wastewater reuse. I am sure that you have been immensely benefited by participating in the programme.

I have no doubt that on your return to your respective states you will try to put into practice what all you have learnt here with the zeal of a missionary to implement wastewater aquaculture for increasing overall fish production in India.

With this note, may I congratulate the organizers as well as the participants for the successful completion of the very very important training programme.

Chapter 16

Some Contributions on Miscellaneous Subjects

In this group the following publications have been selected after due edit of several others:

1. Fertility of hybrids of Indian Carps and preliminary studies in the F_2 Fishery Carp hybrids.
2. Artificial fertilization of eggs and early development of milk fish *Chanos chanos* (Forskal).
3. *Mugil cephalus* Artificial propagation
4. Embryonic development transparent goby (*Gobiopterus chuno*) and several others.

Fertility of Hybrids of Indian Carps and Preliminary Studies on the F_2 Fishery Carp Hybrids

Studies on the hybridization in Indian carps were first initiated following the success of induced breeding of these fishes by hypophysation. A number of interspecific and intergeneric hybrids were produced by crossing cultivated carps belonging to 3 genera and 7 species. Majority of these hybrids had attained maturity and were found to be fertile.

One interspecific hybrid (rohu-calbasu) and one intergeneric hybrid (catla-rohu) were bred by hormone injections and F_2 generations of these hybrids obtained. The percentage of fertilization in both the cases was very high. The morphological characters of the hybrids were observed to be intermediate between those of the parent species. The growth rates, as compared to the parent species, were studied. Another intergeneric hybrid (mrigal-kalbasu) could be stripped after hormone injection and outcrossed with catla, rohu and mrigal producing viable eggs which could be reared to adult stage.

Introduction

Study of hybridization in fishes is assuming increasing importance in both theoretical and applied Ichthyology in recent years only. In fish culture, hybridization is mainly aimed at evolution of a hybrid having desirable cultural qualities far superior to the parent species. Work on hybridization and mass selection of cultivated species have received special attention in USSR, and a number of interspecific and intergeneric hybrids of sturgeons have been produced (Nikoljukin and Shipelvskaya, 1959; Nikoljukin, 1971). In the West, experiments on hybridization have been mainly carried out on trouts (Baldwin, 1956; Buss, 1956; Buss and Wright, 1958) and more recently on sunfishes of the family Centrarchidae and commercial catfishes (Ricker, 1948; Smith, 1961; Childers, 1971). Hybridization in Indian carps was first successfully attempted in 1958, and a number of hybrids were produced (Chaudhuri, 1959). Since then systematic attempts are being made to produce better hybrids by reciprocal crosses, backcrosses and three way crosses.

The present paper deals with the fertility of major carp hybrids and production of subsequent F_2 generation hybrids for their evaluation. The morphological characters as well as growth rate of some of the F_2 hybrids have been dealt with.

Experiments and Results

Production of F_1 Hybrids

With the success of the induced breeding of Indian major carps by hypophysation, it was possible to strip major carps after hormone injection and artificially fertilize the stripped eggs with the sperms from injected males of other species. A number of interspecific and inter- generic hybrids were produced by crosses and reciprocal crosses of carps belonging to 3 genera (*Catla, Labeo* and *Cirrhinus*) comprising eight species (Chaudhuri, 1959, 1971; Hickling, 1968). Of the various hybrids produced, catla-rohu (male parent named first) hybrid was the best having the good qualities of both the parents. The head was smaller than catla and the body broader than rohu. The growth rate of the hybrid was faster than rohu and nearly as fast as catla. Production in terms of body flesh, however, was more than the both the parents. Other hybrids showed much promise were rohu-catla, rohu-calbasu, calbasu-rohu and catla-calbasu.

Fertility of F_1 Hybrids

Majority of the hybrids reached adult stage and maturation of gonads was observed. Catla-rohu hybrid matured in 3 years. A number of adult hybrids (2.2 to 3.4 kg) were cut open and gonads examined. Ripe females as well as oozing males were mostly observed. A few hybrids had ill-formed testes while a single specimen had hermaphroditic character. One of the gonad resembled an immature ovary while the other gonad was an immature testes. In August, 1960, two-year old fully mature specimens of rohu-calbasu, calbasu-rohu, mrigal-rohu and rohu-mrigal hybrids were seen. The males were mostly in oozing condition. However, sufficient number of each hybrid could not be examined for determining the sex-ratio. Among these different hybrids which reached adult size, sterility was observed only in few specimen of cyprinus-rohu hybrid.

F_1 Hybrids

Inter Specific F_2 Hybrids

Rohu-Calbasu

The interspecific hybrid rohu-calbasu showed much promise. The hybrid has a small head and grows much faster than the slow-growing maternal species. The hybrid attained full maturity in about 2 years.

On August 6, 1960, one set of ripe rohu-calbasu hybrid comprising one female and two males were selected for induced breeding experiment. The female weighed 1.2 kg and the two males weighed 1.0 and 0.8 kg respectively. The female completely spawned after administration of two doses of injection of rohu pituitary hormones. A total of 1.2 lakhs of eggs were released, 65 per cent of which were fertilized and 45,000 spawn obtained, of which 40,000 spawn were stocked in a nursery pond and 5,000 reared in the laboratory for further studies. Growth of fry was quite satisfactory.

Survival in the pond was very poor due to presence of predatory aquatic insects. In the laboratory very high survival was achieved. F_2 hybrid showed characters intermediate between rohu and calbasu, with variations in colouration of the body fins, barbels and in caudal spots. Detailed studies of the variations were made. In three and half months the F_2 hybrids attained an average size of 183 mm.

Inter-generic F_2 Hybrids

Catla-Rohu

Although catla-rohu hybrids were observed to attain maturity in 3 years, the first attempt to breed the hybrids was made only on July 16, 1962, when one pair of four-year-old hybrids were selected for induced breeding. The female (4.1 kg) received two injections at an interval of 6 hr and at a dose of 2.5 mg/kg and 6 mg/kg of rohu pituitary gland respectively. The single male (3 kg) received only one injection at the rate of 2 mg/kg. The breeding hapa was fixed in river with thick muddy water. The hybrid spawned successfully and released 6 15 lakh eggs having 76.2 per cent fertilization. Hatching was very satisfactory but many hatchlings died and only 135 lakhs of spawn could be obtained. 50,000 spawn were stocked in one 0 04 ha nursery pond in Killa and 80,000 were stocked in one 0.06 ha pond at Kausalyaganga fish farm. After 3 weeks 15,538 fry were recovered from Killa pond and 13,216 fry from Kausalyaganga pond. The growth study of fry in the rearing ponds was continued. Growth of fry was not uniform. The size-range varied from 90 to 185 mm with an average of 126.47 mm, attained within a growth period of about 100 days.

Study on the comparative growth of catla-rohu F_2 hybrid with that of the parental species catla and rohu was taken up and an experiment was set up in two 0.1 ha ponds at Kausalyaganga fish farm. But study on the replicate pond could not be continued because of large-scale poaching in that pond. The comparative growth study however, was continued in one pond only. From the growth study it is observed that catla-rohu F_2 hybrid grows faster than rohu but slightly slower than catla.

Study on the gut-contents of a number of juvenile specimens of F_2 hybrid (size range 65-142 mm and weight 2-30 g) revealed that the feeding habit is more like rohu. Mud, sand particles, detritus, decaying vegetative matters constituted 90 per cent of the feed. The rest consisted of *Oscillatoria*, Naviculids, *Gomphonema*, *Scenedesmus* and a few other green algae.

The general appearance and morphological characters are intermediate between the two parental species. Further study could not be possible since the ponds were needed for fry rearing work and the F_2 hybrid got mixed up with the F_1 hybrid and other hybrids.

Out-crossing of Intergeneric Hybrid *Mrigal-Calbasu*

Mrigal-Calbasu

The hybrid attained maturity in 2-3 years. One eight year old fully mature female was injected with pituitary hormones and stripped. 60 per cent of the eggs hatched out. The hybrid had a slightly fringed lower lip and two pairs of prominent black barbel resembling that of the material parent. Colour of the body was intermediate between parent species.

In the hybridization experiments carried out by Prof. Chaudhuri at Cuttack, it was observed that the large majority of the interspecific and intergeneric hybrids attained maturity in 2-3 years. These F_1 hybrids could be induced-bred by hypophysation and F_2 generation of hybrids were produced. Even outcrossing by fertilizing eggs of an intergeneric hybrid with the milt of 3 species of carps belonging to three distinct genera was successful.

Hybridization and selective breeding of cultivated fishes is aimed at the evolution of a hybrid or a new strain having qualities especially from the fish culture angle such as, quicker growth, better food conversion, better resistance to diseases and unfavourable environmental conditions, and better nutritional qualities. To produce such strain it is essential to continue investigations for several generations of the hybrids by back crossing and out crossing of subsequent generations. This depends mainly on the reproductive success of the hybrids. Carp hybrids have the advantage that majority of them are fertile which quality would help the fish-culturists to develop a strain of superior quality by repeated back-crossing and out-crossing of the hybrids which show great promise.

[*Reprinted from the Journal of the Inland Fisheries Society Vol. V, December, 1973.*]

References

Baldwin, N. S. 1956. Hybrid trout. *Sylvia*, 12 (3): 1-4.

Buss, K., 1956. Results of species hybridization within the Family Salmonidae. *Progr.FishCult.,\B(4)*: 149-58.

Buss, K. and J. S. Wright, 1958. Appearance and fertility of trout hybrids. *Traits. Amer. Fish. Soc.* 87',: 172-81.

Chaudhuri, H., 1959. Experiments on hybridization of Indian Carps. *Proc. Indian Sci. Congr.*, 46.

1971. Fish hybridization in Asia with special reference to India. Seminar/study tour in USSR on Genetic Selection and Hybridization of Cultivates Fishes. 19 April-29. May, 1968. Lectures. *Rep.* Genetic Selection and Hybridization *FAO/UNDP (TA)* (2926).: 360 p.

Childers, W. F., 1971. Hybridization of fishes in North America (Family Centrarchidae): *Ibid.* Hickling, C. F., 1962. *Fish culture*, Faber and Faber, London: 295 p. 1968. Fish hybridization. *FAO Fish. Rep.*, 44 (4): 492 p.

Nikoljukin, N. I., 1971. Hybridization of Acipenseridae and its practical significance. Seminar/study tour in USSR on Genetic Selection and Hybridization of cultivated fishes. 19 April-29 May, 1969. Lectures. *Rep. FAO/UNDP (TA)* (2926;: 360 p.

Nikoljukin, N. I. and G. P. Shpilevskaia, 1959. Back crosses of hybrid sterlet and beluga in comparison with parental forms. *Dokl. Akad. Nouk. SSSR*, I25: 641-9.

Ricker, W. E., 1948. Hybrid sunfish for stocking small ponds. *Trans. Amer. Fish. Soc.*, 75: 84-96. Smith, S. B., 1961. Selectivity and hybridization in management and fish stocks. *Can. Fish. Cult.*, 29: 25-30. Soward, C. L., 1959. Experiments in hybridizing several species of trout. *Progr. Fish Cult.*, 21 (4): 147-50.

Artificial Fertilization of Eggs and Early Development of Milk Fish *Chanos chanos* (Forskal)

Hydrated eggs obtained from a female milkfish, *Chanos chanos*, were artificially fertilized with the milt collected from a male injected with acetone-dried pituitaries of salmon. The fertilized eggs (1.1–1.25 mm in diameter) developed normally in sea water in basins and petri dishes at a salinity of 30-34 per cent and successfully hatched in 25-28.5 hours at a temperature of 26.4–29.9°C. The yolk was completely absorbed in about 2.5 days and during this period many postlarvae died. A few larvae were reared up to 5 days but all died on the 6th day. Attempts were made to feed the postlarvae with freshly hatched trochophore larvae of oysters obtained from eggs artificially fertilized in the laboratory could not be ascertained.

Introduction

The milkfish *Chanos chanos* (Forskal) is an important food fish and is extensively cultured in brackishwater ponds in Southeast Asia, particularly in the Philippines, Indonesia, and Taiwan. Since it does not attain maturity or breed in captivity, the only source of its seed is the coastal waters from where milkfish fry are collected during the breeding season. While the milkfish industry is rapidly panding, the present supply of fry is inadequate to meet the growing demand. Thus, the urgent need to develop a dependable source of seed is now being realized as fishery scientists in a number of countries are engaged in research in inducing maturation and spawning in both pond-grown and wild milkfish by hypophysation.

Our knowledge on milkfish, especially on parameters such as spawning habits and spawning grounds, nature of eggs and the characters of early larvae, is meager. Since Delsman's description (1929) of probable milkfish eggs, there was hardly any information until Senta *et al.* (1976) reportedly collected a few eggs from a spawning ground near Batbatan island off Antique province in the Philippines.

Preliminary experiments carried out in Taiwan to induce maturation in tank-reared milkfish (Liao and Chang, 1976) have been partially successful. Earlier attempts made in the Philippines to gather information on mature milkfish from the wild and induce them to spawn by hormone injections were not successful (Angeles, 1968; Inland Fisheries Project Technical Report, 1974; Delmendo and Angeles, 1975). More recently, experiments carried out at the Oceanic Institute in Hawaii (Nash and Kuo, 1976) and at the SEAFDEC Aquaculture Department, Philippines (Vanstone, *et al.*, 1976a) to induce spawning in adult milkfish captured from the wild by injections of semipurified salmon gonadotropin (SG-G100) resulted in the release of hydrated eggs from the injected females. These eggs, however, were not fertilized.

The first breakthrough in inducing spawning in wild milkfish by hormone injections and successful hatching of artificially fertilized eggs has recently been achieved at the Pandan and Tigbauan breeding stations of the SEAFDEC Aquaculture Department (SEAFDEC Progress Report, 1977). While the details of the induced breeding experiments conducted by us are being published elsewhere, the present article gives an illustrated account of the artificial fertilization of eggs and the embryonic and larval development of milkfish.

Artificial Fertilization of Eggs

On May 13, 1977 at 0100 hr. three sabalos were caught from the fish trap near Tigbauan. These were transferred to the transport cage and brought to the SEAFDEC Research Station at Tigbauan, Iloilo. During sexing, two of them were found to be females and one a male. The male was spent and it was not oozing any milt. One of the females weighing 8.6 kg appeared recently spawned as evidenced by the presence of a few hydrated eggs near the genital pore. A good number of eggs were then collected from the ovaries by inserting a cannula through the genital pore. The eggs thus collected were transferred to moist Petri dishes. Milt was also collected from a running male weighing 7.0 kg by using a cannula. This fish was caught from the same fish trap on May 12 and was injected at 2000 hr of the same night with 3 ml of physiological solution containing 25 mg of acetone-dried salmon pituitary gland. The "dry method" (Chaudhuri *et al.*, 1966) was used in artificially fertilizing the extruded eggs with milt. After about 10 min, the fertilized eggs were diluted with a little water; after another 10 min they were diluted further and the excess milt was washed away. This procedure was repeated several times and the eggs were transferred to a basin with sufficient water. Artificial fertilization was successfully done at 0500 hr. At about 0800 hr, the same female was killed by punching the heart to bleed the fish after which the ovaries were taken out. The ovaries were almost spent and only a few thousands eggs were obtained. These were fertilized by sperm obtained through squeezing the testes excised from a dying male.

In the morning of June 22, 1977 two sabalos, one male and one female were caught from the same fish trap at Tigbauan. On examination the female (6.8 kg) was found to be recently spawned and was oozing hydrated eggs. The male (6.2 kg) was partially spent.

At 0930 hr artificial fertilization of the eggs was done following the method described above. Both fishes were killed to obtain eggs and sperms.

Incubation of Eggs

The fertilized eggs were kept in aerated seawater in basins. The salinity and temperature during the incubation period ranged from 30-34 ppt and 26.4 to 29.9°C, respectively. At 30 ppt the developing eggs settled at the bottom of the Petri dishes but when salinity increased to 34 ppt or above, they were bouyed up to the subsurface and the surface.

The stages in the development of eggs starting from early cleavage were examined through a microscope. Photomicrographs of important stages of segmentation, blastulation, differentiation of embryo, and hatching were taken.

Larval Rearing

The newly hatched larvae were reared in aquaria, basins and beakers of seawater. These were supplied with feed after 21/2 days. The feed consisted mainly of the trochophore larvae of oysters obtained by artificially fertilizing oyster eggs with sperms from ripe males. Occasionally, the milkfish larvae were also supplied with a small quantity of *Chlorella.* Water temperature and salinity were recorded at regular intervals throughout the period of larval and postlarval development. Photomicrographs of the newly hatched larva and subsequent stages were taken.

Results

The development of the fertilized eggs was studied in detail and the characters of the developing embryo, larval and postlarval stages of milkfish were noted. The fertilized eggs obtained on June 22, 1977, developed only up to the early gastrula stage after which they started to disintegrate. This may have been due to the delay in the artificial fertilization of the hydrated eggs as these have remained in the abdominal cavity too long.

Fertilized Eggs

The fertilized eggs were spherical, nonadhesive and transparent. The yolk had a light yellow tinge and no oil globule. Furthermore, it was granular and appeared to have fine segmentation. The sizes of the fertilized eggs ranged from 1.1 to 1.25 mm with a mean diameter of 1.13 mm. Each egg had a very narrow perivitelline space. Eggs fertilized on June 22, 1977 had a mean diameter of 1.2 mm.

Embryonic Development

Cleavage

The blastodisc appeared in about an hour after fertilization and the first cleavage dividing the blastodisc into two blastomeres occurred 75 min after fertilization. The second cleavage, perpendicular to the first followed within 10 min. In another 15 min the 8-cell stage was reached. The fourth cleavage, which appeared parallel to the second, was effected 10 min later and the 16-cell stage was obtained. The 32-cell stage followed in the next 15 min and the sixth cleavage in another 20 min. The 128-cell stage was reached in 2 hr and 55 min. As successive cleavages occurred, the blastomeres further decreased in size. The morula stage was reached 4 hr and 10 min after fertilization.

Formation of Embryo

The blastula is formed in about 1,1/2 hr. The blastoderm cells start spreading over the yolk mass and at about 8 hr after fertilization, yolk invasion progressed considerably. The egg reached the late gastrula stage 9 hr after fertilization when yolk invasion was two-thirds complete. In another 45 min, gastrula was completed. The yolk plug and blastophore became visible and the embryonic streak was faintly indicated.

Differentiation of Embryo

In about 10 per cent hr after fertilization the embryonic rudiment became distinct with 5-6 somites. Within 40-45 min the somites have increased to 9-10 in number and the Kupter's vesicle appeared. Head and tail were clearly differentiated 13 hr after fertilization and 19-20 somites could be counted.

The optic and auditory vesicles were visible 14,1/2 hours after fertilization and the embryo became C-shaped. In another hour more myotomes appeared, the embryo developed further and looked like a girdle over the yolk. Also at this stage brain differentiation progressed considerably. At 18,1/2 hr after fertilization, the embryo further elongated and the tail started to separate from the yolk. The unpaired finfold appeared and the optic lens was indicated. In another 1,1/2 hr the head separated completely from the yolk and the heart beat was discernible. The tail became further elongated, the yolk was reduced and the optic lens became distinctly visible. The embryo developed further and twitching movements of the body could be observed approximately 24 hr after fertilization. In another 4 hours the yolk was further reduced and the tip of the tail extended nearer to the head. At 28 hr after fertilization the embryo was fully differentiated and ready to hatch.

Hatching

The elongated tip of the tail struck against the head end of the egg shell causing the latter to break; the head came out first. The larva shook off the shell and emerged completely from the egg case.

Larval Development

Newly Hatched Larva

The milkfish embryo hatched 281/2 hours after fertilization at water temperature ranging from 26.4-29.9°C. The second batch of fertilized eggs took only 25 hours to hatch. The newly emerged larva had unpigmented eyes and no fin buds; the mouth was not formed and the anus was situated posterior to the yolk mass. The yolk sac was broad and projected anteriorly near the head end of the larva.

The newly hatched larva was 3.2 mm in length (preserved specimens 2.8-2.9 mm.) The yolk was about 2.15 mm long and 0.5 mm wide. Black pigmented cells were present in the finfold except the tip of the caudal region. There were pigments sparsely scattered in the yolk mass and on the head and body of the larva.

The larva swam slowly up to the surface of the water and then gradually dropped down and remained suspended in the column of water in an oblique position with head down and belly up. Occasionally the larva went up and came down in the characteristically inclined posture.

2-hour Old Larva

The 2-hour old larva was about 3.4 mm long. The length and maximum width of the yolk was 2.3 mm and 0.45 mm, respectively. The width of the body was 0.65 mm with preanal length of 2.6 mm. About 34-35 preanal mytomes could be observed. Melanophores were scattered on the yolk and a few could be seen on the head. The distribution of pigment cells on the finfold was almost the same as in the newly hatched larva.

10-hour Old Larvae

The 10-hour old larva was about 3.8 mm in length. Preanal length was 2.8 mm. The greatly reduced yolk was 2.1 mm long and 0.35 mm wide (maximum). The larva still remained suspended in the water column in the oblique position and occasionally went up and down.

24-hour Old Larva

The day-old larva measured about 4.0 mm in length. The yolk sac was reduced considerably. Faint pigmentation in the eyes commenced. The hindgut was clearly visible. Melanophores were scattered on the head and trunk and also on the dorsal finfold.

30-hour Old :arva

The 30-hour old larva was about 4.3 mm long and its preanal portion was 3.2 mm in length. About 34 myotomes could be counted up to the anus. The mouth was indicated and the yolk further resorbed. The pectoral finbud had also appeared..

48-hour Old Larva

The two-day old larva was about 5.0 mm in length. The yolk was very much reduced. The eyes were fully pigmented and pectoral fins elongated. The mouth was fully formed and the esophagus was distinctly visible. The pigment cells on the unpaired finfold had almost disappeared. There was pigmentation on the dorsal part of the body, and denser on the caudal peduncular region. There were a few dense melanophores over the posterior portion of the gut.

56-hour Old Larva

The 56-hour old larva was similar in size to the 48-hour old larva. The remnant of the yolk sac was seen only as a slender streak. The mouth was open and the intestine fully formed. The optic lens was shiny.

There was a row of melanophores along the dorsal edge of the myotomes and another row over the intestine along the ventral edge of the myotomes. Only a few pigment cells were present in the unpaired finfold near the beginning of the caudal portion of the finfold.

The fry made slow directed movements with occasional jerks and swam on the surface as well as at the bottom of the water column.

Postlarval Development

72-hour Old Larva

The 3-day old larva measured the same length of about 5 mm. The larva slowly moved up and down and occasionally rested obliquely in the water column or crept at the bottom. Sometimes it nibbled on the bottom. The larva, photographed and preserved, had a speck of food in the intestine. One row of melanophores was present along the dorsal edge of the myotomes and another row along the ventral edge. Only 32 preanal myotomes could be observed.

81-hour Old Larva

The 81-hour old larva swam slowly on the water surface and occasionally went down. No further differentiation in the body or in the unpaired finfold could be observed. The dorsal as well as ventral rows of melanophores were distinct.

141-hour Old Larva

The differentiation in the 6-day old larva was not appreciable. A faint indication of the anal rays was present. The dorsal row of melanophores was scattered and considerably reduced in number; the pigments were denser posteriorly. The ventral row of melanophores was distinct and very dense over the posterior end of the hindgut.

Larval Rearing

For brevity of manuscript some of the stages development and the descriptions of egg/embryo/larva are not produced.

Since the mouth and the intestine were fully formed in the 56-hour old larvae and the yolk sac was almost absorbed, these were given newly hatched trochophore larvae of oysters as feed. The eggs of oysters were artificially fertilized in the laboratory by sperms obtained from male oysters. A few mL of *Chlorella* were also introduced along with the trochophore larvae. However, there were larval mortalities during this critical period when the yolk was completely absorbed and the larvae started to feed.

The milkfish larvae were also fed with trochophore on the fourth day. But the majority of the 4-day old larvae died. The dying ones were found to be infested with ciliates which got into the aquaria and basins where the larvae were reared along with the trochophore larvae. A few milkfish larvae still alive on the fifth day were transferred to culture dishes and Petri dishes.

On the sixth day only 2 larvae were alive. These were kept at a salinity of 29-30 ppt and 24-26°C temperatures. However, both died when they were 141 hours old.

Discussion

The success in artificial fertilization of milkfish eggs has bridged a big gap in our knowledge on the nature of its eggs and larvae. The study supports the observations made by Delsman (1926; 1929) about 50 years ago on the characters of milkfish eggs and larvae.

Available information on the maturity, spawning habits, eggs and larval development of milkfish is very scanty. Tampi (1957) was the first to make detailed observation on the milkfish gonads. He found the average diameter of a mature intraovarian egg to be 0.8 mm. Delsman (1929) found 0.7 mm as the average diameter of a mature milkfish ovarian egg. Vanstone *et al.* (1976a) and Nash and Kuo (1976) injected mature females with ovarian eggs 0.8 mm in size and obtained hydrated eggs 1.2 mm in diameter. In our experiments on induced breeding of milkfish, routine examination of the ovarian eggs of female spawners was done by collecting egg samples by a cannula inserted through the genital pore. Females having ovarian eggs of 0.7 mm in diameter or more were ordinarily selected for hormone injection.

Fertilized Eggs

Our observations on the fertilized milkfish eggs agree mostly with those described by Delsman (1926) and Chacko (1950). Delsman (1926, 1929) in his description of milkfish eggs, however, did not mention any perivitelline space. We observed a very narrow perivitelline space which is hardly discernible in many eggs. Another interesting feature that we observed, especially, in the second experiment was the prominent marking patterns visible in the egg shell itself giving it a segmented appearance. At first, we thought the segmented appearance was due to abnormalities in the hydrated eggs left in the abdominal cavity for a relatively long time. However, similar markings were also observed in the hydrated (preserved in Bouin's fluid) and developing eggs in the first experiment. The markings were very faint in most of the developing eggs and could not be clearly observed in the shells of embryos about to hatch. Such characteristic markings are also present on the surface of the egg shells of the silver mullet *Mugil curema* (Anderson, 1957), yellow fin menhaden (Reintjes, 1962) and *Mugil cephalus* and *Liza troschelii* as observed by one of us (Chaudhuri).

Buoyancy of Eggs

Our observations on the buoyancy of milkfish eggs were not conclusive. While the unfertilized eggs quickly settled at the bottom of the containers, the fertilized eggs were observed to go down slowly at a salinity of 30 ppt. When salinity was increased to 34 ppt or above, however, the eggs either stayed at the subsurface or floated at the surface of the water.

Delsman (1929) described the milkfish eggs as pelagic. Senta *et al.* (1976) made similar observations. According to the latter, live milkfish eggs floated to the water surface when placed in a glass jar. The observations of different authors, however, are varied with regard to the buoyancy of marine fish eggs. The grey mullet *Mugil cephalus* has pelagic eggs, slightly smaller (0.93-0.95 mm) than milkfish eggs and having a big oil globule (0.38 mm). Liao (1975) mentioned that the fertilized grey mullet egg is bouyant and stays afloat near the surface of water with slight aeration but settles down slowly in still water. Tang (1964) observed that the eggs were suspended in the circulating water but sank to the bottom in standing water, while Yashouv (1969) reported the sinking of mullet eggs towards the end of incubation. According to Kuo *et al.* (1973), the majority of eggs which sink are unfertilized. The bouyancy of eggs, however, depends on several factors. May (1974), for example, found that the eggs of the sciaenid fish *Bairdiella icistia* fertilized in lower salinities were larger and more bouyant than eggs fertilized in higher salinities. Furthermore, Blaxter (1969) observed that the pelagic eggs of most marine fishes sink when put in salinities below a certain threshold.

Incubation Period

The incubation period for eggs depends largely on temperature. According to our observations, milkfish eggs hatched in 25 to

281/2 hours at a temperature range of 26.4-29.9°C. Delsman (1929) observed that milkfish eggs from catches made in the morning hatched in the evening of the same day. Taking into account the short incubation period of the pelagic eggs of related species, he concluded that the eggs had been shed 24 hours prior to hatching (Schuster, 1960). Delsman and Hardenburg (1934; cf Schuster, 1960), stated that spawning took place in the evening from 8 to 10 hours and the eggs developed in about 24 hours. Senta *et al.* (1976) likewise found that most of their collected eggs hatched in the evening. The two recently spawned females observed by us appear to have spawned sometime before midnight since they were caught in the trap shortly after midnight. May (1970) stated that the incubation period in *Bairdiella icistia* eggs depended both on temperature and salinity. The duration of hatching tended to be greater at lower salinities and temperatures. Grey mullet eggs hatched in 59 to 64 hours at 20.0-21.0°C (Tang, 1964). Liao (1975) reported that the hatching of grey mullet eggs took place in 34 to 38 hours at 23-24.5°C temperature and 49 to 54 hours at 22.5-23.7°C at the salinity of 30.1-33.8 ppt. Chaudhuri (1968) mentioned that the incubation period of *Mugil cephalus* in India is 48 hours at 25.5°C-23.5°C. Kuo *et al.* (1973) hatched grey mullet eggs in 36 to 38 hours at 24°C and in 48 to 60 hours at 22°C compared that of the mullet, the period of incubation in milkfish is generally much shorter.

Newly Hatched Larva

Our observations on the characters of newly hatched milkfish larva agree well with the description of Delsman (1929) except on one point. While he stated that the head does not project in front of the yolk in the newly hatched larva, we, on the contrary found that the yolk mass does not reach as far anteriorly as the tip of the head. The length of the newly hatched larva is 3.2 mm as compared to that of Delsman's which is about 3 mm. A great variation in the size of newly hatched larvae have been observed in many fish species. In the grey mullet the size of newly hatched larvae varies from 2.2 to 3.5 mm. While Tang (1964) obtained 2.2 mm long larva, Kuo *et al.* (1973) reported the size as 2.65 mm, and according to Liao (1974) newly hatched grey mullet larvae measure from 2.8 to 3.5 mm in length. The newly hatched larvae of milkfish, as well as the grey mullet are more or less at the same stage of development, having unpigmented eyes, scattered melanophores and without findbuds or mouth parts.

Kuo *et al.* (1973) observed that newly hatched mullet larva when inactive remains suspended in water column in an inclined position with the ventral side oriented towards the surface; when in motion, it jerks, rights its position, darts rapidly upward then sinks to its resting position. We have observed similar movements in the newly hatched milkfish larva.

Larval Rearing

The larval development of milkfish that we have observed is generally in agreement with those described by Delsman (1929). The growth of the larvae was initially very fast and continued for 2 days during which the yolk was steadily resorbed. From the 3rd day until the 6th day when all larvae died, there was practically no further growth. Similar observations were made by Kuo *et al.* (1973) on mullet. They also observed that the 2nd growth period coincided with the onset of larval feed (5th-8th day) and intensification of feeding (9th-12th day).

The mouth was fully formed in 2-day old milkfish larva and complete resorption of yolk was effected in 2,1/2 days. Delsman (1929) had also made similar observations. In grey mullet, formation of the mouth occurred on the 3rd day (Liao *et al.*, 1971) and 2nd to 3rd day (Kuo *et al.*, 1973) and the yolk was completely absorbed by the 5th day. Both Kuo *et al.* (1973) and Liao *et al.* (1971) mentioned two "critical" periods in mullet larvae when very high mortalities occur. While the former indicated the 2nd to 3rd day coinciding with the opening of the mouth and the 8th to 12th day preceding its second growth phase, the latter reported the two critical periods as the 3rd day and 11th to 13th day, respectively. We observed the first critical period to be the 2nd to 3rd day which coincided with the complete resorption of yolk and full formation of mouth and intestine. Majority of the larvae died during this period. Senta *et al.* (1976) could rear only one larva beyond the critical period from the collection of eggs which hatched.

This problem of heavy mortality in rearing marine fish larvae has baffled fishery scientists for decades. Attempts were made to overcome this difficulty by providing suitable food for the larvae in adequate quantities. May (1970) made a review of various available larval food and discussed the major food types used by fish breeders. Mullet breeders had to face the difficult task of feeding mullet larvae to obtain higher rates of survival. Concerted efforts are being made along this line during the current decade. Liao *et al.* (1971) mentioned the quality, size, density and mobility of the food as the important factors for developing larval rearing techniques. They used eggs and trochophore larvae of oysters for feeding the mullet larvae during the first few days. The density was maintained as high as 400-500 organisms/mL. This was followed by rotifers and copepods as the larvae grew and *Artemia salina* was provided to the bigger larvae. "Green water" containing *Chlorella* and other algae was tried initially without much success. Kuo *et al.* (1973) used veliger larvae of both oysters and sea urchin followed by day-old nauplii of *Artemia salina.* The survival rate was improved by supplementary feeding with marine plankton from the 3rd day onward. Prepared feeds such as cooked egg yolk have been successfully fed to many marine fish larvae but these generate problems of water fouling due to decay of leftover food. In our experiments, the milkfish larvae were provided with trophore larvae of oysters from the 3rd day but how far they utilized these feeds could not be ascertained.

Besides the nonavailability of the right type of feed, there are other causes of mortality such as high temperatures and high salinities which should be carefully avoided while rearing larvae. Higher temperatures generally increase the metabolic rate and accelerate yolk absorption, and hasten death by starvation while high salinities may increase larval mortality by causing osmotic or ionic changes in the interior milieu (May, 1975).

Due to wholesale mortality of the larvae when about 6 days old, it was not be possible to rear them to the stage (10-15 mm) when similar larvae are collected by fry gatherers from the coastal waters. However, one interesting point about the milkfish larvae is that, unlike their nearest relatives, namely, the 10-pounder *Slops saurus* (Family: Elopidae) and the tarpon *Megalops cyprinoides* (Family: Megalopidae), which have the characteristic ribbon-shaped leptocephalus stage (Alikunhi and Rao, 1951), the milkfish larvae do not pass through such stage (Delsman, 1929).

In the foregoing, we have discussed our observations on milkfish eggs and larvae and compared these with the observations of other workers. In the absence of adequate information on milkfish eggs and larvae, comparison has been made with the other important food fish, namely, the grey mullet *Mugil cephalus* whose eggs and larvae are akin to those of *Chanos chanos* and about which considerable information is available. Mullet rearing has been a problem of the last decade. It is hoped that the information available on the rearing of mullet larvae will be of great help in solving the problem of feeding milkfish larvae and developing the technique of commercial production of milkfish seed.

[A joint paper by H. Chaudhuri, J. Juario, Jurgenne, H. Primavera, R. Mateo, R. Samson, Erlinda Cruz, E. Jarabejo and J. Canto (Jr.)]

References

Alikunhi, K. H. and Rao, S. N. 1951. Notes on the metamorphosis of *Elops saurus* Linn and *Mega lops cyprinoides* (Broussonet) with observations on their growth. *J. Zool. Soc. India*, 3:99-109.

Anderson, W. W. 1957. Early development, spawning, growth, and occurrence of the silver mullet *Mugil curema* along the South Atlantic coast of the United States. U.S. Fish and Wildlife Service, *Fish. Bull.* No. 119, 57:397-414.

Angeles, H. G. 1968. A preliminary report on the observations and possibilities of induced spawning of mullet and milkfish. *Indo-Pac. Fish. Counc. Occas.* PAP. 71(8)1-11.

Blaxter, J. H. S. 1969. Development of eggs and larvae. Hoar, W. S. and D. J. Randall (eds.), Fish physiology. New York, Academic Press, Vol. 1, pp. 177-252.

Chacko, P. I. 1950. Marine plankton from waters around Krusadai island. *Proc. Indian Acad. Sci.*, 31 (3): 162-76.

Chaudhuri, H. 1968. Breeding and selection of cultivated warm-water fishes in Asia and Far East–a review. FAO Fish. Rep. (44) 4:30-66.

Chaudhuri, H., J. Juario, R. Samson and L. Tiro 1977. Notes on external sex characters of *Chanos chanos* (Forskal) spawners. *Fish. Res. J. Philipp.* July-Dec. 1976, 1(2): 76-80.

Chaudhuri, H., S. B. Singh and K. K. Sukumaran. 1966. Experiments on–large-scale production of fish seed of the Chinese grass carp, *Ctenopharyngodon idellus* (C.& V.) and the silver carp, *Hypophthalmichthys molitrix* (C. and V.) by induced breeding in–ponds in India. *Proc. Indian Acad Sci.* (b) 63:80-95.

Delmendo, M. N. and A. G. Angeles. 1975. Preliminary observations on the spawning of milkfish *Chanos chanos* (Forskal). Fisheries Forum, Philippine Council for Agricultural Research, FNRCA, Pedro Gil St., Manila, Jan. 28, 1975:1-22.

Delsman, H. C. 1926. Fish eggs and larvae from the Java Sea, Treubia, 8(3-4):389-412.

Delsman, H. C. 1929. Fish eggs and larvae from the Java Sea. Treubia, 11 (2):275-286.

Delsman, H. C. and J. D. F. Hardenberg. 1934. De indische zeewissen en zeewisserij. Batavia, Visser and Co., 388 p. (Cited by Schuster, 1960).*

Inland Fisheries Project Technical Report No. 5. (1974): January 1 through June 20, 1976, University of the Philippines College of Fisheries, pp. 9-20.

Kuo, C-M., F. H. Shehadeh, and K. K. Milisen. 1973. A preliminary report on the development, growth and survival of laboratory reared larvae of the grey mullet *Mugil cephalus* L, J. Fish. Biol., 5:459-470.

Liao, I–C. 1957. Experiments on induced breeding of the grey mullet in Taiwan from 1963 to 1973. *Aquaculture*, 2:1-24.

Liao, I-C, Y-J. Lu, T-L. Huang and M. C. Lim. 1971. Experiments on induced breeding of the grey mullet *Mugil cephalus* Linnaeus. *Aquaculture*, 1:15-34.

Liao, I-C, and Y-S. Chang. 1976. A preliminary report on the gonadal development of adult milkfish *Chanos chanos* reared in tank. Proceedings, International Milkfish Workshop Conference, Tigbauan, lloilo, Philippines, May 19-22, 1976, pp. 121-132.

May, R. C. 1970. Feeding larval marine fishes in the laboratory: A review, *Calif. Mar. Res. Comm.*, CalCOFI Rept., 14:76-83.

May, R. C. 1974. Factors affecting buoyancy in the eggs of *Bairdiella icistia* (Pisces: Sciaenidae). *Mar. Biol.* 28:55-59.

May, R. C. 1975. Effects of temperature and salinity on fertilization, embryonic development, and hatching in *Bairdiella icistia* (Pisces: Sciaenidae), and the effect of parental salinity acclimation on embryonic and larval salinity tolerance. U.S. National Marine Fisheries Service, Fish. Bull. 73(1):1-22.

Nash, C. E. and C.M. Kuo. 1976. Preliminary capture, husbandry and induced breeding results with the milkfish *Chanos chanos* (Forskal). Proceedings, International Milkfish Workshop Conference, Tigbauan, lloilo, Philippines, May 19-21, 1976. pp. 139-159.

Reintjes, J. W. 1962. Development of eggs and yolk-sac larvae of yellowfin menhaden. U.S. *Fish and Wildlife Service,* Fish. Bull. No. 202, 62:93-102.

Schuster, W. H. 1960. Synopsis of biological data on milkfish *Chanos chanos* (Forskal), 1975. *FAO Fish. Biol.* Synop. No. 4, Rome, FAO, FB/00/S4. 58 pp.

SEAFDEC Progress Report 1977. Background information on Milkfish Research Breakthrough at the SEAFDEC Aquaculture Department, Tigbauan, lloilo, Philippines, No. 1, May, 1977.

Senta, T., S. Kumagai and L. Ver. 1976. Occurrence of milkfish eggs in the adjacent waters of Panay Island, Philippines. Proceedings, International Milkfish Workshop Conference, Tigbauan, lloilo, Philippines, May 19-21, 1976, pp. 167-180.

Tampi, P. R. S. 1957. Some observations on the reproduction of the milkfish *Chanos chanos* (Forskal). *Proc. Indian* Acad. Sci., 46:256-273.

Tang, Y- A. 1964. Induced spawning of striped mullet by hormone injection *Jap. J. Ichth.* 12:23-28.

Vanstone, W. E., A. C. Villaluz and L B. Tiro, Jr. 1976a. Spawning of milkfish *Chanos chanos* in captivity. Proceedings, International Milkfish Workshop Conference, Tigbauan, lloilo, Philippines, May 19-22, 1976. pp. 222-226.

Vanstone, W. E., A. C. Villaluz, P. E. Bombeo and R. B. Belicano. 1976b. Capture, transport and domestication of adult milkfish *Chanos chanos*. Proceedings, International Milkfish Workshop Conference, Tigbauan, lloilo, Philippines, May 19-22, 1976. pp. 206-221.

Yashouv, A. 1969. Preliminary report on induced spawning of *M. cephalus* (L.) reared in captivity in freshwater ponds. Bamigdeh 21:19–24.

1. Experiments in Artificial Propagation and (Larval) Development of *Mugil cephalus* Linnaeus in India

Editor's Note

Experiments on induced breeding of *Mugil cephalus* L., by pituitary injections were carried out in 1961 and during 1969-73 near the Chilka Lake. Homoplastic pituitary glands were generally used for hypophysation. The effective dosages varied from 8-16 mg/kg of fish in 1961 and 2-8 glands per fish in other years and dry method of stripping was followed. Success in hypophysation of *M. cephalus*, was achieved in 1961 when a single hatchling was obtained out of a few developing eggs. During 1969-73, out of 50 sets injected 31 responded and eight spawners yielded 26,050 hatchlings. The embryonic and larval development were studied. Majority of the hatchlings died when 3-4 days old but a few thousands could be reared upto 10 days and they were gradually acclimatised to fresh water, while a few could be reared for more than a month, one was reared for 325 days. Fluctuations and sudden fall in water temperature and lack of proper food appeared to be responsible for high mortality of larvae.

Introduction

Mugil cephalus L., constitutes one of the most important commercial fisheries in the coastal waters and estuaries in India. It is cultured both in brackish and freshwater ponds. The young ones of this species along with those of a few other grey mullets are collected from the estuaries of both east and west coast of the country during November to March for stocking in ponds. Natural supply of fry is unpredictable and invariably comprises a mixture of economic and uneconomic species. However, the quantity of seed of *M. cephalus* collected from the natural sources is inadequate to meet the growing demand of seed for aquaculture in the country. With a view to evolve a dependable technique to breed *M. cephalus* and for large-scale production of its seed, experiments were initiated by the Central Inland Fisheries Research Institute, Barrackpore, India in 1961 when, for the first time in the world, success was achieved in breeding *M. cephalus* by injection of fish pituitary hormones (Barrackpore 1962, IPFC 1963). The experiments carried out in India on induced breeding of *M. cephalus* by hypophysation and on rearing of fry in 1961 and 1969-73 have been described in the paper.

During October to January maturing specimens of the grey mullet migrate from the to sea for spawning (Jhingran, 1958). Enroute to the spawning ground these spawners are intercepted in the coastal areas near the lake mouth and caught in large numbers by fishermen.

For catching mullet spawners, a simple type of drag net, locally called *Chhata jal* (length 10-12 m and width 1.2 m), operated by two persons is used. The fishes on their spawning migration are usually chased by dolphins and sometimes by sharks and other predators and are driven towards the shore. The fishermen awaiting for the opportunity catch them by swiftly casting the net and dragging it in a semicircle in ankle deep water on the shore. The catch of *M. cephalus* usually comprises 0.3-0.75 kg males and 0.75-2.1 kg females. When the requisite number of good spawners were collected they were transported to the camp laboratory by an outboard motor boat in 1961 and by jeep in subsequent years. Three females and six males could be transported successfully in a fry carrier or plastic pool (350 litre capacity). The camp was set up near the fishing site to avoid the problem of excessive handling and transportation of brood fish.

Breeding Hapas and Holding Tanks

The breeding hapas (cloth container) and plastic tanks were used for holding spawners before and after injection.

During 1961, breeding hapas were fixed in the outer channel. The hapas were tied to bamboo poles inside a bamboo fencing so as to prevent entry of predators and isopods. But, in 1969 and 1970-71, these were fixed either in the sea or in pools near the lake mouth which got connected with the open sea during high tide. Hapas were also fixed in a creek connected with the lake where salinity was fluctuating (4 to 16 per cent). During two experiments in 1970-71, breeding hapas fitted to wooden frames were fixed in the pools with the help of two anchors. In 1971-72 and 1972-73, hapas were fixed in circular, portable plastic tanks (9,000 1) filled with sea water pumped in from the sea.

Experiments on Induced Spawning

The experiments were carried out from first week of December to second week of January. The technique of hypophysation was same as developed for Indian major carps (Chaudhuri, 1960; 1963).

The dosages varied according to the condition of the fish and was expressed in number of glands per fish and also mg/kg weight of fish. The weight of glands injected could be made from the average weight of glands of male and female donors recorded earlier (Table 1).

Table 1: Sexwise Range of Size, Body and Pituitary Gland Weight in *M. cephalus*

Year	*Sex*	*Size Range in (mm)*	*Weight Range in (g)*	*Weight Range of Pituitary Glands in (mg)*
1961	Male	360–413	440–690	5.0–6.8
1974	Male	324–374	300–500	2.8–6.0
1961	Female	495–550	1260–1760	12.8–15.4
1974	Female	401–561	750–1880	7.0–16.0

Dry method of stripping was adopted as the injected female fish failed to spawn in the hapas after administration of two homoplastic pituitary extract injections at an internal of 6-8 hours. A third injection was administered when the fish failed to respond after the initial two injections. Ripe males without injection were also successfully used.

The eggs remained for 15-20 minutes for water hardening and they were released in trays glass jars, plastic pools and also in hapas fixed in plastic tanks and in natural depression on the sea coast and outer channel of the lake for incubation. The larvae were reared in the same surroundings as eggs, but they were kept indoors except for hapas fixed in large tanks and outer channel or sea. Water circulation and aeration facilities were provided as per availability. The larvae were fed on artificial feed (boiled yolk, milk powder, fine G. O. C. powder etc.) earlier and in later years with plankton collected from the lake mouth area.

Results Experiments in 1961

Preliminary induced breeding experiments were conducted simultaneously on *M. cephalus* and *M. macrolepis* during 1961 when extracts of pituitary gland were injected @ 8-16 mg/kg/body weight of the recipient fish in 2-3 doses. The eggs were easily stripped from injected females and mixed with freely flowing milt from males but fertilisation was not successful in lake water presumably due to low water salinity (5 per cent). It was observed that sperms of *M. cephalus* died as soon as they came in contact with lake water but sperms of *M. macrolepis* were active only for a short while. However, in sea water they were active for about 10 minutes and successful fertilisation occurred but the developing eggs died and one hatchlings was produced.

The results of 1961 experiments, however, indicated that it could be possible to get successful fertilisation on these mullets if experiments were conducted in sea water with facilities for circulation and aeration of water which is essential for proper development and survival of hatchlings.

Experiments During 1969-73

After a gap of 8 years a second attempt was made during December, 1969 and thereafter in subsequent years during December-January in 1970-71, 1971-72 and 1972-73.

In 1969, 14 sets of *M. cephalus* spawners were injected and on two occasions eggs were fertilized but the development did not proceed beyond cell division stage. In 1970-71, 14 sets were injected out of which successful fertilization was obtained in 5 sets and two sets of spawners yielded hatchlings. Out of 12 sets of experiments conducted in 1971-72, hatchlings were produced on four occasions only even though successful ovulation took place in 11 sets. In 1972-73, 10 sets of experiments were conducted and 9 sets gave positive results but only 3 sets yielded hatchlings.

Effective Dosage

While the minimum effective dosage was 2 (donors–one male and one female *M. cephalus*) glands per fish, the maximum dosage was 8 glands per fish for inducing ovulation in *M. cephalus.* Synohorin was used along with pituitary extract in 17 experiments mainly during 1969, 1970-71 and pituitary extract of Indian major carps was tried on one occasion, both failed to induce successful ovulation.

Stripping and Fertilization

The time taken for the injected female to become ready for ovulation usually varied from 4½ to 12 hours after the last injection at a water temperature 16.0 to 27.5°C. The abdomen of the recipient spawners distended considerably and dialation of the genital opening before ovulation was remarkable. In some cases, blood clot was observed in the genital opening just before the specimens were ready to ovulate. The distended abdomen became very loose and soft to touch and the eggs were released freely without or with slight pressure on the abdomen. Percentage of fertilization was in general poor. It varied from 5 to 50 per cent only.

Immediately after fertilization the eggs were observed floating at the surface but subsequently sank to the bottom in still water. It was rather a difficult task to seggregate the good eggs from the bad ones since both bad and good eggs lay at the bottom in still water in the hapa/tray.

Embryonic and Larval Development

Preliminary observations of larval development were made in the camp laboratory on fertilized eggs during 1970. The average size of the fertilized eggs was 0.8 mm. Each egg contained one large, yellow shining oil globule of 0.34 mm in diameter and a narrow peri-vitelline membrane. Blastoderm was formed within 3 hours of fertilization. Gastrulation was completed after 14 hours and 40 minutes of fertilization.

The embryo was fully formed 21 hours after fertilization and pigmentation started 1 hr later. 33 hours old embryo was densenly pigmented with dark stellate chromatophores which were concentrated on dorsal surface of the embryo. About 16 myotomes could be counted at this stage. The tail was free from the yolk mass and the twitching movements of the embryo at an interval of about 2 minutes was seen. Rapid twitching movements at an interval of 50 60 seconds were observed after 37½ hours of fertilization. The movement was vigorous in the head region and this resulted in the emergence of head first and the tail became free later from the egg case after 3-4 rapid movements of hatchling. The newly hatched larva was comma-shaped and had a large yolk mass with prominent oil globule. The beginning of the mouth formation was indicated as a ventral depression in the head region in one day old larva but the formation of mouth was completed only in a 3 day old hatchling.

Rearing of Larvae

In 1970-71, 1,500 hatchlings were obtained from one injected female which were reared in enamel trays and in plastic pools in sea water. Many hatchlings died on the third day after hatching. The rest were transported in plastic bags under oxygen packing to the Central Inland Fisheries Research Substation, Cuttack at a distance of 116 km. The larvae were then released in big enamel trays and in glass jars (10 1 capacity) containing sea water. No food was supplied until the larvae were four days old. Subsequently they were fed with plankters, *viz.*, rotifers, nauplii of copepods and cladocerans collected from freshwater ponds. Plankton was filtered through a close meshed cloth before feeding the larvae. In one tray the larvae were fed with mashed boiled egg yolk. Mortality among the larvae, however, continued and all of them died within 6 days after hatching.

In 1971-72, out of 4 sets of spawners, over 4,350 healthy hatchlings were obtained. Attempts were made to rear them simultaneously in lass jars and also in outdoor and indoor plastic pools fitted with aerators, near the experimental site. Plankters collected from the lake mouth area were supplied to the hatchlings after sieving. Two batches of larvae survived for 5 days but died within 6 days after hatching while hatchlings of other two batches did not survive for more than 2 to 3 days. Mass mortality appeared to have occurred due to fluctuation of temperature leading to considerable fall of water temperature in plastic tank at night. Out of the hatchlings which survived for 5 days at the experimental site, 700 hatchlings (3 days old) were transported in sea water media to Cuttack under oxygen packing.

They were reared in enamel trays provided with aerators. Gradual freshening of the sea water was continued from 4 days onwards by daily addition of freshwater. Rotifers, copepods and cladocerans collected from the freshwater ponds were supplied to the larvae after filtration. *Chlamydomonas* sp. cultured in the laboratory was also provided as food. Majority of the larvae died when 7 days old, only one larva survived for 13 days.

Over 20,200 hatchlings were produced from 3 sets of spawners in 1972-73. The first batch produced 6,000 hatchlings died within 8 days. The second batch comprising 100 hatchlings died on the third day. The last batch of 14,100 hatchlings survived for the maximum period. Larvae were reared in plastic pools containing sea water at the experimental site for a month. Plankters collected from the lake mouth area were sieved, those measuring 32 to 64 were supplied to the larvae in the morning and evening.

With the increase in size of the larvae, correspondingly bigger size of plankters were provided. Majority of the larvae survived for 10 days. Mortality started thereafter and continued till only 14 larvae survived on the 30th day of rearing. These were transported to the Research Substation, Cuttack and released in a plastic pool containing sea water for further rearing. 8 more larvae died on the 32nd day of rearing. By way of replenishment, freshwater was added to the plastic pool daily to acclimatise the fry to a purely freshwater condition. 4 more larvae died on the 40th day of rearing The remaining two larvae were transferred to a big plastic pool (9001 capacity) and were fed regularly with freshwater zooplankters consisting of rotifers, cladocerans and copepods. One of the fry which had reached a length of 72 mm and weight of 6 g died when it was 160 days old, due to attack of leeches. The surviving one could be reared upto 325 days when it died as a result of heavy fungal infection.

Discussion

The experiments conducted in 1961 demonstrated that mature specimens of the grey mullet could be collected for artificial propagation and that they easily respond to stripping after hormone injections resulting in ovulation, fertilization and hatching of eggs. Experiments conducted by Tang and his associates in 1963 (Tang, 1964) was a step forward. While he could hatch eggs in aquaria with circulation of sea water and also under natural conditions near the sea coast, the hatchlings did not survive beyond four days. This success aroused interest in artificial breeding of mullets in various parts of the world and successful breeding experiments were carried out in Israel and Hawaii. Research in the field of breeding and rearing was intensified in Taiwan. In India, subsequent experiments conducted during 1969-73 showed further progress towards production of hatchlings and rearing of larvae.

With regard to dosage, majority of the workers calculated it in terms of number of pituitary glands either per fish or per kg body weight of fish. While in our experiments, dosages were calculated in mg of pituitaries per kg weight of recipient fish but in subsequent years, dosages were calculated in terms of the number of pituitary glands. Distinction was, however, made between glands from male and female donors since male glands were much smaller than those of females (Table 1).

Tang (1964) and Liao *et al.* (1971; 1972) obtained successful results by injecting pituitary hormones in combination with Synahorin. No significant advantage in the use of Synahorin was observed by us. Further study, however, is needed for determining the role of Synahorin in the precipitation of spawning in mullets. Shehadeh and Ellis (1970) attained success in inducing spawning in *M. cephalus* by injection of salmon pituitary extract and Synahorin. Shehadeh (1973) reported success in induced spawning by using partially purified salmon gonadotrophin. While the present authors obtained negative result in injecting a female mullet with Indian major carp pituitary extract. Yashov (1969) achieved success with extract of common carp glands.

In the experiments described in this paper, generally two injections were given to the female brood fish at an interval of 6-8 hours and stripping was done in another 4,1/2 to 12 hours. In Taiwan (Ling, 1970), the usual interval between the first and successful ovulation was achieved within 48 hours after the first injection. Yashouv (1969) gave three injections at intervals of 7 hours and spawning occurred 17 hours after the third injection (31 hours after first injection). It is interesting to note that mullets from Chilka Lake area of India took only 10–20 hours for ovulation after the first injection. This variation in the total period required for ovulation could probably be attributed to the state of gonadal maturity and water temperature.

Ling (1970) observed that the size of hatchlings of *M. cephalus* varied from batch to batch (2.1-3.1 mm). Similar variation in size of hatchlings were also observed by us. This is also true with regard to the size of eggs and oil globule. In 1961, the average size (diameter) of the fertilized eggs and oil globule was observed to be 0.48 mm and 0.29 mm respectively. But in 1970-71, the same was found to be 0.8 mm and 0.34 mm respectively. According to Ling (1970), the diameters of fertilized eggs in Taiwan varied between 0.930.95 mm and that of oil globule was around 0.33 mm. As regards the nature of egg, Ling (1970) mentions, "fertilized eggs are buoyant and tended to stay afloat near the surface when the water was in slight motion, but some of them may settle down slowly in still water". According to Yashouv (1969), the eggs floated at the beginning but sank to the bottom when embryos were formed Our observations showed that the eggs immediately after fertilization floated but subsequently sank to the bottom in still water. Buoyancy of the eggs perhaps depended on the salinity of water. In Taiwan, salinity 6f water varied from 30.1 to 35.8 per cent (Liao *et al.* 1971), whereas it varied between 4.0-26.0 per cent in the present experiments.

Experiments conducted so far in India and in countries mentioned above demonstrated that induced breeding of the mullet by hypophysation was not difficult, but the real problem lay in hatching and rearing of the larvae. High mortality of hatchlings is a frequent occurrence especially during the critical period when the mouth is formed and the larvae start feeding when 3 days old. Although much progress has been made during the last decade, yet a lot more has to be achieved to make the technique a dependable one. Liao *et al.* (1971; 1972) observed high dissolved oxygen content and minimum variation in water temperature and salinity as essential factors for success in incubation and hatching, while our present observations indicate that salinity does not play a major role in incubation since eggs were found to develop and hatch successfully in outer channel water wherein salinity fluctuated from 14.0 to 26.0 per cent. For rearing larvae high dissolved oxygen content in water is necessary. Fluctuations and sudden fall in water temperature have been observed to be detrimental to the survival of the hatchlings.

Supply of the right kind of food to the larvae is very important. In Taiwan (Liao *et al.*, 1971; 1972), better results were obtained by feeding mullet larvae with oyster larvae, rotifers, larval stages of copepods, diatoms and unicellular green algae, ("green water"). Artificial food such as egg yolk, wheat flour, milk powder, ground pig liver etc., individually or in combination were fed. In our experiments, the initial attempts made in rearing larvae did not give satisfactory results. In 1972-73, it was arranged to rear the fry at the experimental site and feeding them with natural feed collected from the Lake mouth which is an important source of food organisms of mullet fry. Jhingran and Natarajan (1970) have observed congregation of large number of *M. cephalus* fry in the lake mouth every year in winter (Nov.-Jan.). The supply of food from this source ensured better results.

[A joint paper by H. Chaudhuri, R.M. Bhowmick, G.V. Kotwa, M.M. Bagchi, R.K. Jana and S.N. Gupta in J.I.F.S., 1977]

References

Barrackpore, 1962. Annual Report 1961-62. Central Inland Fisheries Research Institute, Barrackpore, West Bengal, India.

Chaudhuri, H., 1960. Experiments on induced spawning of Indian major carps by pituitary injections. *Indian J. Fish.*, 7: 20-48.

______1963. Induced spawning of Indian carps. *Proc. nat. Inst. Sci. India (B)*, 29 (4): 478-87.

Hendrickson, J. R., 1969. Artificial culture of Milk-fish and mullets. Some limitations on the art at present. *Symposium*, Pacific Science Associatian, Kuala Lumpur.

IPFC, 1963. *Procs. Indo-Pacif. Fish. Coun. 10, Soeul, Sect-1;* 40 and 70.

Jhingran, V. G., 1958. Observation on the seaward migration of *Mugil cephalus*, Linnaeus from the Chilka for breeding. *Curr. Sci.*, 27: 181-82.

Jhingran, V. G. and A. V. Natarajan, 1970. A study of the fisheries and fish populations of the Chilka Lal:e during the period 1957-65. *J. Inland Fish. Soc. India*, I: 49-126.

Liao, I. C. *et al.*, 1971. Experiments on induced breeding of the grey mullet, *Mugil cephalus* Linnaeus. *Chin. Amer. Jt. Comm. Rural Reconstr. Fisheries Ser.*, No. II: 10-29.

______1972. Preliminary report on the mass propagation of grey mullet, *Mugil cephalus* Linnaeus. *Ibid., Ser.*, No. 1?: 1-14.

Ling, S. W., 1970. A brief review of the work done on the induced breeding of *Mugil cephalus* in Taiwan. J. *Inland Fish. Soc. India*, I: 1-12.

Shehadeh, Z. H. *et al.*, 1973a. Induced spawning of grey mullet *Mugil cephalus* L. with fractionated salmon pituitary extract./. *Fish. Biol.*, 5: 471-79.

Shehadeh, Z. H. and J. N. Ellis, 1970. Induced spawning of the stripped mullet, *Mugil cephalus* L. *J. Fish. Biol.* 2: 293-303.

Tang, Y. A., 1964. Induced spawning of stripped mullet by hormone injection *Jap. J. Ichthyol*, 12 Q): 23-28.

Yashouv, A., 1969. Preliminary report on induced spawning of *Mugil cephalus* (L) reared in captivity in freshwater ponds. *Bamidgeh*, 21 (1): 19-24.

2. Notes on the Embryonic Development of the 'Transparent Goby', *Gobipterus chuno* (Hamilton)

The contents is as under:

1. Introduction
2. Breeding Season and Fecundity
3. Embryonic Development
   (*i*) Unfertilized Egg
   (*ii*) Fertilized Egg
   (*iii*) Early Segmentation
   (*iv*) Formation of the Embryo
4. Hatching and Hatchling
5. Discussion
6. Summary
7. References

This article is the maiden production of Prof. Chaudhuri as research assistant immediately after joining the Central Fisheries Deptt. at Barrackpore. The inspiration he derived from the successfull research probably gave a thrawt to the research career.

Introduction

In July, 1948, Mr. A. David of this Research Station, drew my attention to certain specimens of tiny fish with well-developed eggs in his collections from the Kulti river. On detailed examination I found them to belong to a Sicydiaphine Goby, *Gobiopterus chuno*. Mature specimens of this species were subsequently collected by me in September, 1948, from Kulti and an account of the embryonic development of *Gobiopterus chuno* which was worked out is given in this paper.

The minute transparent goby, *Gobiopterus chuno* (Hamilton) that is common in brackish water fields near the Kulti outfall, in the River Hooghly and also in the Pulta settling tanks and filter beds, was found to be mature during June, July and October, 1948. Since fertilized eggs could not be collected from these natural habitats, artificial fertilization was attempted in the laboratory successfully on the 29th September, 1948, at Barrackpore, by stripping eggs from the ripe female into a petri-dish with a little water and teasing the dissected-out testis of the ripe male, in the dish.

This minute goby is abundant in fresh-water tanks of the Jobra Fish Farm, Cuttack. It was found to breed in these tanks almost throughout the year. By keeping ripe specimens of either sex in the laboratory aquaria the process of deposition of eggs was fully observed during November, 1949. Artificial fecundation of ova also was successfully repeated and the different stages were re-examined for confirmation. The hatchlings lived up to 12 hours in the laboratory.

Our knowledge of the life history of Indian Gobiids is still very meagre, though the contributions of Bhattacharya (1916), Raj (1916), Aiyar (1935), Jones (1937) and Alikunhi *et al.* (1948) have helped in elucidating the embryonic and larval life of species like *Gobius ostereicola, Acentrogobius neilli, A. viridipunctatus* and *Olossogobivs giuris. Gobiopterus chuno* is usually abundant in many carp nurseries under observation in Orissa, and the role of this ' weed fish' in the pond economy has to be correctly assessed for profitable carp culture. An intimate knowledge of the life habits of the species is essential for any such assessment and the following account of its early development will form a useful link in our knowledge of the species.

Pillay and Sarojini (1950) have recently published their observations on the larval development of *Gobiopterus chuno.*

Breeding Season and Fecundity

As already pointed out, gravid females have been collected during January to November, indicating that the species breeds almost throughout the year. The adult males are easily distinguished from the females by the presence of a distinct fingerlike genital papilla; the corresponding structure in the female being a stouter prominence. The body being always of glassy transparency, the ovaries in the gravid females are clearly seen through by the naked eye. The ovarian eggs are transparent; some scattered chromatophores are present in the ovarian wall and in some specimens the entire ovary has a light pinkish tinge. 140 to 150 eggs mature at a time. The ova are closely packed in the ovary and the immature eggs and early oocytes too are generally present in considerable numbers in the ripe ovaries. The maximum length of the adult fish, so far collected by me is only 25 mm. For the size of the adult fish the eggs are relatively large. Deposition of eggs was observed by keeping ripe specimens in glass aquaria in the laboratory. The fish was swimming about mostly near the bottom; the genital papilla appeared highly tumid and extended. Eggs were slowly coming out in ones or twos and since the fish was moving near the bottom, as soon as the egg was coming out of the genital papilla, dirt and particles of debris began to stick to it, indicating that the egg is sticky in nature. The genital papilla was often touching the bottom of the aquarium, and the eggs as they came out, adhered to the bottom.

Embryonic Development

Unfertilized Egg

The unfertilized mature egg is more or less spherical in shape and has an average diameter of 0.54 mm. The egg membrane is slightly swollen. There is a distinct micropylar opening around which numerous fine radiating adhesive filaments are present. The egg gets attached to some substratum by means of these filaments. The yolk which often has a yellowish tinge (when preserved) is coarsely granulated and has several oil globules of different size.

Fertilized Egg

Immediately after fertilization the zona radiata begins to swell up considerably and within five minutes it forms an oval capsule enclosing the egg. The capsule continues to swell up and within ten minutes becomes characteristically pear-shaped. The outer surface of the zona radiata appears finely striated longitudinally. The average dimensions of these egg capsules are as follows:

Total length of the egg capsule: 1-12 mm.

Maximum width of the egg capsule: 0-73 mm.

Diameter of the fertilized egg: 0-45 mm.

A streaming movement of the egg protoplasm now takes place. The egg appears in two distinct halves, the lower half, nearer to the attached portion of the egg capsule, of accumulated protoplasm in the form of a finely granular blastodisc and the upper half forming the bulk of the egg, representing the yolk.

Early Segmentation

40 minutes after fertilization the cap like blastodisc undergoes the first cleavage to form the two blastomeres. The second cleavage takes place at the end of the first hour while the 8-celled stage is reached at the close of the next half hour. Two hours after fertilization, the 4th cleavage results in 16-cells. Subsequent divisions rapidly follow and in three hours and fifteen minutes after fertilization, the blastodisc is transformed into a cluster of small cells forming the blastoderm.

Formation of the Embryo

The crown of blastoderm cells now begin invasion of the yolk, spreading over the latter as a thin layer. About five hours after fertilization, the embryo is indicated as a median thickening of the blastoderm. 40 minutes later the head and tail ends of the embryo are distinguishable. Two myotomes are visible at the hind portion of the embryo which is slightly elongated. 17 hours after fertilization, the embryo gets further elongated and measures 0-60x0-17 mm. The rudiments of the optic cups appear, and five myotomes are differentiated. The head of the embryo is directed towards the base of the capsule. Two and a half hours later, the embryo becomes roughly C-shaped and measures 0-66x0-22 mm. The auditory vesicles have appeared and 10 myotomes are visible. The head and tail ends of the embryo are beginning to get free from the yolk mass which is steadily being used up. A narrow space has appeared between the head and the yolk, locating the position of the heart. Within another two hours, further differentiation takes place and the embryo measures 0.72 mm. x 0-30 mm. Tail is further elongated and the lens has appeared in the optic cup. 13 myotomes have been differentiated and the embryonic fin fold is clearly seen around the tail. Rudiment of the heart has appeared but has not begun pulsating. The embryo begins to execute faint side to side twitching movements of the tail. Between the 21st and 24th hours after fertilization, it affects a change in its orientation within the egg capsule, by a complete reversal of the position of the head, which accordingly becomes nearest to the distal end of the egg capsule. In the 27½ hours old embryo, the heart begins pulsating and two minute concretions appear in each of the auditory vesicles. The dorsal and ventral fin folds are better differentiated. The pulsations of the heart are regular and rapid, at the rate of about 108 beats per minute. 17 myotomes are seen.

Circulation of blood is fairly well established when the embryo is about 38 to 42 hours old, and the rate of heart beats increases to 155 per minute. The hind portion of the alimentary canal is indicated. The tail end is elongated and curved forwards. The embryo moves frequently within the egg capsule. In the 45 hours old embryo, the eyes begin to get slightly pigmented. The rudiments of the pectoral fins appear as bud-like prominences. In another 9 hours, the eyes get fully pigmented and dark. The yolk is reduced to a small rounded mass. The auditor vesicles are now closer to the eyes. The alimentary canal is well differentiated. The total length of the embryo is now about 1.62 mm., the post-anal portion is longer than the pre-anal. Pectoral rudiments are larger and on the dorsal side, the embryonic fin fold is seen to commence from the nape.

In the 62 hours old embryo, the rudiment of the air bladder is visible as a small oval structure on the dorsal aspect of the yolk mass. The buccal invagination is clearly indicated. The eyes are dark and shining. The pectoral rudiments are larger and flap-like. Blood is now red in colour. The embryo makes constant vigorous movements and when about 84 hours old, it is ready for hatching.

Hatching and Hatchling

The rupture of the egg capsule takes place near its distal end. The head portion of the embryo presses on the distal part of the capsule where the maximum strain is felt owing to the vigorous movements of the embryo. During hatching the head portion first comes out of the capsule.

The hatchling is very slender and measures 2.2 mm. in length. The eyes are fully pigmented and the mouth and anal openings have appeared. The anus is very anterior in position and is situated at the level of the 7-8th myotome.

A small yolk mass still remains. The dorsal embryonic fin fold commences at about the level of the hind portion of the yolk mass. Ventrally there is a small pre-anal fin fold. The tip of the notochord is straight and the caudal fin is rounded. A single chromatophore is present, ventrally midway between the anus and the tip of the tail. Two or three chromatophores are present on the dorsal aspect of the air bladder also.

Discussion

Like many other gobiids, *O. chuno* appears to be a perennial spawner. For the tiny fish, the ovarian eggs are larger than those of *Acentrogobius neilli* which grows to more than three times the length of *O. chuno.* The shape of the egg capsule is characteristic and is easily distinguished from that of *A. neilli, A. viridipunctatus* or *G. giuris.* The period of incubation is longer than in *A. neilli* and shorter than in *A. viridipunctatus,* but this is found to fluctuate with the temperature of the water. The differentiation of structure is typical of Gobiids. The change in the orientation of the embryo soon after it begins movement within the egg capsule is interesting. While Aiyar (1935) does not specially mention this change in position of the embryo in *A. neilli,* seem to indicate that such a change does take place in that species; the embryo had not commenced movement. In *A. viridipunctatus* the reorientation appears to take place at an earlier stage (Jones, 1937); while in *Boleophthalmus boddaerti* such a change is indicated more or less at the same stage as in *A. neilli* (Jones, 1937). Alikunhi *et al.* (unpublished) has also observed that in G. *giuris* a similar reorientation of the embryo takes place within the egg capsule. The relative density of the yolk and protoplasm during the different phases of embryonic differentiation and hatching requirements are perhaps responsible for this reorientation in the position of the embryo.

The slender hatchling is a little longer than that of *A. neilli* and is in more or less the same stage of development as the latter. The scarcity of pigment in the hatchling is in conformity with the condition in the adult which is of an almost glassy transparency.

Summary

Gobiopterus chuno is common in brackishwater fields and is often found in rivers and freshwater ponds. It is found to breed almost the whole year round. The early development of this species was worked out by artificial fertilization of eggs. The eggs, with an average diameter of 0.54 mm. are sticky and get attached to objects by means of adhesive filaments. After fertilization the zona radiata swells up into a pear–shaped egg capsule and measures about 1.12 mm. in length. The yolk is pale yellowish in colour and coarsely granulated. The period of incubation lasts 80 to 96 hours. The hatchling, 2.2 mm. long, has a single chromatophore mid-ventrally behind the anal region and a small patch over the air bladder. The eyes are fully pigmented, the mouth is fairly well formed, the air bladder is present and pectorals are functional.

Conclusion

Various advancements and successes in the field of Aquaculture.

It may be viewed that the Fishery Scientists of India have worked ceaselessly in the field of development of aquaculture and achieved useful research findings in National and International fields. They have initiated active waves of development and management issues to revolutionize the protein revolution in India.

A few such achievements which emanate from the research success since 1940 may be noted as below:

1. Enactment of Marine Fisheries Regulation Act in some Coastal States.
2. Emergence of West Bengal and Andhra Pradesh as the foremost Centres of Aquafarming.
3. Farming of Nile Tilapia fed by treated effluents of breweries.
4. Introduction of several exotic fish and their successful breeding.
5. On shore Commercial pearl production. Standardised by Central Marine Fisheries Research Centres.
6. Setting up of SPF Shrimp brood Stock and SPF Shrimp naupli Project Centre in Andamans by MPEDA.
7. Hiralal Chaudhuri best farmer award Instituted by "Fishing Chimes, Jaysree Charitable Trust".

8. Passing in the Parliament of Coastal Aquaculture Authority Bill which became an Act. A Coastal Aquaculture Authority has been Constituted.
9. Approval of Fish Farmers day by Central Government in 2000 which is being Celebrated every year on 10th July.
10. Extension of large brackish water and Coastal Marine Farmers by several newly floated Companies.
11. Farming of Sea weed through self-help groups.
12. Standardization of Technology for Calm production by CMFRI.
13. Continuation of research on commercial production of Mariculture also on Sea-Cucumber, Snail and Azolla etc.
14. Modernization of Training Programme by CIFA/CIFRI for spread of improved farming technologies.
15. Immense progress made Shrimp farming technologies in the fields of seed production hatchery etc.
16. Emergence of hatchery integrated Fish Farming system.
17. Progress made in Ornamental fish production and commercialization under FFOA and NCDC programme.
18. Initiation of various welfare schemes for upliftment of poor fisherman such as pension scheme, personal accident Insurance scheme, Savings-cum-relief scheme, Land housing scheme etc.
19. Advances in aquaculture and Marine Biotechnology to meet human protein requirement.
20. The works of Aquaculture Scientist during six decades from 1940 to 2000 have initiated diversification of Aquacultures in India for nutritional Security and employment for poor in rural and Semiurban areas which centre round several fish species.
21. Setting up of hatcheries for giant fresh, water prawn seed production in non-coastal states.
22. Vast improvement of technology of hatchery.
23. Establishment of Central Institute of Fisheries Education Universities and College on Animal Husbandry, Veterinary and fishery in various States.

24. Improvement of techniques of Seed-production of several cat fishes and enhancement of production.
25. Revolution of Scenario in Poly-farming and integrated farming system.
26. Popularization of Paddy-cum-Fish farming system all over the country.
27. Improvement of Cage culture system.
28. Emergence of Fish Farmers Development Agencies as a viable mechanism for spreading fish production technologies.
29. And very many other issues.

(*Courtesy*: *Fishing Chimes*).

References

Aiyar, R. G. 1935. Observations on the Development of *Acentrogobius Neilli. (Gobius Neilli* Day). *Zool. Anz., Bd.* Ill, 83-92. Alikunhi, K. H., Rao, G. L. and Jacob, P. K. 1947. Bionomics and Development of *Glosso- gobius giuris* Hamilton. *Proc. 34th Ind. Sci. Congr.* (*Delhi*); (3); 179.

Bhattacharya, D. R. 1916. Stages in the Life-history of *Gobius, Petroscrites* and *Hemirhamphus. Mem. Ind. Mus.*, 5 (4); 381-392.

Jones, S. 1937. Observations on the breeding habits and development of certain brackishwater fishes of Adyar, Madras. *Proc. Ind. Acad. Set.*, 5 (6); 261-289.

Pillay, T. V. R. and Sarojini, K. K. 1950. On the Larval Development of the Indian Transparent Goby, *Gobiopterus chuno* (Hamilton) with observations on its Bionomics. *Proc. Nat. Inst. Sci. Ind.*, 16 (3); 181-187.

Sundararaj, B. 1916. Notes on the Fresh-water Fish of Madras. *Rec. Ind. Mus.*, 12 (6); 249-294.

Willey, A. 1910. Notes on the Fresh-water fisheries of Ceylon. *Spol. Zeyl.*, 7 (26); 88-105.

[Published in the Proceedings of the National Institute of Sciences in India, 1951]

Some Common Latin Words Used in Biological Science

A, ab.. off, from, away

A, ah.. not, without

Ac... to, forward

Acantho.. spine, therm

Acet.. vinegar

Acetabul.. vinspar cup

Aci, acu.. a point, needle

Acro.. topmost, tip

Acti, actine.. ray

Ad.. tom at, toward, into

Adelo.. unknown, secret

Adeno.. gland

Adr, adro.. thick, stout

Agni, agnus.. Iamb

Ala.. wing

Alb, albine,. white

Alg, algis.. poan

Alia.. other, different

Alis.. pertaining to

Alti.. tail

Alveo.. hollow

Ambi.. around

Amby.. blunt

Ameb, ameba.. change

Ana.. up, again, backward

Andro.. man

Angui.. snake

Angusti.. narrow

Ante.. before

Antho.. flower

Anti.. opposed to against

Aphan.. hidden, invisible

Aphel.. smooth

Apo.. away from, apart

Arbor.. tree

Arche.. ancient

Archi.. first
Arctic.. northern
Areni.. sand
Argo.. shining, bright
Argyr.. silver
Arthro.. joint
Articulus.. joint
Asc, asco.. bag, bladder
Asc, azule.. dry, parched
Aspi.. shield
Astro.. star
Atri, Atrius.. vestibule
Aura.. gold
Auris.. ear
Austral.. southern
Auto.. self
Auxo.. grow, enlarge
Badis, basac.. walk
Bates.. one who haunts/walks
Batho, bathy.. deep, high
Belon.. needle, dirt
Belu.. beast, moneter
Benacus.. a deep lake
Bi, bis.. two, twice
Boreal.. north
Botry.. a bunch of grapes
Brachium.. arm
Brachy.. short
Brady.. slow
Branchium.. gill
Brevi.. short
Broncho.. windpipe
Bucca.. cheek
Byas.. fine thread
Byth.. depth of sea
Caerule.. blue
Cal, calo.. beautiful
Cald, calid.. hot, warm
Calix.. cup
Calypte.. cover, voil
Campus.. sea monster
Cand, candid.. white brilliant
Capill, capilus.. hair
Capit, caput.. head
Carchar.. jagged
Cardia.. heart
Carni.. flesh
Carpic.. wrist
Carry-O, Karyo.. nut, nucleus
Cata.. down, downward
Catr, cetra.. shield
Cauda.. tail
Cauli.. a stem, stalk
Ceno.. recent, common
Centra.. against
Cephal-e.. head
Ceps.. head
Cera, cerato, kera.. horn
Cerebro.. brain
Cero-o.. tail

Cervi-x.. neck
Chalyb.. steel
Chan-o.. gape, oven
Chao.. empty space
Cheeta.. bristle
Chela.. claw
Chic-a.. a blade of grass
Chili, chilo.. thousand
Chilo.. lip
Chiro.. hand
Chloro.. green
Choano.. funell
Choric.. skin, membrane
Chromo.. time
Chryso.. gold
Chthon.. land
Cilium.. eyelid, fine hair
Circum.. around
Cirrho-O.. orange coloured, tawny
Cirrus.. curl of hair
Clade.. branch, sprout
Clavicle.. a key
Cleido.. key, clavicle
Cleithro.. bar, bolt, key
Cleva.. club
Clino.. bed
Clipeus, cly.. shield
Cloisto.. closed
Coccin ,. scarlet
Coli.. swell
Com, can.. with, together
Cool-O.. hidden
Corn, cornus.. horn
Corpus.. body
Crassi.. thick
Crista.. crest
Cteno.. comb
Cucullus.. hood
Cyanc.. dark blue
Dactylo.. finger
Dar-o, dir.. neck, old, hide
Dasi, dasy.. hairy
De.. from, away, out, down
Del, deloa.. visible
Derm-o.. skin
Di.. double, two, twin
Dia.. through, across
Diadrom.. wandering
Digitus.. finger
Dino.. terrible, whirling
Dis.. apart
Dolicho.. long
Dondro.. tree
Dorsa, dorsum.. back
Dromo.. to run
Dropen.. Sickle
Dys.. bad, badly
E, ex.. cut, without, from
Ec, ek.. out of, off

Echibno.. spine, hedgehog

Ecto.. outside, outer

Ef.. out, from away

Eido.. form, image

Eluthero.. true

Em, en.. in, into

Encaphalo.. brain

Endo, ento.. within

Eo.. dawn

Ephippi.. a saddle

EPi.. upon, to, cover

Erythro.. red

Eso (oeso).. within

Etes.. one who dwells

Ethmo.. a sieve

Etymo.. truth

Eu.. gocoo, wall, true

Eum.. the place where

Eury.. wide, broad

Exter-O.. outer

Extra.. on the outside

F ale, falx.. sickle

Fasciatus.. banded, bundle

Femus.. thigh

Fer.. to carry, bear

Ferrugin.. rust colored

Filii, filum.. thread

Flab.. yellow

Fluvia.. river

Foli.. leaf

Font, fontan.. fountain

Fusca.. brown

Galb.. yellow

Gallban.. Greenish yellow

Gastro, gastero.. stomach

Gero, ger.. to bear

Gladius.. sword

Glans.. Gland

Glauco.. grey

Glossa, glotto.. stomach

Glypno, glypto.. engraved

Gnatho.. jaw

Gompho.. angles

Gonic.. angled

Gracil.. glender

Grandi.. large

Gravi.. heavy

Gymno.. naked

Gyno.. female

Gyro.. round

Hali, halo.. sea, salt

Halpo.. single, simple

Hasta.. spear

Helec, helo.. marsh

Helio.. sum

Hemato, haemo.. blood

Hepato.. lives

Hetero.. different, other

Hippo.. horse

Hispid.. hairy

Holo.. whole
Homo.. like, same
Homo.. man
Hormon.. excite
Hyalo.. glass, transparent
Hydro.. water
Hygro.. moist
Hyo.. 'U'/'Y'-shaped
Hyper.. over, above, excessive
Hypo.. under, beneath
Hypso.. high
Ichthy-o.. fish
Ilio.. intestine
Infra.. below, beneath
Infundibulum.. funell
Integri.. whole
Inter.. between, among
Intra.. within
Labium, Labrus.. lip
Latero.. side
Lati, laty.. wide, broad
Leio.. smooth
Leo.. equal
Lepido, Lepis.. scale
Lepto.. fine, slender
Leuce, Leuko.. white
Limno, Limne.. lake
Lingua.. tongue
Litho.. rock, stone
Littor.. seashore, shore
Longi.. long
Lopho.. crest
Lucid.. shining
Luni.. moom
Lute, Luti.. yellow
Lute.. yellow
Macro.. large, long
Macula.. spot
Malaco.. soft
Manne.. dwarf
Manus.. hand
Mastig-o.. whig
Maxilla.. jaw
Medi.. middle
Mees, mes.. middle
Mega, megalo.. large
Melano.. balck
Mers.. part, thigh
Meta, met.. after, behind
Micro.. small
Mio.. less
Molli.. soft
Monime.. fixed
Mono.. single
Multi.. many
Myo.. muscle, mouse
Myso.. muck
Nasi, nass.. nose
Natau.. swim, flow
Neoto.. swim

Nephre.. kidney
Neuro.. nerve
Nigro.. black
Noo.. new
Noto.. back
Nudi.. naked
Ob.. against, in the way of
Ochr.. pule yellow
Oculus.. eye
Odenta.. tooth
Oiko.. House
Oligo.. few, less
Onco, onono.. barb, hook
Onycho.. claw
Operculu.. cover, lid
Ophi-o.. snake
Opistho.. behind
Opthalmo, Opte.. eye
Ora, os.. smooth
Ornitho.. bird
Ortho.. straight
Os, ossi.. bone
Ost, ostop.. bone
Oto.. year
Ovus.. egg
Oxy.. sharp, pointed
Pachy.. thick
Pallid.. pale
Palso.. ancient
Palvi.. smell
Pan, pant-a.. all
Para.. beside, by the side of
Pector.. breast
Pedi, pes.. foot
Pelagos.. sea
Pelt.. shield
Per.. through, by mean of
Peri.. around, about
Perisso.. out numbered
Phago.. to eat
Pheo.. dusky grey
Phora.. to carry, bear
Photo.. light
Phyllos.. leaf
Physa, physo.. bladder
Phyto.. plant
Pilus.. hair
Pinna, pluma.. feather
Pisei.. fish
Plagio.. sideways
Platy.. flat
Pleco, pleoto.. twisted
Pleo.. more, swis
Pleuro.. lateral
Pliesto.. most
Plio.. more
Ploto.. float, sail
Pneumo.. lung
Podo, pes, pug.. foot
Poecili.. many colored

Pogono.. beard

Poly.. many

Porphyr.. purple

Post.. after

Potano.. river

Pre.. before

Primi.. first

Pro.. before

Procto.. anus

Profund.. deep

Proto.. first

Proxim.. near

Psamno.. sand

Psoudo.. false

Pters.. wing

Ptilo, pterulo.. feather

Pulchellus.. beautiful

Pulmo.. lungs

Purpur.. purple

Pyrrho.. orangeeree

Rachi, rhaci.. spine

Racti.. straight

Ramph.. curved bork or hook

Remus.. oar

Reni.. kidney

Reti.. network

Retro.. backwards

Rhado.. rags, red

Rhin, rhino.. Shark

Rhino.. none

Rhiso.. root

Rhyncho.. beak

Rostro, rostrum.. ancut, beak

Rubicund.. very red

Rubra.. red

Rugos.. wrinkle

Rumi, ramus.. branch

Sagitta.. arrow

Saleni.. moon

Sanguini.. blood

Sapro.. rottan

Sarco.. flesh

Sauro.. lizard

Saxa. rock

Saxatil.. leaving among rocks

Scapho.. host

Scat-o.. dung

Scelio.. crooked

Schiate, ochizo.. split, cleave

Scolux.. word

Scypho..cup

Seier-o.. hard

Semi.. one - half

Serra-t.. saw

Sestro, sesto.. sieve

Sets, seti.. bristle

Sol.. sun

Soleno.. pipe, channel

Soma-to.. body

Soutum.. shield

Sphee.. wadge
Sphyro.. hammer
Spini.. spine
Splanchno.. viacera
Spondylo.. vertebra
Squams.. scale
Stage.. cover, roof
Stema-to.. mouth
Steno.. narrow
Steonano.. crown
Sterni.. breast
Stetho.. cheat
Stinums.. spot, brand
Stole.. a long robe
Stropto.. twiword
Sub.. under
Sulca.. groove
Super, Supra.. above
Sym, sys.. together, with
Tarai.. ankle
Telos.. complete
Tergum.. back
Terti.. third
Tetra.. four
Thermos.. heat
Tiana.. across
Topca.. rough
Trachea.. windpipe
Tracholo.. neck, throat
Triono.. hair
Troono.. wheel
Tropho.. food, nourish
Tropo.. twin
Tule, to-o.. and, for, distant
Turbell.. a little crowd
Tycha.. chance
Ult.. beyond
Unguis.. claw
Uni.. one
Uro, ours.. twil
Ventro.. belly
Virid.. green
Vittat.. striped
Vora.. to eat
Vulgur.. common
Xantho.. yellow
Xeno.. stranger
Xero.. dry
Xipno, zipho.. sword
Xygo.. yoke, write
Xylo.. wood

Some Common Greek and Latin Prefixes and Common Words

Greek

A, AH–WITHOUT

ALLOS–OTHER

AMPHI–BOTH

ANA–UP, UPON, BACKWARD

ANDRE–MALE

ANISO–UNEQUAL

ANTI–OPPOSED TO, AGAINST

APO–AWAY FROM, APART

ARCHAEO, ARCHI–ANCIENT

ARTHO–JOINT

AUTO–SELF

BRACHIO–ARM

CATA–DOWN

CEPHALO–HEAD

CERA TO, KERAS–HORN

CHELE–CLAW

CHILO–LIP

CHIRO–HAND

CHLORO–GREEN

COEL–HOLLOW

CYANO–BLUE GREEN

DACTYLO–FINGER

DERM, DERMATOS–SKIN

DIA–THROUGH, ACROSS

DI–DOUBLE, TWIN

DIPLO–DOUBLE

DYS–BADLY, BAD

EC, EK–OUT OF, OFF

ECHINO–SPINE

ECTO–OUTSIDE

EN, EM–IN, INTO

ENDO–INSIDE

ENTO–WITHIN

EOS–DAWN

EPI–UPON

ERYTHROS–RED

EU–GOOD, WELL, TRUE

EURYS–BROAD

GLOSSO -TONGUE

GNA THO–JAW

GYMNO–NAKED

GYNO, GYNE–WOMAN, FEMALE

HALO–SALT

HAPLO–SINGLE, SIMPLE

HELIOS–SUN

HEMI–HALF

HETERO–OTHER

HOLOS–ENTIRE

HOMO–SAME

HYPER–ABOVE, BEYOND, GREATER THAN

HYPO–UNDER, BELOW

ISO–EQUAL, SIMILAR

KARYON–NUCLEUS

KRYPTOS–HIDDEN

LEPTO–FINE, BLENDER

LEU CO–WHITE

LUTEA–YELLOW

MACRO–LARGE, BIG, LONG

MALAKOS–SOFT

MEGA, MEGALO–LARGE

MEION–SMALLER, LESS

MELANO–BLACK

MERO–PART, PORTION

MESO–MIDDLE

META–AFTER, BEHIND

MICRO–SMALL

MIKTOS–MIXED, MIXING

MONO–SINGLE, ALONE

NOTOS–BACK

ODONTO–TOOTH

OIKOS (ECO)–HOUSE

OLIGO–SMALL, FEW

OPHIO–SNAKE

ORTHO–STRAIGHT

OSTEO–BONE

OURA, URO–TAIL

PARA–BY THE SIDE OF

PELAGOS–HIGH SEA

PERI–AROUND, ABOUT

PHAGO–TO EAT

PHOTOS–LIGHT

PHYSO–BLADDER

PHYTON–PLANT

PODOS, PES–FOOT

POLY–MANY

PRO–BEFORE

PROCTOS–ANUS

PROTO–FIRST

PTERA–WIND

RHIS, RHINO–NOSE

RHODON–RED

RHYNCHOS–SNOUT

STENOS–NARROW

STOMOS–MOUTH

SYM, SYN–WITH, TOGETHER

TELE, TELO–DISTANT, FAR

TELOO–COMPLETE

THERMOS–HEAT

TOPOS–PLACE

TROPHEIN–TO NOURISH

XEROS–DRY

ZYGO–UNITE, YOKE

Latin

A, AB–OFF, FROM, AWAY

AD–TO, AT, TOWARDS, INTO

ANT, ANTE–BEFORE

BI, BIS–TWICE

CEREBRO–BRAIN

CO, COM, CON–WITH, TOGETHER

CONCHO–SHELL

CORPUS–BODY

DE–FROM, AWAY, OUT, DOWN

DEMI–HALF

E, EX–OUT OF, WITHOUT

EXTERO–OUTER

EXTRA–ON THE OUTSIDE

FILUM–THREAD

INFRA–BELOW, BENEATH

INTER–BETWEEN, AMONG

INTRA–WITHIN

LATERO–SIDE

LEPIS, LEPIDOS–SCALE

LIGNUS–WOOD

LITTUS, LITUS–SHORE

LUX, LUCIS–LIGHT

MACULA–SPOT

MANUS–HAND

OB–ON THE WAY OF, AGAINST

OS–MOUTH, BONE

PER–THROUGH, BY MEANS OF

PLEION–MORE

PLEISTOS–MOST

POST–AFTER

PRE–BEFORE

SUB–UNDER SUPER

SUPRA–ABOVE

UNI–ONE

Index

R

S

T

U

W